A series of student texts in

CONTEMPORARY BIOLOGY

General Editors:
Professor E. J. W. Barrington, F.R.S.
Professor Arthur J. Willis

The Diversity of Green Plants

Second edition

Peter R. Bell,
M.A.

Professor, Department of Botany, University College, London

Christopher L. F. Woodcock,
Ph.D.

Zoology Department, University of Massachusetts, Amherst, Mass.

Edward Arnold

First published 1968
by Edward Arnold (Publishers) Ltd.,
41 Bedford Square
London, WC1B 3DQ
Reprinted 1969
Second edition 1971
Reprinted (with corrections) 1972
Reprinted (with additions) 1975
Reprinted 1976, 1978, 1980

Boards Edition ISBN: 0 7131 2309 5
Paper Edition ISBN: 0 7131 2310 9

Printed in Great Britain by
Whitstable Litho Ltd., Whitstable, Kent

Preface to the Second Edition

The Diversity of Green Plants has been written in the conviction that never before has the science of botany had so much to offer the enquiring student. Apart from the fundamental importance of photosynthesis and other aspects of plant metabolism in the continuance of animal life, plants are often particularly suitable for the study of basic problems of growth and development by the newest biophysical and biochemical techniques. It seems likely that certain aspects of plant growth will before long be resolvable in terms of the interaction of recognizable macromolecules.

How is this new knowledge to be fully appreciated? In our view by providing the botanist at the beginning of his training with a basic knowledge of plant morphology and evolution. Not only does an awareness of the inter-relationships of the organisms used in current research facilitate an understanding of the significance of the results; it also provides the opportunity for recognizing new lines of profitable enquiry. Our purpose, therefore, is to promote our science in all its aspects by presenting a concise account of the structure and reproduction of the varied groups of autotrophic plants, both living and extinct, proceeding from the simple to the complex.

Although we have adopted a form of classification as a framework for our account, we have thought of the autotrophic plants as a whole. Each is the morphogenetic expression of inherited information, much of it presumably common to all autotrophs. The fossil record does, however, give undeniable evidence of an evolutionary progression. We have therefore discussed the likely origin of each level of organization and its possible selective advantages, but unwarranted phylogenetic speculation has naturally been avoided. We have referred throughout to experimental work relevant to the problems of growth and form, and to the detection of evolutionary affinities, but detailed consideration of these topics has had to be omitted as beyond the scope of this book.

The Diversity of Green Plants is intended primarily for undergraduate students. We hope, however, that it will also be useful to all those seeking information about the relationships of the autotrophs commonly used in research. Fuller accounts of the plants we describe can be found in the standard works to which our superscripts refer.

ACKNOWLEDGEMENTS

We are grateful to Professor Dr. K. Mühlethaler for Fig. 1.1, to Professor Dr. G. Drews for Fig. 1.2, to the Cambridge Instrument Co. Ltd. for Fig. 7.22, to Dr. H. G. Dickinson for Fig. 8.24, and to the Biology department of the Northern Polytechnic, London, N.7, for the use of its slide collection.

In addition to the authors and publishers cited, we acknowledge with thanks permission to reproduce figures from the following: the Councils of the Linnean Society of London (Figs. 7.47, 8.14, 8.15), of the Royal Society (Fig. 5.21), and of the Royal Society of Edinburgh (Fig. 3.8); the Trustees of the British Museum (Natural History) (Figs. 2.24, 2.25, 2.40, 2.43, 3.15, 3.16, 3.18, 3.22, 3.23, 3.24, 3.27, 8.5); and the University of Michigan Press (Figs. 2.41, 2.42, 2.44). We have also been fortunate in being able to call on the excellent draughtsmanship of Drs. J. G. Duckett (Figs. 4.2, 4.8b, c, 4.9, 4.10, 4.14, 4.16, 4.17, 4.18, 4.19, 4.20, 4.25 (in part), 4.28, 4.29, 5.29b, 5.30, 6.24, 6.31 (in part), 7.11, 7.25b, c, 8.31) and H. G. Dickinson (Figs. 7.27, 8.22, 8.32), and of Elizabeth Harrison (Figs. 2.19, 2.47, 4.6, 4.7, 4.12, 5.3, 5.9, 5.11).

Our colleagues in various places gave us invaluable criticism and advice, but we take full responsibility for any errors that may remain in the text.

London and
Amherst, Mass.
1971

P.R.B.
C.L.F.W.

Contents

I

The Principles Governing the Evolution of Autotrophs

SIGNIFICANCE OF AUTOTROPHIC NUTRITION

The living state is characterized by instability and change. A living cell consumes and releases energy by means of numerous chemical reactions, called collectively metabolism, taking place within it. Metabolism is synonymous with life. Even the apparently inert cells of seeds show some metabolism, but admittedly only a fraction of that which occurs during germination and subsequent growth. At normal temperatures, only in a dead cell, provided it remains sterile, does metabolism stop.

To maintain the dynamic state of a living cell it must be provided with sources of energy. The most usual sources are chemical and frequently consist of sugars. Together with these sources of energy, a cell also requires those materials for the repair and maintenance of its structure which it is unable to make for itself. These vary with the nature of the cell, but examples are certain metals which are essential components of important enzymes, the nitrogen of the proteins, and certain complex molecules called vitamins. All these nutritional requirements of the cell must be met; if not, the dynamic state we call life ultimately ceases.

It is a remarkable property of a large part of the Plant Kingdom that the cells, or in a multicellular plant at least some of the cells, are able to utilize radiant energy in their metabolism. The energy, absorbed by the pigment chlorophyll, is subsequently used to assimilate carbon dioxide from the air, and to convert it into sugar within the cells. This property, photosynthesis, releases the organisms concerned from the necessity of an external source of carbohydrate, and their nutritional demands are consequently relatively simple. Plants which are able to utilize radiant energy in this way are termed autotrophs; those which are unable to use radiant energy and which must depend upon carbohydrates supplied from without are termed heterotrophs.

So far as is known, autotrophs utilize for photosynthesis principally that radiant energy falling within the wave-band called 'visible light'. Consequently, autotrophs can exist only in an environment continuously or intermittently illuminated. Some small autotrophs can, it is true, also exist as heterotrophs, so we may distinguish between obligate autotrophs, such as higher plants, normally unable to utilize sugar supplied externally, and facultative autotrophs (or facultative heterotrophs) which can adapt their metabolism to the nature of their environment.

Chlorophyll is a complex pigment, existing in a number of slightly different forms. The molecule is in part similar to that of the active group of the blood pigment haemoglobin, but chlorophyll contains magnesium instead of iron. The photosynthetic pigment is not free in the cells, but is always associated with lipoprotein membranes, which seem generally to

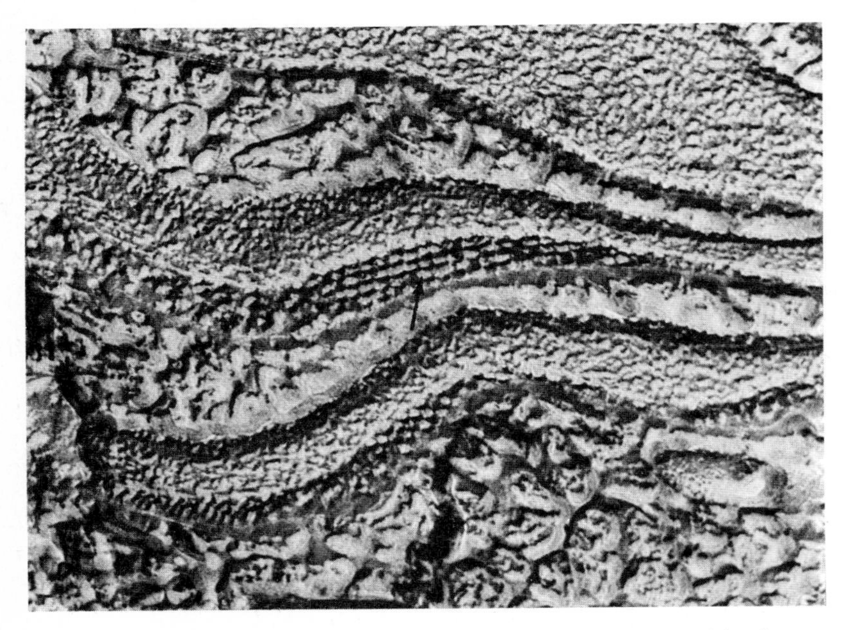

Fig. 1.1 A replica of part of the chloroplast of spinach prepared by freeze-etching. The arrays of larger particles (indicated by the arrow), associated with membranes, may be the sites at which light energy is transformed into chemical energy. (×94,000).

have a granular substructure (Fig. 1.1). Pure chlorophyll is green, and absorbs particularly in the red part of the spectrum. In cells its colour may be masked by accessory pigments. Although the function of these is not precisely known, in certain instances part of the energy they absorb undoubtedly contributes to photosynthesis. Chlorophyll, however, is the pigment principally concerned. It is present in all autotrophs, and absent from all obligate heterotrophs, including the whole of the Animal Kingdom.

The possession of chlorophyll and the consequent ability to utilize the sun's energy clearly bestow a great advantage on autotrophs. There is no necessity to forage for a supply of carbohydrates, and Mereschkowsky's image[62] of the restless lion and the placid palm as contrasting the consequences of heterotrophic and autotrophic nutrition respectively, although poetic, is nevertheless apt. Autotrophs, of course, as all living organisms, remain dependent upon a supply of water and minerals, but such motility as occurs is in relation to light rather than mineral sustenance. Green unicellular organisms, for example, often show marked positive phototaxis.

THE STRUCTURE OF THE AUTOTROPHIC CELL

Two fairly distinct kinds of cellular organization occur amongst the autotrophs as a whole. In the first, termed ***procaryotic***,[25b,75] the cell possesses no distinct nucleus, although a region irregular in outline and of differing density occurs at the centre of the cell. This is referred to as a nucleoid, and the genetic material is believed to lie therein. In the electron microscope this region appears fibrillar rather than granular, and the fibrils may indicate the site of the deoxyribonucleic acid. The protoplast of such cells is bounded by a membrane and this membrane here and there invaginates into the cytoplasm (Fig. 1.2). Evidence is accumulating that these membranous invaginations are the sites of photosynthesis, and they disappear or become very reduced if the cells are grown in the dark. This simple kind of autotrophic cell is found in both the photosynthetic bacteria and the Cyanophyta, perhaps representative of the most primitive kind of photosynthesizing organism. Remains very suggestive of autotrophic procaryotes have been found in Canadian rocks believed to be about 2×10^9 years old. Geochemical evidence of photosynthesis, together with what are possibly remains of bacterial cells, comes from even more ancient rocks in South Africa, estimated to be at least 3×10^9 years old.[72a]

In the cells of all other autotrophic plants the nucleus, the photosynthetic apparatus, and the membranes bearing the enzymes responsible for the electron transport chain of respiration are separated from the

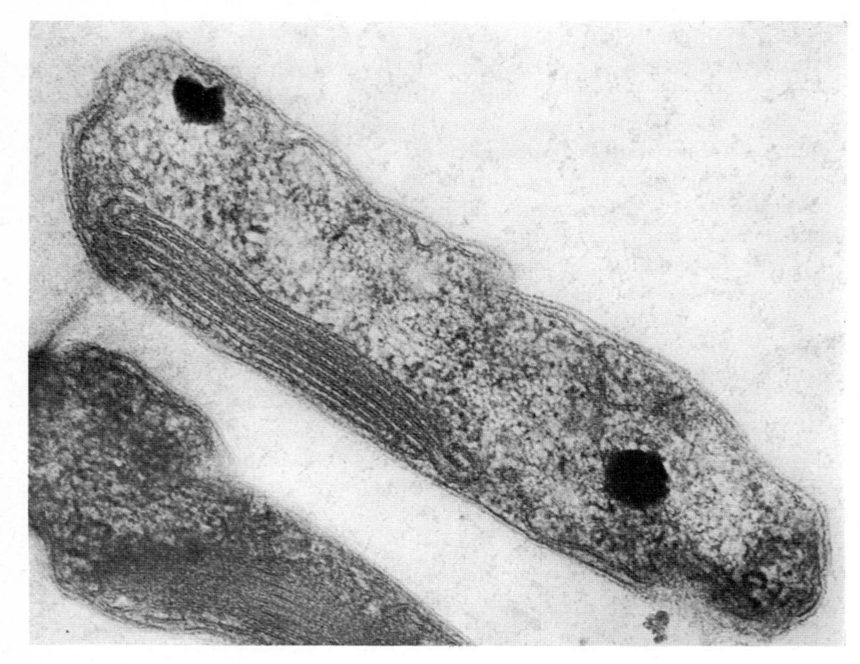

Fig. 1.2 Median section of a photosynthetic bacterium, *Rhodopseudomonas viridis*. The chromatophore is formed from an invagination of the cell membrane (plasmalemma). The folding of the membrane gives rise to a stack of lamellae, similar to those seen in the chloroplasts of higher plants (×95,000).

remainder of the cytoplasm by distinct envelopes. Such cells, termed ***eucaryotic***, have evidently been capable of giving rise to much more complicated organisms than the procaryotic. The photosynthetic apparatus, which consists of numerous granular lamellae running parallel to one another, is contained in one or more ***plastids*** (or chromatophores). The envelope of the plastid consists of two unit membranes, the inner of which invaginates into the central space and generates the internal lamellar system. The respiratory enzymes are contained within another distinct organelle, the ***mitochondrion***. Both plastids and mitochondria contain nucleic acids, and possess a procaryote-like biochemistry. There is some evidence that they possess genetic systems of their own, partly independent of that of the nucleus.

The simplest eucaryotic autotroph is thus a single cell containing a single nucleus, a plastid and a mitochondrion. This condition is represented in *Micromonas pusilla* (Fig. 1.3), a minute alga ubiquitous in the sea. Electron micrographs of this organism undergoing fission show how the plastid and mitochondrion divide at the same time.[58]

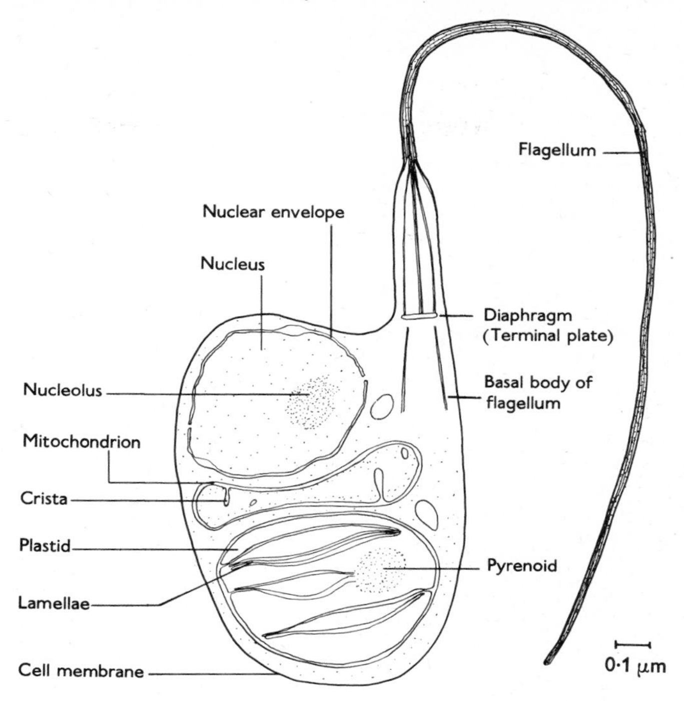

Fig. 1.3 *Micromonas pusilla.* (From electron micrographs by Manton, 1959, *J. Mar. Biol. Ass. U.K.*, **38**, 319)

THE EVOLUTIONARY CONSEQUENCES OF AUTOTROPHIC NUTRITION

It seems beyond doubt from the fossil record of life and from the biological and geological inferences that can be drawn from it that life began in water. The early forms of life were probably also autotrophic. The plants that evolved in this aquatic environment, and still in the main exploit it, have certain biochemical and structural features in common, and are collectively termed algae. Our treatment of the autotrophic plants consequently begins at this level of organization (Chapters 2 and 3).

At some stage, possibly in the Silurian period (Table 1.1) or even earlier, vegetation began to colonize the land. These early colonists, and consequently the whole of our existing land flora, almost certainly emerged from that group of aquatic plants today represented by the green algae (Chlorophyta).[20] The Chlorophyta and the land plants (a term which means plants suited to life on land and not merely plants growing on land) have the same photosynthetic pigments, and basically the same photosynthetic apparatus. Moreover, at least one green alga (*Cladophorella*; see p. 38) which grows

Table 1.1 The Geological Table

ERA	PERIOD		AGE (in 10^6 years)	First Authentic Appearance
QUATERNARY	Pleistocene and Recent		0–1	
TERTIARY (or Cenozoic)	Pliocene, Miocene	Upper Tertiary	1–28	
	Oligocene, Eocene, Paleocene	Lower Tertiary	28–60	
MESOZOIC	Cretaceous	Upper	60–135	Angiosperms*
		Lower		
	Jurassic	Upper	135–180	
		Middle		
		Lower (Lias)		
	Triassic		180–225	
PALAEOZOIC	Permian	Upper	225–270	Cycad- and *Ginkgo*-like plants, *Glossopteris*
		Lower		
	Carboniferous	Upper (Pennsyl-vanian)	270–350	Conifers
		Lower (Mississippian)		Pteridosperms
	Devonian	Upper	350–400	Ferns, Sphenopsids
		Middle		
		Lower		Vascular plants: Lycopods and Psilopsids
	Silurian		400–440	Triradiate spores
	Ordovician		440–500	
	Cambrian		500–600	
PRE-CAMBRIAN	Proterozoic		600–1,500	Calcareous algae
	Archaeozoic		1,500–3,000+	Fungi, algae, bacteria (see p. 3)

* See pp. 335–339 for a discussion concerning the first appearance of the angiosperms.

on damp mud is covered on its upper surfaces by a material which, judging from its resistance to acids and oxidizing agents, closely resembles cutin. This perhaps indicates the way in which the cuticle, ubiquitous in land vegetation, was derived.

Any consideration of the evolution of an autotrophic land flora must thus necessarily take into account the physiological features of the green algae, and how these may have been modified in the transition to terrestrial life.

Recent research into algal environments is yielding much information relevant to this problem. It is commonly found, for example, that from 5 to 35% of the light striking the surface of a lake or sea is reflected, the actual amount lost depending upon the angle of incidence. The light penetrating the water is then gradually absorbed as it advances, so that up to 53% of the radiation passing the surface may be dissipated as heat in the first metre.[69] Consequently, in warm and temperate regions, the rate of photosynthesis of submerged plants is normally controlled by the amount of light reaching them, and not by the amount of carbon dioxide in the water. We can see at once that the first colonists of land, emerging on to bare mineral surfaces, would almost certainly have had to contend with intensities of illumination strikingly higher than those experienced by their aquatic ancestors. This would have provided opportunities for greatly increased photosynthesis.

Another discovery of recent research, also very relevant to the problem of the colonization of the land, is the surprising extent to which algae release photosynthesized materials into the surrounding water. In Windermere, for example, up to 35% of the total carbon fixed is continuously lost in this way.[34] Losses of this order are clearly possible only from aquatic plants. As vegetation advanced from marshes, or from littoral belts subject to periodic inundation, on to relatively dry substrata, much more of the fixed carbon must have been conserved within the plant body.

The emergence of green plants from seas or lakes must therefore have involved striking changes in carbohydrate metabolism. The increased assimilation of carbon, resulting from the increased intensity of illumination on land, and a diminishing loss of photosynthesized carbohydrates by outward diffusion as plants reached relatively dry areas, could have resulted in embarrassingly large, and possibly toxic, quantities of carbohydrates in the cells. The physiology must therefore have changed simultaneously to meet this new situation.

We have no evidence that adaptation to terrestrial life was accompanied by any reduction in the amount of chlorophyll in the chloroplasts (the term used for plastids that are unambiguously green), or in the efficiency of the photosynthetic apparatus. Evolution at this critical period seems rather to have been concerned, not with changes in the photosynthetic process, but with the development of mechanisms for utilizing, and removing in a non-metabolic and osmotically inactive form, the increased supply of carbohydrates. It is in this light that we should interpret the thick cell walls, extensive lignification, and resinous deposits of the early land plants (Chapters 4–6). Cellulose, lignin, resin and phlobaphene are all derived from sugars. All are osmotically inactive, and their formation could be regarded as a kind of detoxication mechanism protecting the metabolism from an excess of soluble carbohydrates.

Natural selection would, of course, have ensured that those forms survived in which the cells containing the condensed carbohydrates, and the materials derived from them, assisted the functioning of the plant as a whole.

We can envisage how this led to the evolution of xylem, the principal lignified tissue of land plants. Although largely dead, xylem is of paramount importance in the structure of plants since it provides both a skeleton supporting the plant in space, and an effective system for the transport of water and solutes. Viewed biochemically, it also represents the removal by condensation of considerable quantities of carbohydrates from the metabolism of the plant. Cutin and sporopollenin, condensation products of a fatty nature, are also given essential roles in land plants. They serve to lessen the loss of water by evaporation from the living cells, and thus help to solve another of the physiological problems plants encountered as they left aquatic environments or saturated atmospheres.

In the course of evolution many complex and bizarre forms of growth have appeared in land plants, but the material from which they are fashioned has remained predominantly carbon, extracted from the atmosphere. This diversity is evidently related to the tetravalent nature of carbon, and the great range of compounds that can be formed from it. Had not the photosynthetic fixation of this versatile element arisen on the Earth's surface, plant life, and the animal which depends upon it, would have been impossible. Indeed, it is difficult to conceive of any alternative form of life appearing in its absence.

THE MOBILITY OF PLANTS

Although the earliest plants probably soon acquired motility of the kind seen today in *Chlamydomonas* and *Euglena* (Chapters 2 and 3), this appears to have been rapidly lost in the evolution of higher forms. Autotrophic nutrition, together with the ability to manufacture a skeleton and a vascular system from the assimilated carbon, eventually made the large and firmly anchored terrestrial plant a practical possibility. Such an organism is, of course, immobile, and thus suffers a serious disadvantage, not shared by the higher animal, at times of natural catastrophe, such as volcanic eruption or fire. Plants, however, very frequently possess a remarkable mobility, or at least a ready transportability by agencies such as wind and water, in their reproductive bodies. Fern spores, for example, have been caught in aeroplane traps in quantity at 5,000 ft and even higher, and the hairy spikelets of the grasses *Paspalum urvillei* and *Andropogon bicornis* have been encountered at 4,000 ft above Panama.[8] Lakes, seas and the coats and feet of animals also play their part in distributing plants. The immobility of the individual is thus frequently compensated for by the mobility of the species, and devastated areas and new land surfaces become colonized with amazing rapidity and effectiveness.

THE LIFE CYCLE OF AUTOTROPHS

A life cycle, involving segregation and recombination of the genetic material, appears as basic to the evolution of plants as to that of animals. In one part of the cycle the nucleus contains a single set of chromosomes (and is consequently termed ***haploid***), and in the other two sets of chromosomes (and is termed ***diploid***). The cycle is seen at its simplest in the unicellular autotrophs of aquatic environments (Chapters 2 and 3), where haploid individuals in certain circumstances behave as gametes and fuse, so forming a ***zygote***. The zygote, which contains a diploid nucleus, either undergoes meiosis at once, or only after some delay, in which case the diploid condition can be thought of as having an independent existence. Either the haploid or the diploid generation, or both, may be multicellular. The multicellular plant is called a ***gametophyte*** if it produces gametes directly, and a ***sporophyte*** if it produces, following meiosis, individual cells which either behave as gametes immediately or which develop into gametophytes. Each generation may also multiply itself asexually. These various possibilities are summarized in Fig. 1.4.

A life cycle is thus basically a nuclear cycle, and it is not necessarily accompanied by any morphological change. In the alga *Dictyota* (p. 91), for example, the gametophyte and sporophyte are identical, and it is necessary to observe the manner of reproduction in order to identify the place in the cycle which any individual occupies. Such a life cycle is termed ***isomorphic*** (or homologous). Frequently, however, the two generations of the cycle have different morphologies, one often being less conspicuous than the other, and sometimes even parasitic upon it. These cycles are termed ***heteromorphic*** (or antithetic). Although the algae show both isomorphic and heteromorphic life cycles, those of land plants are exclusively heteromorphic. Occasionally there may be a morphological cycle without a corresponding nuclear cycle, as in the apogamous ferns (see p. 220), but this is regarded as a derived condition.

Gametes are always uninucleate, and, when motile, usually naked cells. In the simplest form of sexual reproduction, termed ***isogamy***, the two gametes involved in fusion are free cells and morphologically identical. Nevertheless, detailed investigations continue to show that gametes from the same parent rarely fuse. Consequently some measure of self-incompatibility, and hence physiological differentiation between the parents, appears always to be present.

Isogamy was probably the most ancient condition, and this appears to have been succeeded by ***anisogamy***. Here the gametes, although still free

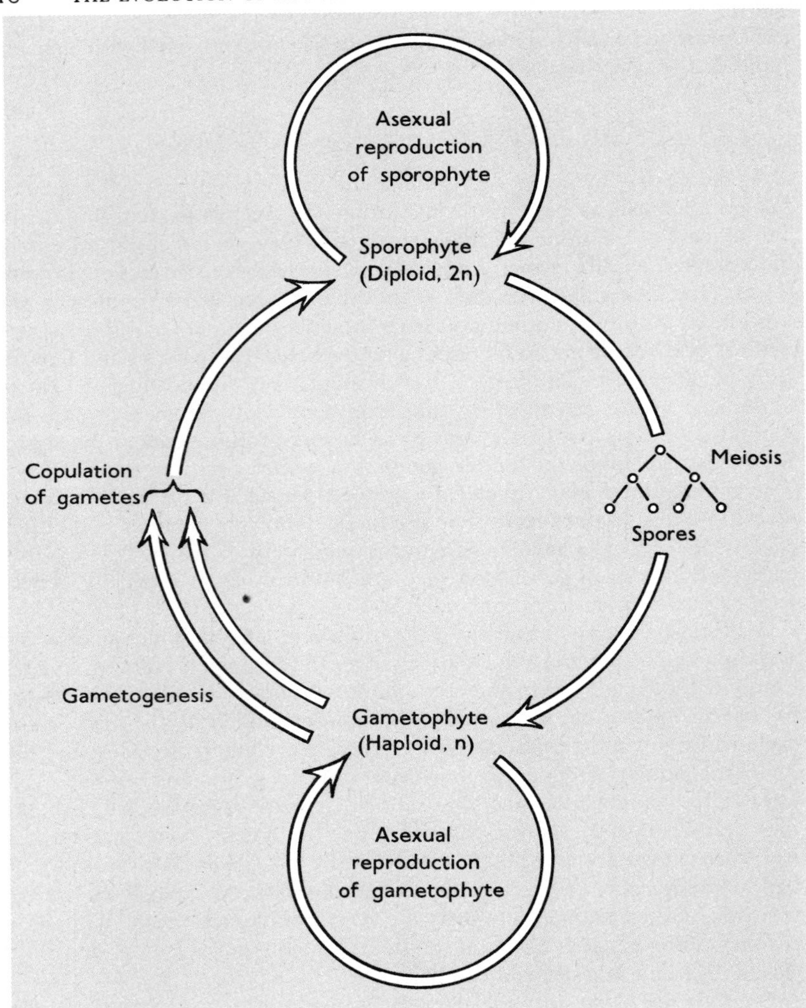

Fig. 1.4 The life-cycle of autotrophic plants generalized. The large circle represents sexual reproduction. Only relatively few species display all the reproductive potentialities shown.

cells, are morphologically dissimilar, but usually differ in little more than size. The larger, which may also be less mobile, is called the female. The extreme form of anisogamy is ***oogamy***, in which the female gamete is now a large, non-motile cell, filled with food materials. This cell may either float freely in water, as in the alga *Fucus*, or be retained in a chamber, as in some algae and all land plants. The chamber bears various names according to the group of plants being considered. Since the progression from isogamy is accompanied in many algal groups by an increase in somatic complexity, it appears almost beyond doubt that this morphological progression is also a phylogenetic one.

In several instances of sexual reproduction it has been shown that one or both gametes produce traces of chemical substances, termed gamones, which cause the appropriate gametes to approach each other. In the ferns malic acid may play this role. The male gametes certainly accumulate around a source of this substance placed in water. Malic acid is one of the components of the Krebs cycle, and may be liberated in small amounts from newly opened archegonia.

THE LIFE CYCLES OF THE TRANSMIGRANT FORMS

The transition to a terrestrial environment clearly presented a number of problems in relation to sexual reproduction. Although all land plants are oogamous, and are presumably derived from oogamous algae, fluid was still necessary in the initial land plants to allow the motile male gametes to reach the stationary female. This problem appears to have been met first by the egg becoming enclosed in a flask-shaped structure, the ***archegonium***, the neck of which provides a collecting chamber for the male gametes, and second by the male gamete becoming a highly motile cell. The male gametes of the lower archegoniate plants (Chapters 4, 5 and 6), termed ***antherozoids*** (or spermatozoids), are remarkable cytological objects. Each is furnished with two or more highly active flagella, and both the cell and nucleus have an elongated snake-like form, well suited for penetration of the archegonial neck. Dependence upon water is thus reduced to the necessity for a thin film in the region of the sex organs at the time of maturity of the gametes.

The archegonium is common to all the lower land plants, but its origin remains tantalizingly obscure. It may have appeared immediately before the colonization of the land, possibly as a consequence of morphogenetic tendencies seen today in association with the eggs of the Charales (p. 61) and certain red algae (p. 96). If so, these antecedents of the transmigrants left no similar descendants among living algae. Whatever the exact time of the evolution of the archegonium, however, there are no compelling reasons for regarding it as having been evolved more than once. Consequently we are not justified in regarding the migration of autotrophic life on to land as having occurred more than once in geological time, since the fossil record indicates that the most primitive forms of land plants were probably all archegoniate.

If the transmigrant forms were archegoniate, what was the nature of their life cycles? This is largely a matter of conjecture. However, as will be seen in later chapters, except for one approach to isomorphy (in the living Psilopsida), the lower archegoniate plants possess markedly heteromorphic life cycles in which the conspicuous generation is either the gametophyte (Bryophyta), or the sporophyte (Lycopsida, Sphenopsida, ferns). The transmigrants possibly had an intermediate position, with more or less isomorphic cycles. Nevertheless, heteromorphic cycles were probably very soon developed as terrestrial vegetation diversified and exploited particular features of the new environment. The cycle in which the sporophyte was the dominant generation clearly had the greater evolutionary potential, since, with the exception of the Bryophyta, it is characteristic of all existing terrestrial vegetation.

SEXUAL REPRODUCTION IN LATER TERRESTRIAL VEGETATION

An important step in the evolution of sexual reproduction on land was undoubtedly the emergence in the archegoniate plants of heterospory. This involves the production of spores of two sizes, the larger giving rise to a wholly female gametophyte and the smaller to a male (see pp. 155, 222). In homosporous archegoniate plants, quite considerable growth of the gametophyte is often required before it acquires the ability to produce egg cells. In the primitive heterosporous plants, however, the small female gametophyte formed on germination of the megaspore produces archegonia in a very short time. The microspore also develops rapidly, and antherozoids are soon liberated from the diminutive male gametophyte. The time involved in the gametophytic phase is thus reduced to a minimum.[18] In plants in which the sporophytic phase is dominant, speedy reproduction of this kind has the obvious advantage of accelerating the establishment of new forms and hence the rate of evolution.

In the higher archegoniate plants (Chapter 6) we see how sexual reproduction becomes increasingly independent of water. These archegoniates are exclusively heterosporous, but the megaspore is retained and germinates within a specialized sporangium called an ovule. In some forms (*Cycas*, p. 242; *Ginkgo*, p. 272) fertilization is still effected by flagellate male gametes, but the only fluid necessary is a small drop, immediately above the archegonia, into which the gametes are released. Other higher archegoniate plants escape even from this requirement. The male gametophyte is filamentous, and, as a consequence of its growing towards the female gametophyte, it liberates the male gametes (which now lack any specialized means of locomotion) directly into an archegonium. In a few allied plants (e.g. *Gnetum*, p. 280) modifications of the female gametophyte result in the disappearance of the archegonium, and ultimately we arrive at the embryo sac and the finely ordered cytology that is characteristic of the sexual reproduction of the flowering plants. Comparative morphology and the fossil record indicate that the morphological sequence we have considered here also represents the evolutionary development of sexual reproduction in land plants. Compared with the cytological elegance of fertilization in an angiosperm, the clumsy antherozoid of *Cycas* is thus not only barbarous, but also primitive.

THE CLASSIFICATION OF AUTOTROPHIC PLANTS

Having outlined the general principles which have governed the evolution of autotrophic plants we can now proceed to consider them according to their level of organization and in their natural alliances. We require a general classification to provide a framework for this information. Of the several general classifications available, all of which have some virtues, we have chosen that shown in Table 1.2. This is convenient for our purposes, and we do not wish to imply that the relative ranking given to the higher categories has any special significance. The classification depends firstly upon the presence or absence of a differentiated vascular system (which separates the Tracheophyta from the remainder), and subsequently upon what appear to be the natural affinities of the plants concerned.[27] In the arrangement of the algae we have followed Round.[71]

Table 1.2

Division	Sub-Division	Class	Order
Cyanophyta		Cyanophyceae	Chroococcales
			Hormogonales
Chlorophyta			Volvocales
			Chlorococcales
			Ulotrichales
			Cladophorales
			Chaetophorales
			Oedogoniales
			Conjugales
			Siphonales
			Charales
Chrysophyta		Xanthophyceae	
		Chrysophyceae	
		Bacillariophyceae	
Pyrrophyta		Desmophyceae	
		Dinophyceae	

Table 1.2—*continued*

Division	Sub-Division	Class	Order
Cryptophyta			
Euglenophyta			
Phaeophyta			Ectocarpales
			Laminariales
			Fucales
			Dictyotales
			Cutleriales
Rhodophyta		Rhodophyceae	
		{Bangioideae	
		{Florideae	
		(Sub-Classes)	
Bryophyta		Hepaticae	
		Anthocerotae	
		Musci	
Tracheophyta	Psilopsida		Psilotales
			Psilophytales
	Lycopsida		Lycopodiales
			Selaginellales
			Isoetales
			Lepidodendrales
			Pleuromeiales
	Sphenopsida		Equisetales
			Calamitales
			Sphenophyllales
			Pseudoborniales
	Pteropsida	Filicinae	Cladoxylales
			Coenopteridales
			Ophioglossales
			Marattiales
			Filicales
		Gymnospermae	Pteridospermales
			Cycadales
			Bennettitales
			Caytoniales
			Cordaitales
			Coniferales
			Ginkgoales
			Gnetales
		Angiospermae	
		{Dicotyledonae	
		{Monocotyledonae	
		(Sub-Classes)	

2

The Algae, I

Mainly autotrophic organisms in which the sex organs are unicellular, or, if multicellular, then lacking a wall of sterile cells. Sporangium usually a single cell, or, if multicellular, then all the cells fertile. Mostly aquatic, the thalli showing little cellular differentiation in the somatic regions.

The simplest autotroph imaginable is a single cell floating in a liquid medium, synthesizing its own sugar, and reproducing at intervals by binary fission. Such organisms do in fact exist in both fresh and salt waters. Examples are provided by *Euglena* (see Fig. 3.13), occurring in ponds and streams, and the minute marine *Micromonas* (see Fig. 1.3).

These organisms are examples of algae,[37,71] a group of plants showing the greatest diversity of any major division of the Plant Kingdom. They range from minute, free-floating, unicellular forms to large plants, exclusively marine, several metres in length. Nevertheless, despite this enormous variation in size, the algae remain predominantly aquatic and their structure comparatively simple. In the smaller multicellular species (e.g. *Pediastrum boryanum* (see Fig. 2.20)) the cells resemble each other in appearance and function, and they can be regarded as forming little more than an aggregate of independent units. In the larger, however, there is some morphological and cellular differentiation, although it is far less striking than in most land plants.

The larger algae are found only in the sea, commonly anchored to a rocky substratum. Their restriction to a marine environment is perhaps accounted for by the relative impermanence of inland waters in geological time, and the consequent limiting of the opportunity for the evolution of complex freshwater forms. Although marine algae are in some instances able to withstand inundation in fresh water (e.g. *Fucus*, p. 87), they do not survive indefinitely or grow in these conditions, presumably because fresh waters are unable to supply minerals at a rate adequate to support their metabolism. A large alga present in European seas is *Laminaria* (see Fig. 3.17), some species of which may reach 4 m in length. In the Pacific, off the west coast of America, are the gigantic *Nereocystis* and *Macrocystis*, with thalli commonly exceeding 50 m. Maintaining the integrity of a thallus of this size raises mechanical problems because, although the sea provides considerable supporting upthrust, currents and turbulence cause more sustained tensions and pressures than would similar movements in a gaseous medium. The toughness, and hard rubbery resistance to any kind of distortion found in the larger algae, are thus necessary qualities for the survival of these species. These attributes arise principally from the general properties of the cell walls and of the surface, and not from any specialized strengthening elements.

As would be expected of a group exploiting the aquatic habitat, algae have a number of biochemical characteristics peculiarly their own. Many, for example, accumulate fats and oils rather than starch.[33] The microfibrils of algal cell walls have in several instances been found to contain the polysaccharides mannan and xylan in addition to cellulose.[36] The nitrogenous polysaccharide chitin is found as an outer layer of the wall in *Cladophora prolifera* and possibly in *Oedogonium*. Pectin, a polymer based on galacturonic acid, is a common component of algal cell walls, sometimes forming a distinct outer sheath (e.g. *Scenedesmus*, Fig. 2.19). Colloids such as fucin and fucoidin, unknown outside the algae, occur in the amorphous matrices of the walls of brown algae. Alginic acid, which occurs in quantity in the middle lamellae and primary walls of several brown algae, is extracted commercially and finds a wide range of uses as an emulsifier in industry. Complex mucilaginous polysaccharides rich in galactan sulphates are characteristic of the red algae.[52b]

There is no evidence that the major groups of algae have any close relationship with each other. Nevertheless, there are sufficient morphological, physiological and ecological similarities between these plants to make the term 'alga' a useful one. Earlier the algae were grouped with the fungi and bacteria in the Thallophyta, but current opinion attributes to each algal group the status of a Division. We have followed this practice in Table 1.2.

Study of the structure and reproduction of the algae reveals a number of ways in which these simple autotrophs have increased their morphological and reproductive complexity. We shall in the main be concerned with the illustration and discussion of these trends, and we shall not attempt a complete taxonomic or morphological survey of any group. Nevertheless, in the first of these two chapters devoted to the algae the treatment of the Cyanophyta and Chlorophyta is fuller than that of the remainder of the algae in the second. This is because of the great importance of the Cyanophyta in relation to the evolution of the photosynthetic apparatus, and of the Chlorophyta in relation to that of the land flora.

THE PROCARYOTIC ALGAE

CYANOPHYTA

Habitat	Aquatic, or in moist situations.
Pigments	Chlorophyll *a*, β-carotin, biliproteins, myxoxanthin, myxoxanthophyll.
Food reserves	Glycogen, cyanophycin.
Cell wall components	Hemicellulose, occasionally cellulose, pectin, possibly mucopeptide.
Reproduction	Asexual. Genetic recombination observed, but mechanism unknown.
Growth forms	Unicellular, cellular aggregates, filamentous.

The Cyanophyta (blue-green algae) are amongst the simplest autotrophic plants in existence. Although the cells may occur in aggregates, sometimes even arranged linearly in a filament, there is little co-ordination of activity. The Cyanophyta are thus considered to be capable at most of forming colonies of individuals, rather than true multicellular organisms.

CYTOLOGY The cells, which rarely exceed 10 μm in diameter, are also less differentiated internally than those of any other algal Division. They are procaryotic, and in many features resemble the cells of photosynthetic bacteria (see Fig. 1.2).

Under the light microscope, partly because of the small size of the cells concerned, the photosynthetic pigments appear to be dispersed in the cytoplasm, but electron microscopy has revealed that this view is almost certainly erroneous. The cytoplasm contains a number of parallel lamellae, some and perhaps all of which are invaginations of the plasmalemma. Since the amount of chlorophyll in the cells is related to the development of this lamellar system, it is inferred that the photosynthetic pigments are associated with the lipoproteins of its membranes. In some forms small particles (phycobilisomes), about 30 nm in diameter, can be seen lying along the membranes. These are believed to be aggregations of a biliprotein pigment, phycobilin. Since the cells do not contain mitochondria, the respiratory mechanism in addition to that of photosynthesis may also be located upon the internal lamella system. Clusters of minute membranous cylinders containing air ('gas vacuoles') are present in the cells of some species. These vacuoles are unique in being bounded by a membrane consisting solely of protein. They collapse if the cells are subjected to sudden mechanical shock. Vacuoles similar to those of higher plant cells appear to be lacking.

In many filamentous species the chains of cells are interrupted by occasional conspicuously large cells called heterocysts.[32] Although, since they lack photosynthetic pigments, they appear empty, their fluorescence in deep blue light when stained with acridine orange indicates that they contain considerable amounts of deoxyribonucleic acid. A distinguishing morphological feature of the heterocyst is the presence of a narrow pore at one or both poles. Heterocysts have attracted considerable attention, and it is now known that they fix atmospheric nitrogen[76b], the thick wall possibly excluding oxygen. In a few instances heterocysts have been observed to regain internal structure and pigmentation, and then to germinate. On these grounds some have considered them to be vestigial reproductive cells.

The cell walls of the Cyanophyta, composed principally of hemicellulose, cellulose and pectin, are commonly gelatinous. The gelatinous material frequently serves to hold cell colonies together, a feature well seen in *Nostoc* (Fig. 2.5). The morphological integrity of the more strikingly filamentous forms often depends upon the cells being held in linear sequence by a mucilaginous sheath. 'Branching' occurs where breaks develop in the sheath.

DISTRIBUTION The Cyanophyta are widely distributed, occurring wherever water is available. Some are marine (e.g. *Trichodesmium*, responsible for the colour of the Red Sea). Others are found under such extreme conditions as those of snowfields and hot springs (where they can survive temperatures of up to 85°C). Many occur as green slimes on rocks, damp soils and tree-trunks, and as scums on stagnant water. Some live as endosymbionts in the protoplasts of colourless flagellates and amoebae (*Glaucocystis* is often regarded as a composite organism of this kind), and others as colonies in the tissues of higher plants (e.g. the presence of *Anabaena* in the fern *Azolla* and the gymnosperm *Cycas*). Cyanophyta are also a component of some lichens. These symbiotic associations are probably related to the ability of many Cyanophyta both to photosynthesize and fix atmospheric nitrogen.

REPRODUCTION The observed reproduction of the Cyanophyta is entirely asexual, in its simplest form involving nothing more than cell division. This, correlated with the absence of distinct nuclei, is not associated with the formation of any recognizable mitotic figure or chromosomes. In some species, mainly those that are unicellular, the cell enlarges, and the protoplast meanwhile divides to form many naked daughter cells termed ***endospores***. When these are eventually released they develop a cell wall and become new individuals. In unfavourable conditions, thick-walled cysts (***akinetes***) may be formed. Experiments have shown that these resting spores are very resistant to desiccation and extremes of temperature. In filamentous forms the akinetes frequently germinate to form a short thread of rounded cells, termed a ***hormogonium***. This has some capacity for movement which, since there are no flagella or cilia, seems attributable solely to the streaming of mucilage at the surface. The hormogonia eventually settle and give rise to filaments. Also in filamentous forms whole short side branches may become resting organs (***hormocysts***). As already mentioned, heterocysts occasionally serve as reproductive organs.

The Cyanophyta contain two principal Orders, the Chroococcales, mainly single or aggregated spherical cells, and the filamentous Hormogonales.

Chroococcales Although containing the simplest blue-green algae, some of the Chroococcales are nevertheless colonial forms with a regular and conspicuous symmetry. In *Chroococcus* (Fig. 2.1) single cells are occasionally seen, but more usually, and always in the similar *Gloeocapsa*

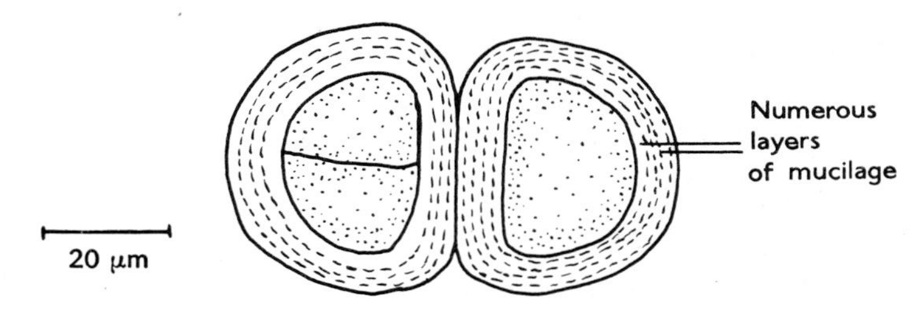

Fig. 2.1 *Chroococcus westii*. (After Boye-Petersen, from Rabenhorst, *Kryptogamenflora*, vol. 14. Leipzig, 1935)

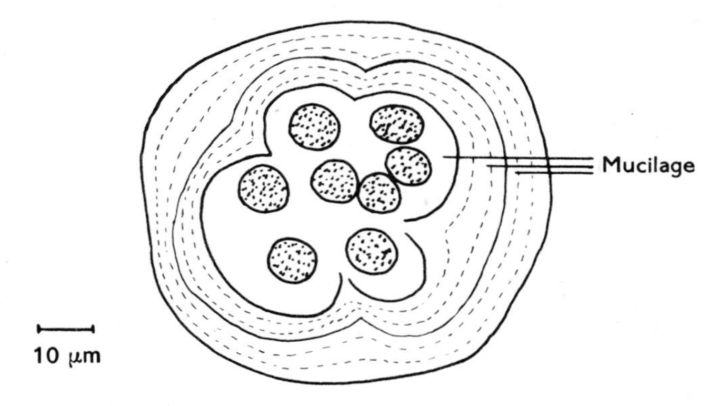

Fig. 2.2 *Gloeocapsa sanguinea*. (After Wille, from Rabenhorst, *Kryptogamenflora*, vol. 14. Leipzig, 1935)

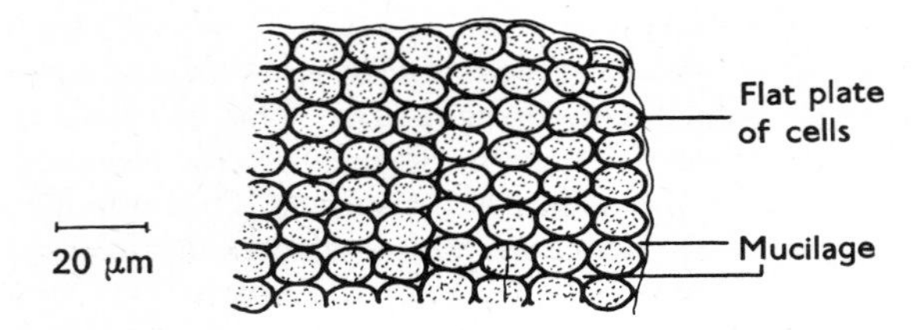

Fig. 2.3 *Merismopedia elegans*. Portion of colony. (After Smith, from Rabenhorst, *Kryptogamenflora*, vol. 14. Leipzig, 1935)

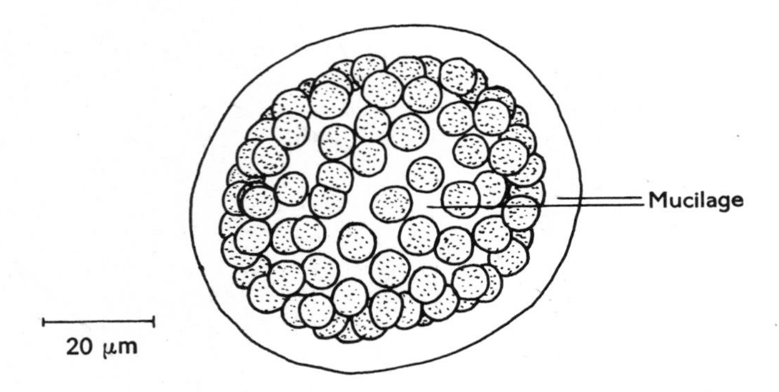

Fig. 2.4 *Coelosphaerium dubium*. Spherical colony. (After Schmula, from Rabenhorst, *Kryptogamenflora*, vol. 14. Leipzig, 1935)

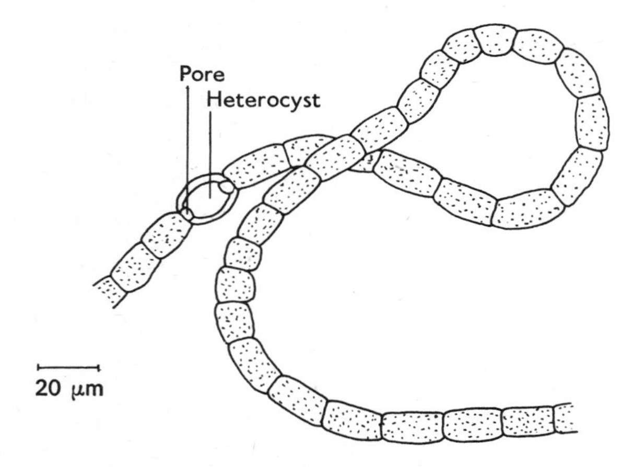

Fig. 2.5 *Nostoc carneum*. Portion of filament with heterocyst. (After Frémy, from Rabenhorst, *Kryptogamenflora*, vol. 14. Leipzig, 1935)

(Fig. 2.2), the cells remain held together after division in a mucilaginous matrix. These aggregates of indefinite size and shape are referred to as ***palmelloid*** forms. In *Merismopedia* (Fig. 2.3) the cells are arranged in regular rows to form a plate, and in *Coelosphaerium* (Fig. 2.4) a hollow sphere. Other geometric arrangements are characteristic of further genera in this Order. Although the cells in these colonial forms appear to be all of similar status and function, there is evidence of polarity, since in vegetative reproduction divisions appear to take place more readily in certain directions than others.

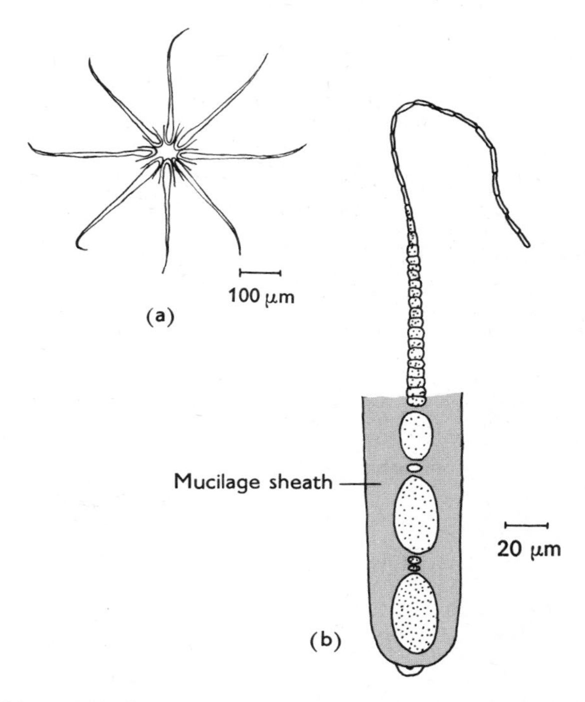

Fig. 2.6 *Gloeotrichia letestui.* (**a**) Star-shaped cluster of filaments. (**b**) Single filament. (After Frémy, from Rabenhorst, *Kryptogamenflora*, vol. 14. Leipzig, 1935)

Hormogonales The simplest filamentous form is *Nostoc* (Fig. 2.5). Here all the cells are of equivalent status and divisions occur sporadically along the filament. In some others, e.g. *Gloeotrichia* (Fig. 2.6), the filament has a basal cell and the remainder diminish in diameter towards the apex. The divisions here take place principally in the central region of the filament. In *Oscillatoria* (Fig. 2.7) the filaments occur individually, but in *Gloeotrichia* the filaments are clustered with their basal cells together, so forming a circular plate. Many genera, e.g. *Scytonema*, have branched filaments, but 'branching' is usually caused by growth of one part of the cell chain being deflected by the other part out of the mucilage sheath

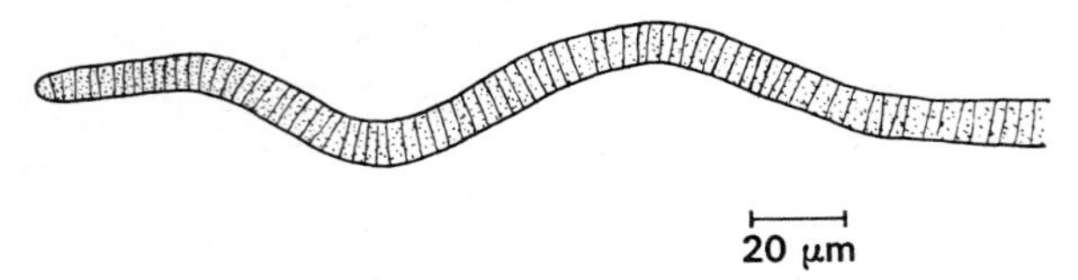

Fig. 2.7 *Oscillatoria meslini*. Terminal portion of filament. (After Frémy, from Rabenhorst, *Kryptogamenflora*, vol. 14. Leipzig, 1935)

(Fig. 2.8). True lateral branching, in which there is a distinct change in direction of growth with respect to the longitudinal axis of the filament, occurs very rarely.

A feature of certain Hormogonales, as yet not readily explained, is a slow mobility. Although this is particularly noticeable in respect of the hormogonia, it is also shown by normal filamentous cultures. *Anabaena*,

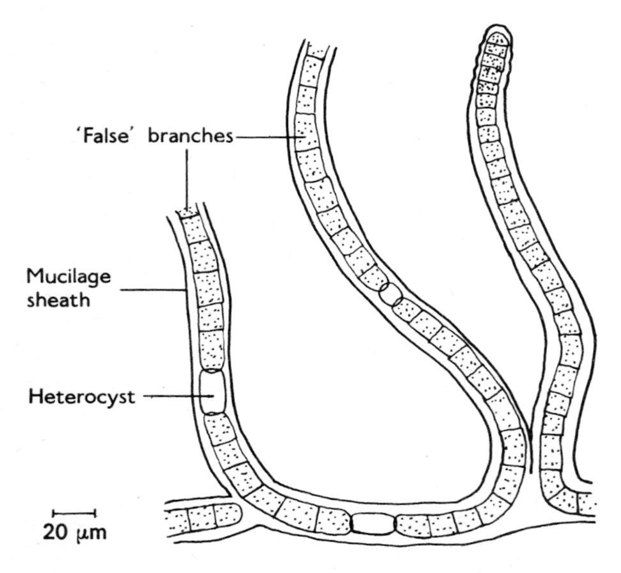

Fig. 2.8 *Scytonema tolypothricoides*. 'False' branching. (After Frémy, from Rabenhorst, *Kryptogamenflora*, vol. 14. Leipzig, 1935)

for example, grown in an agitated medium is uniformly dispersed, but if the shaking is stopped the filaments soon group themselves into a few tight bunches. This is probably a consequence of the chance cohesion of filaments, followed by the gliding of the filaments along each other within the coalescent mucilage. The oscillatory movements characteristic of *Oscillatoria* appear to arise from changes in curvature propagated along the filament, but are otherwise unexplained.

The relationship of the Cyanophyta

Except in respect of the photosynthetic pigments, the Cyanophyta have little in common with any other Division of the algae. The organization of the cell, however, is very similar to that found in the photosynthetic bacteria, and the view that these bacteria and the Cyanophyta had a common ancestor now has considerable support.[28]

THE EUCARYOTIC ALGAE

The cells of all algae other than the Cyanophyta are eucaryotic, and in respect of organization have much in common with those of higher plants.

A striking feature of many of the unicellular eucaryotic algae, and of the zoospores and gametes of the multicellular forms, is the presence of flagella. An unexpected and remarkable discovery of electron microscopy is that flagella from all eucaryotic organisms have a common basic structure, providing a characteristic picture in transverse section (Fig. 2.9). Nine pairs of microtubules, each pair orientated tangentially, are arranged concentrically near the periphery of the flagellum. Usually two more microtubules lie symmetrically at the centre. The movement of the flagella is probably caused by the paired microtubules sliding over one another. The mechanism is not, however, entirely understood since the flagella of some mutants (and of the male gametes of some diatoms), in which the internal structure of this organ is defective, appear to move normally. The microtubules of the flagella are probably identical in nature with those of the mitotic spindle, but different from those which lie near the surface of the cytoplasm in most cells.

There are two broad classes of flagella: the 'whiplash', which is long and smooth, and the 'flimmer', which is shorter and possesses two rows of minute hairs (Fig. 2.9b). In a number of species it has been shown that these hairs are formed within the cell, possibly close to the nucleus. How they arrive at the surface of the flagellum is not certainly known, but the endoplasmic reticulum is probably involved in conducting them to their final position.[52a] In some species the Golgi apparatus is implicated.

In some groups of algae the flagella are scaly and in others (all brown algae) spiny. These and other characteristics of the flagella, such as their

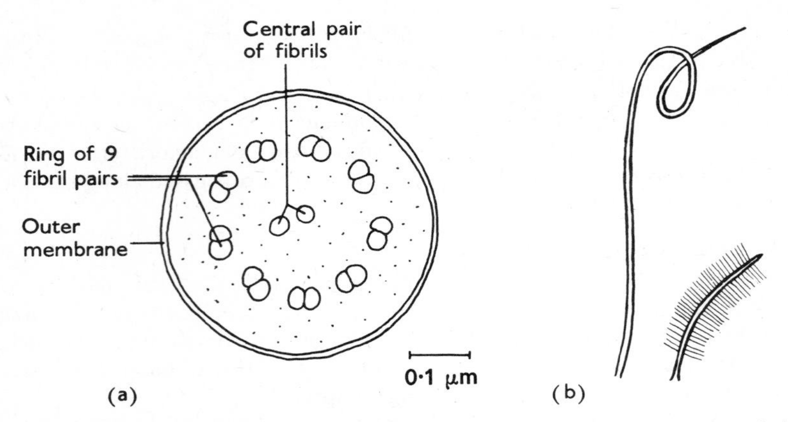

Fig. 2.9 Structure of flagellum. (**a**) Transverse section, showing characteristic 9+2 fibrillar structure. (**b**) 'Whiplash' and 'flimmer' flagella. Diagrammatic.

number, placement, and method and kind of insertion, provide considerable assistance in identifying the relationships of the algae.[59]

CHLOROPHYTA

Habitat	Aquatic (mainly freshwater), or terrestrial in moist situations.
Pigments	Chlorophylls *a* and *b*, β-carotin (α-carotin in Siphonales, γ-carotin in Charales), xanthophylls.
Food reserves	Starch, fat.
Cell wall components	Cellulose, hemicelluloses.
Reproduction	Asexual and sexual (isogamy, anisogamy and oogamy).
Growth forms	Flagellate, coccoid, colonial, filamentous, siphonaceous.
Flagella	2 or 4, whiplash; anterior, equal.

The Chlorophyta (green algae) have always attracted attention because, in respect of metabolism, photosynthetic pigments and ultrastructure, they show so much in common with the vascular plants and the bryophytes. We shall here consider representatives of nine commoner Orders, including the somewhat aberrant Charales. The classification depends upon the degree of development of the thallus, and, where present, the nature of the sexual reproduction. The shape of the chloroplast (chromatophore) also shows some taxonomic variation.

Volvocales This Order contains unicellular and colonial plants. They are widely distributed, and common in fresh water. In those species that are

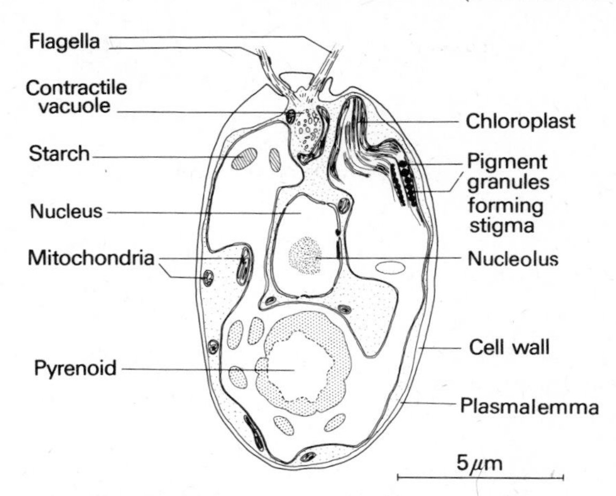

Fig. 2.10 *Chlamydomonas reinhardi.* Longitudinal section of cell. (From electron micrographs by Ursula W. Goodenough)

free-swimming, two or four whiplash flagella of equal length are attached at the anterior ends of the cells.

Unicellular forms

Because of its common occurrence, the genus frequently used to illustrate the unicellular state is *Chlamydomonas* (Fig. 2.10), of which about 400 species have been described. The cells, which rarely exceed 30 μm in major diameter, contain a single basin-shaped chloroplast, towards one side of which is a conspicuous pyrenoid. This is a proteinaceous body which acts as the centre of starch formation, and is a common feature among the Chlorophyta. A granular stigma ('eye-spot'), associated with carotenoid pigments, also occurs within the chloroplast. Although possibly photosensitive, this property is not confined to the stigma, since phototaxis persists in mutants in which this structure is lacking. The movement of *Chlamydomonas*, and of flagellates generally, does not result from random threshing of the flagella, but from regular sequences of contractions (Fig. 2.11). The two vacuoles within the cell discharge their contents at short intervals.

REPRODUCTION *Chlamydomonas* multiplies asexually by zoospores: the parent flagella disappear, and the protoplast divides to form two to eight daughter cells, which secrete cell walls and flagella before being liberated. Sexual reproduction differs in detail from species to species, indicating the possible evolutionary lines along which oogamous reproduction arose from anisogamous. In all species, a parent cell withdraws its flagella, and the protoplast divides to produce 64 or fewer motile gametes, which are usually

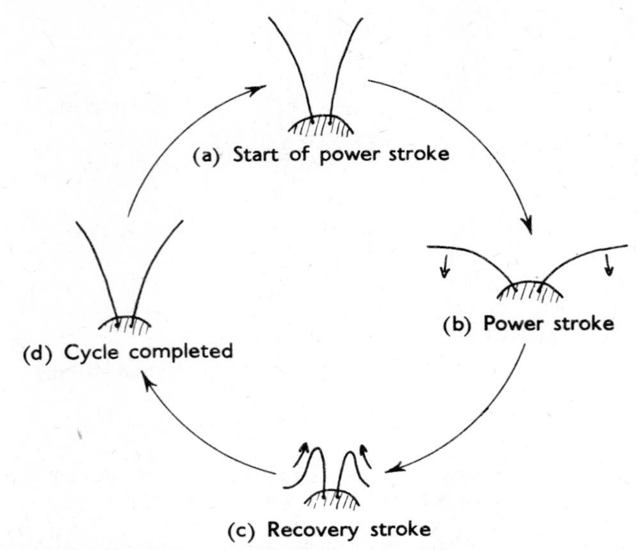

Fig. 2.11 The beating movements of paired flagella.

naked. Fusion of similar daughter gametes (isogamy), the most primitive method of combining nuclear information, occurs in only a few species of *Chlamydomonas* (Fig. 2.12) and other lowly organisms. The initial stage in

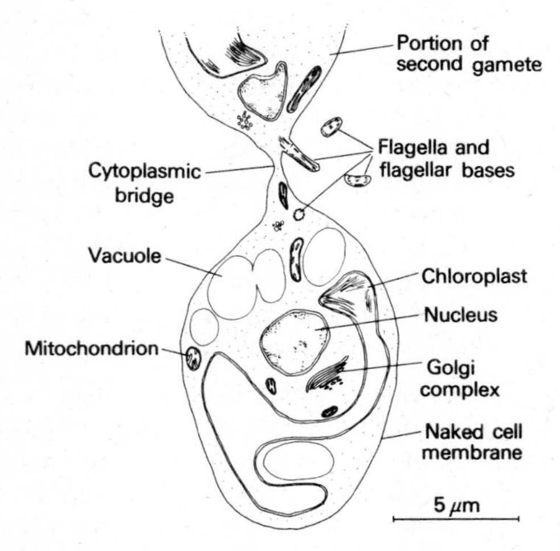

Fig. 2.12 *Chlamydomonas reinhardi.* Beginning of fusion of gametes. (From electron micrographs by Ursula W. Goodenough)

sexual differentiation is indicated in *C. moewusii* and other species in which plus (+) and minus (−) strains exist. Fusion can take place only between gametes of different strains; the species is therefore physiologically heterothallic, though not morphologically. The process of fusion has been carefully studied in *C. moewusii* in which mature cells act directly as gametes, and pairing occurs immediately on mixing cultures of different strains. The juxtaposed flagella extend laterally, and the cells become accurately aligned before a protoplasmic bridge forms between the anterior ends. During this stage, the pair continue swimming, though only one pair of flagella, that of the + strain, remains active. Once the protoplasmic bridge has formed, about 5 min after the beginning of mating, phototaxis becomes negative and continues until the cells come to rest, round off, and secrete a resistant wall. Meiosis presumably occurs sometime before the zygote germinates to release 4 zoospores. Reproduction is anisogamous in *C. braunii*; the − strain produces 4 macrogametes and the + strain 8 microgametes. The oogamous state is reached in *C. coccifera* (Figs. 2.13–2.15) in which one parent produces a single macrogamete with no flagella.

From an examination of the mode of life of *Chlamydomonas*, representative of the simplest nucleate green plants, it is clear that physiological processes of a complexity unequalled in the most advanced blue-green

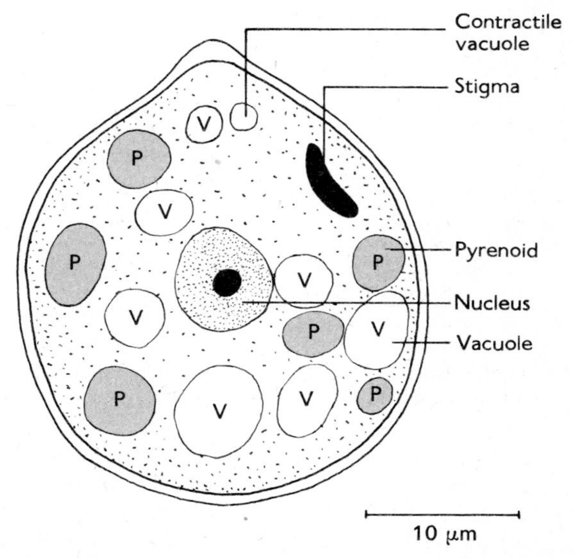

Fig. 2.13 *Chlamydomonas coccifera*. An early stage in the differentiation of the non-motile female gamete. (After Goroschankin, *Flora*, **94**, 420 (1905))

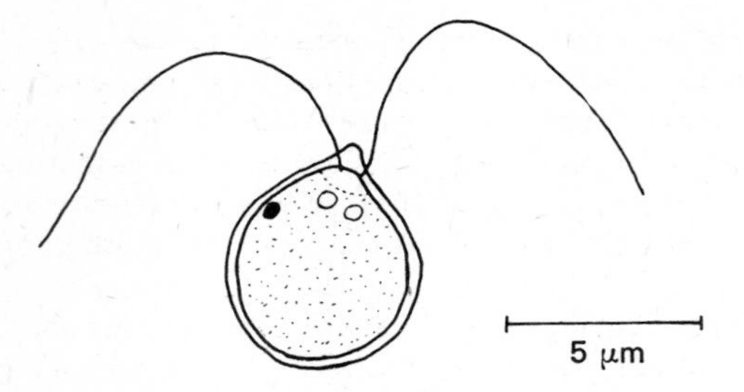

Fig. 2.14 *Chlamydomonas coccifera*. Male gamete. (After Goroschankin, Flora, **94**, 420 (1905))

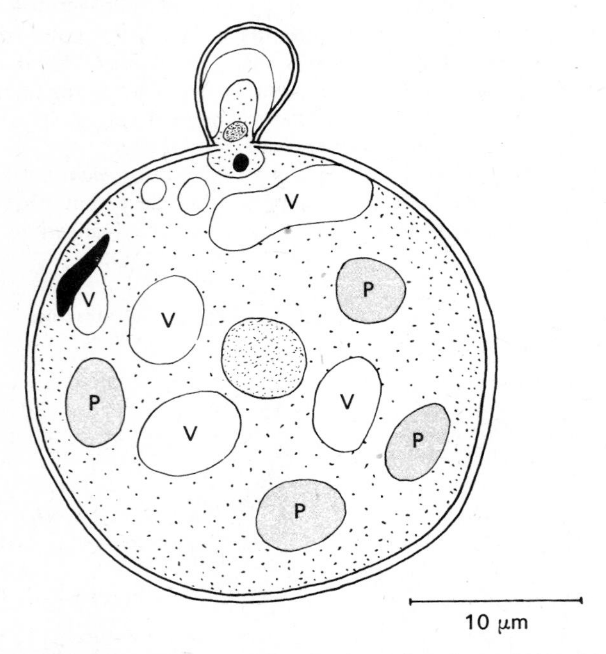

Fig. 2.15 *Chlamydomonas coccifera*. Fusion of male and female gametes. (After Goroschankin, Flora, **94**, 420 (1905))

algae must take place. Possibly the attainment of the nucleate state is an advance in organization without which others, such as the development of flagellate motility and sexual reproduction, are impossible.

Two genera which are related to *Chlamydomonas* are *Dunaliella*, which has no cell wall and is often pigmented bright orange, and the colourless *Polytoma*. Both these organisms could well be assigned to the Animal

Kingdom, while others are borderline cases between algae and fungi.[66] Some taxonomists place all unicellular motile organisms in a separate group, the Flagellatae. However, all forms with a cell wall and one or more chromatophores are readily recognized as plants, and can be assigned to an appropriate algal Class, depending upon the nature of their pigments.

Colonial forms

The Volvocales also contains motile colonies, composed of identical cells, each similar in morphology to *Chlamydomonas*. *Gonium* (Fig. 2.16), for example, has a flat plate of 4 or 16 cells (depending on the species) which are regularly arranged, and held together in a gelatinous matrix.

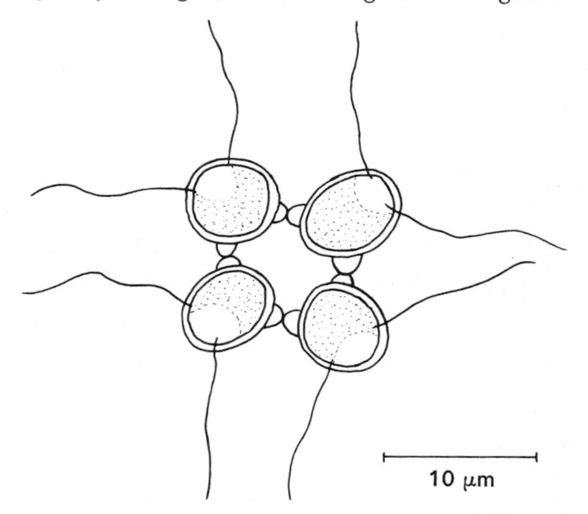

Fig. 2.16 *Gonium sociale*. (After West and Fritsch, *A Treatise on the British Freshwater Algae*. Cambridge, 1927)

The 16 cone-shaped cells of *Pandorina* (Fig. 2.17), however, are arranged in a sphere. In both asexual reproduction (in which a single cell gives rise to a new spherical colony) and gametogenesis, all the cells of these simple colonies become involved simultaneously.

Evidence for both coordination, and division of labour between cells, is seen in *Volvox* (Fig. 2.18). The number of cells in the comparatively large, spherical colonies of this genus is always a power of two, indicating co-ordinated and simultaneous cell division. The flagella beat in unison, producing a steady, rolling motion during which one side of the colony remains in front. It is significant that the cells of this portion have stigmata larger than those of the rest.

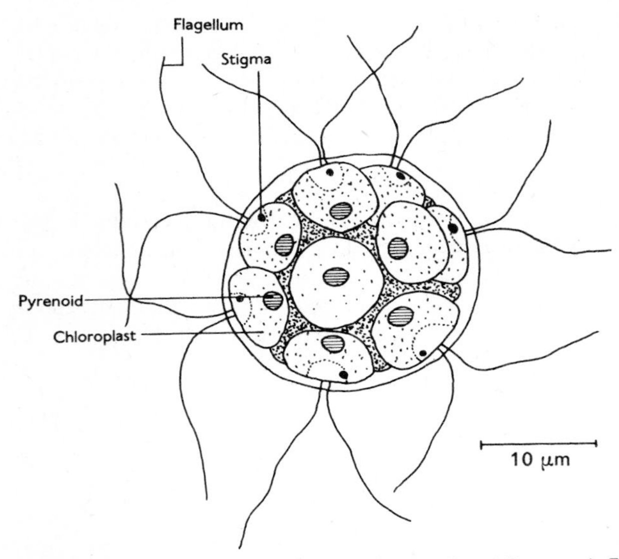

Fig. 2.17 *Pandorina morum*. Adult coenobium. (After West and Fritsch, *A Treatise on the British Freshwater Algae*. Cambridge, 1927)

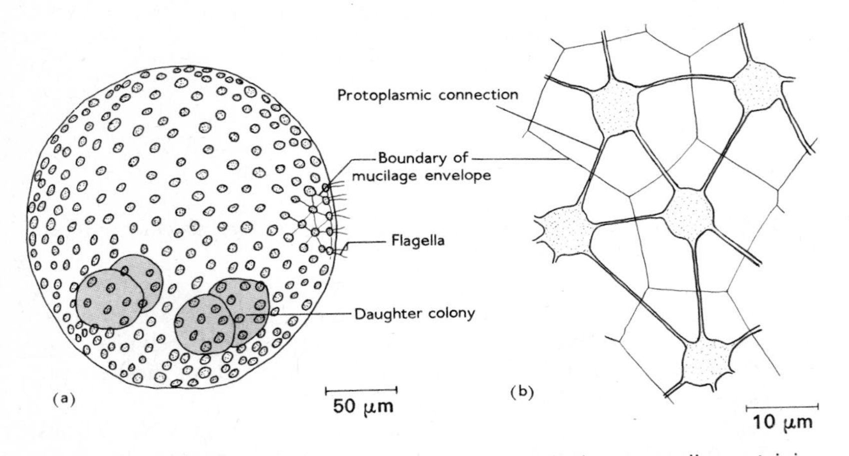

Fig. 2.18 *Volvox* sp. (**a**) Young coenobium reproducing asexually containing four daughter colonies. (**b**) Surface view of colony.

REPRODUCTION Asexual reproduction takes place by the enlargement and repetitive longitudinal division of a vegetative cell in the colony, so that eventually a new sphere of cells is liberated into the central cavity of the mother colony. Initially, the cells of the daughter colony are orientated inversely in relation to those of the mother colony, but at a certain stage of development of the new sphere it invaginates into itself and the orientation of the cells becomes reversed. This is frequently associated with the rupture of the mother colony, allowing the daughter colony to escape, often already with a grand-daughter colony within it.

Sexual reproduction in *Volvox*, in which both monoecious and dioecious species occur, is truly oogamous. The mature oogonia are large flask-shaped cells. Each is developed from a single vegetative cell, and remains in position in the colony up to and during fertilization. The same is true of the antheridia. The sexual cells may be sporadic in the colony, or, where the colony has a distinct anterior and posterior, confined to the posterior region. After fertilization the zygote (oospore) is liberated and soon germinates, presumably undergoing meiosis, to produce one or more biflagellate swarmers. Each of these divides and gives rise to a spherical colony. Those first formed are small and consist of only a few cells, but repeated asexual reproduction leads to the mature size characteristic of the species.

Volvox and related genera differ from the simpler Volvocales in possessing in their colonial thalli somatic cells which play no part in reproduction. These cells consequently perish when a mother colony breaks open to liberate a daughter colony, or when the integrity of a colony is destroyed by the liberation of a number of zygotes. Because of this evidence of organization within the thallus, many have considered the term 'colony' inappropriate in these instances, replacing it by 'coenobium'. In the simpler algae the progression from colony to coenobium is, of course, a gradual one, and, although the coenobial nature of *Volvox* is doubted by none, the status of many transitional forms is obscure.

Chlorococcales This heterogeneous Order contains those lowly green algae in which motility is confined to zoospores and gametes. Simple binary fission of individual vegetative cells does not occur. Apart from this feature, the Chlorococcales show much diversity, both in form and behaviour, indicating few interrelationships. Although most members of the Order occur in fresh waters, a few live in the ocean and moist places on land; others are endophytic in the intercellular spaces of higher plants, symbionts with lower animals (as, for example, in *Hydra*), or constituents of lichens. Genera illustrating increasing morphological complexity are *Chlorococcum* (unicellular), *Scenedesmus* (colonial) (Fig. 2.19), *Pediastrum* (coenobic) (Fig. 2.20), and *Hydrodictyon* (a colonial form with multinucleate cells) (Fig. 2.21). The unicellular alga *Chlorella*, a plant very widely used in experiments on the physiology of photosynthesis, is also a member of the Chlorococcales. Photosynthesis was originally discovered by Priestley and Ingenhousz in 1779 with this organism, but it was not

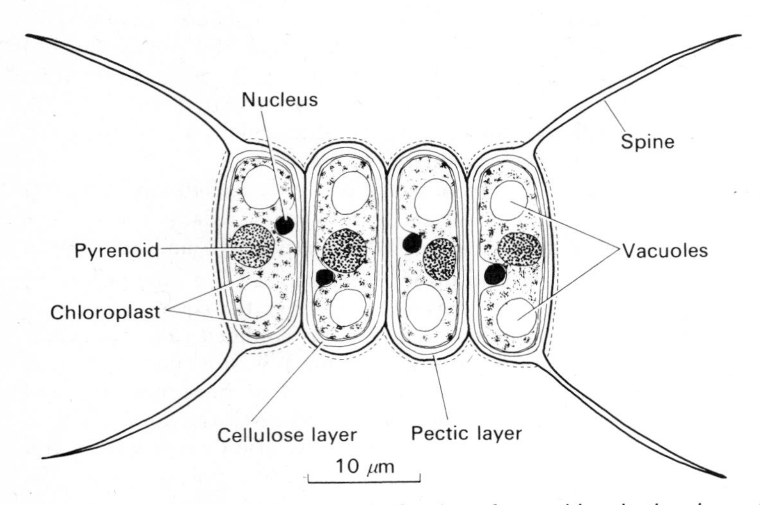

Fig. 2.19 *Scenedesmus quadricauda*. Section of coenobium in the plane of the spines, based on photo- and electron micrographs. The middle layer of the three-layered wall is micellar in structure, but its chemical nature is not certain. The spines consist of bundles of micelles emerging from the middle layer.

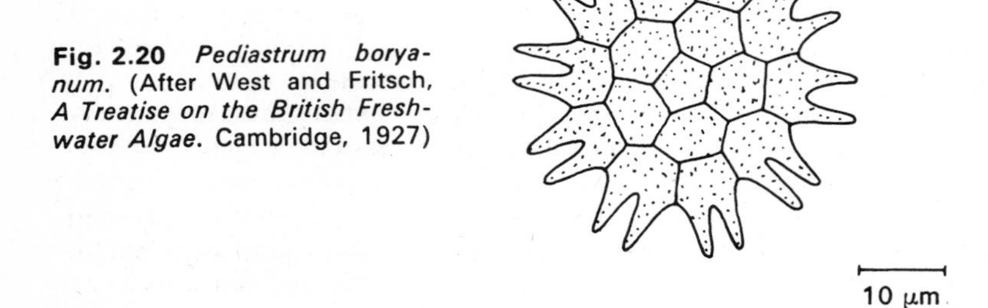

Fig. 2.20 *Pediastrum boryanum*. (After West and Fritsch, *A Treatise on the British Freshwater Algae*. Cambridge, 1927)

until 1900 that *Chlorella* was obtained in pure culture. A unicellular form very similar to *Chlorella*, but lacking pigmentation, is *Prototheca*.

REPRODUCTION Asexual reproduction takes place by spores, either motile (zoospores) or non-motile (***aplanospores***). Sexual reproduction is either isogamous or anisogamous, the gametes being flagellate and resembling the zoospores. The life cycle of *Hydrodictyon* is particularly complex. Oogamy, however, is unknown in this Order.

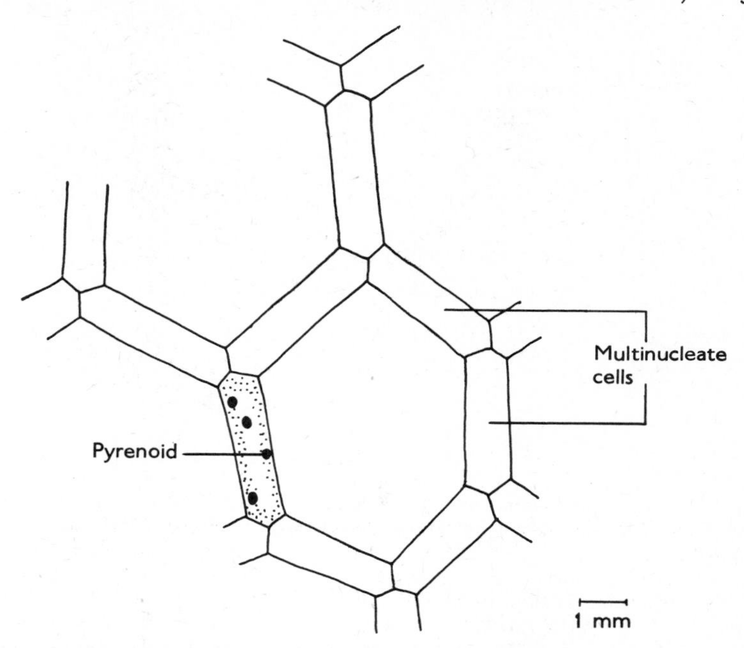

Fig. 2.21 *Hydrodictyon reticulatum*. Portion of young colony. (After West and Fritsch, *A Treatise on the British Freshwater Algae*. Cambridge, 1927)

Ulotrichales The Ulotrichales show considerable advances in organization over the algae described so far, both in vegetative complexity and in the elaboration of the life cycle. All the various forms, including the macroscopic, such as *Ulva*, can be regarded as being derived from a filamentous system. In the simpler forms sexual reproduction is isogamous, and the zygote undergoes meiosis on germination. However, in many of those genera where the thallus shows greater morphological complexity, the zygote germinates without a reduction division, and consequently yields a diploid organism. This generation ultimately produces zoospores, and it is not until these settle and enter a resting stage that meiosis occurs. The haploid generation, alone able to reproduce sexually, is thereby restored. These forms thus show a life cycle in which the two generations are morphologically identical (see p. 9).

Ulothrix

This genus is representative of the simple filamentous forms which lack branching, and in which all the cells are of equal status. Vegetative division is intercalary and all the cells are involved in reproduction, except, in young

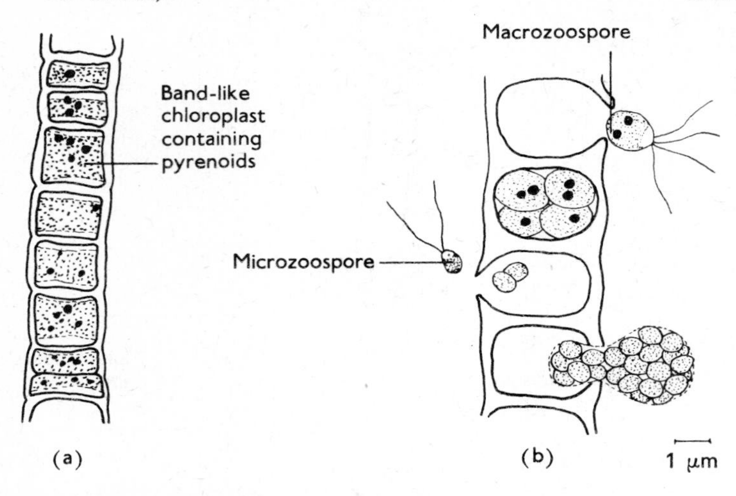

Fig. 2.22 *Ulothrix zonata*. (**a**) Vegetative filament. (**b**) Formation and release of zoospores. (After West and Fritsch, *A Treatise on the British Freshwater Algae*. Cambridge, 1927)

filaments, the basal attachment cell which is usually present. The vegetative cells, which are often wider than long, are uninucleate and have a single peripheral chloroplast (Fig. 2.22). Fragmentation of the filament is a common method of vegetative propagation, but this is caused principally by accidental breakage; simultaneous dissociation of the filament into segments has rarely been observed.

REPRODUCTION In asexual reproduction, 1 to 32 zoospores (the number depending upon the species) are produced in each cell by division of the protoplast. The mature zoospores are pear-shaped and quadriflagellate (macrozoospores), or narrowly ovoid and biflagellate (microzoospores) (Fig. 2.22b). They are liberated through a pore in the wall of the parent cell, and are surrounded by a mucilaginous sheath as they emerge. The free zoospore closely resembles a unicellular member of the Volvocales, devoid of its cell wall. After settling, the commoner macrozoospore attaches itself by the posterior end (to which the stigma has now shifted), and grows out laterally, producing a holdfast cell on one side, and new vegetative cells on the other.

Sometimes, after the initial division in the parent cell, aplanospores with resistant walls are produced instead of zoospores.

In sexual reproduction, gametogenesis resembles the production of zoospores in asexual reproduction. The gametes, however, are uniformly biflagellate, and 8, 16, 32 or 64 (the number again depending upon the species) are produced in each gametangium. *Ulothrix* is physiologically heterothallic, gametes from the same filament being unable to unite. The zygote is mobile for a short while, but then secretes a resistant wall and enters a resting period. Germination commences with a reduction division,

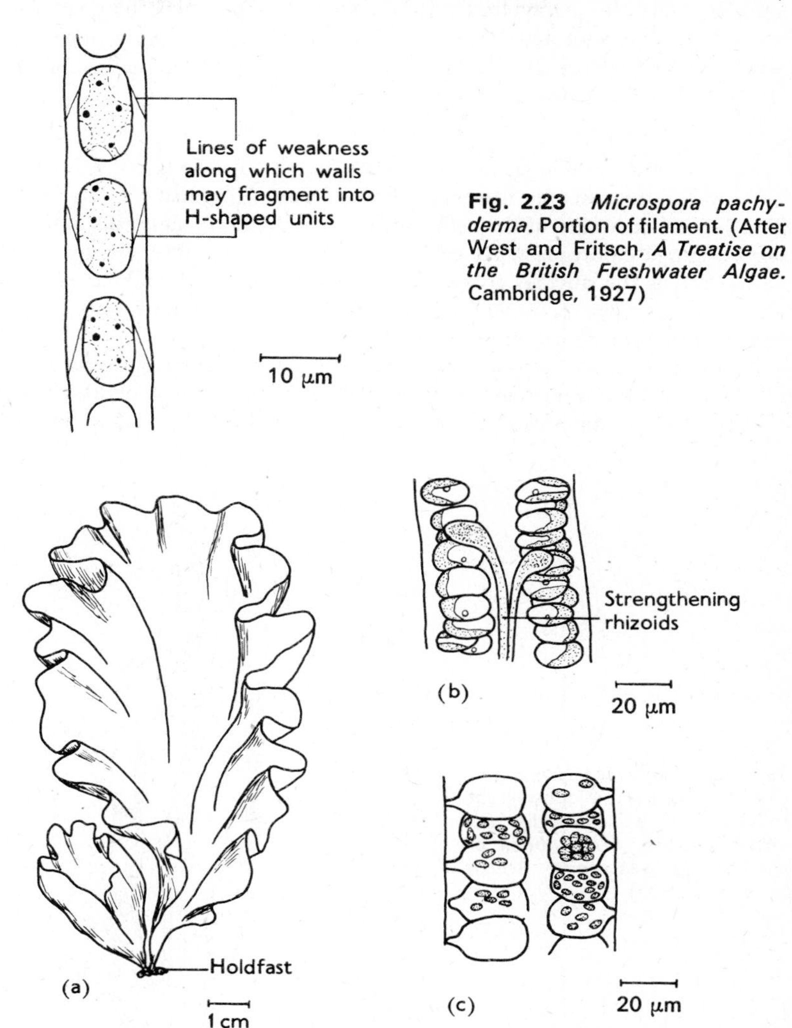

Fig. 2.23 *Microspora pachyderma*. Portion of filament. (After West and Fritsch, *A Treatise on the British Freshwater Algae*. Cambridge, 1927)

Fig. 2.24 *Ulva lactuca*. (**a**) Habit. (**b**) Transverse section of vegetative thallus. (**c**) Transverse section of thallus producing zoospores. (**a** and **c** after Newton, *A Handbook of the British Seaweeds*. British Museum, 1937), **b** after Thuret, from Fritsch, *The Structure and Reproduction of the Algae*, **I**. Cambridge, 1935)

followed by the formation of aplanospores. The diploid state is thus represented only by the zygote, and has no prolonged existence.

Forms related to *Ulothrix* are *Microspora* (Fig. 2.23), in which the cell walls disjoin into characteristic H-shaped pieces when the filament fragments, and *Cylindrocapsa*, where the cells have conspicuously thick walls and a dense chloroplast. *Cylindrocapsa* is also outstanding in showing well-developed oogamy.

Ulva

At first sight, there is little to relate the macroscopic foliaceous thallus of *Ulva* to the microscopic filaments characteristic of so many Ulotrichales. The general structure of the cells, however, is the same, and each cell contains a single chloroplast. At cell division each chloroplast divides at the same time as the nucleus. *Ulva*, in fact, is the first multicellular plant in which it has been possible to follow unambiguous chloroplast division with the electron microscope.[53a]

Young *Ulva* plants always begin their development as simple filaments. Division in three dimensions results in a flattened expanse of tissue, two cells thick, expanding from a narrow stalk and holdfast (Fig. 2.24). Some of

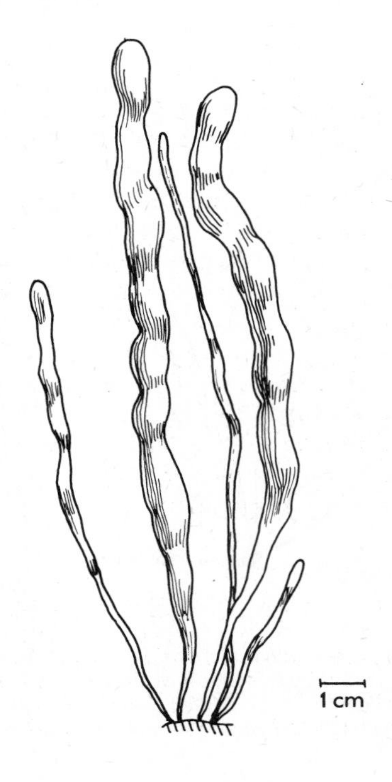

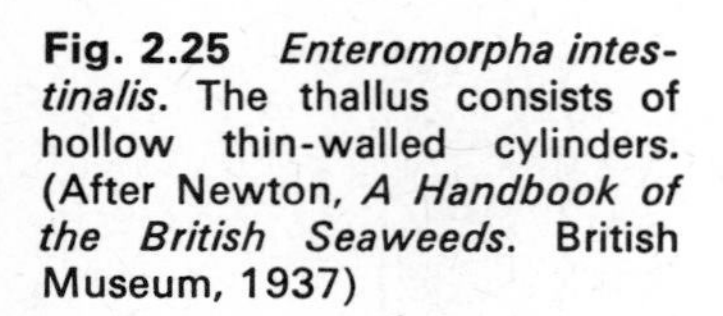

Fig. 2.25 *Enteromorpha intestinalis*. The thallus consists of hollow thin-walled cylinders. (After Newton, *A Handbook of the British Seaweeds*. British Museum, 1937)

the vegetative cells produce multinucleate rhizoids which grow between the two cell layers, strengthening the thallus. This occurs particularly in the stalk.

The marine intertidal zone is the characteristic habitat of both *Ulva* and the closely related *Enteromorpha* (Fig. 2.25). Both species can, however, tolerate wide variations in salinity, and may be found far up tidal rivers.

REPRODUCTION Asexual reproduction occurs as in *Ulothrix*, most of the vegetative cells taking part. After the zoospores have been discharged, the parent thallus remains as a bleached framework of empty cells.

Sexual reproduction is again similar to that of *Ulothrix*, and *Ulva* is also physiologically heterothallic. The zygote, however, does not undergo reduction division on germination, but grows instead into a diploid thallus identical with that of the haploid plant. Presumably reduction division occurs during the production of zoospores in the diploid generation, but this remains to be demonstrated.

Enteromorpha differs from *Ulva* in having a peculiar tubular thallus, and anisogamous sexual reproduction.

The origin of the isomorphic life cycle

There are no intermediate forms in the Ulotrichales indicating how an isomorphic life cycle, such as that of *Ulva*, might have originated. It is possible that a relatively simple mutation prevented meiosis at the zygotic stage, and that the inhibition remained effective until after many cell generations. Since we have no reason to believe that chromosome number in itself determines the kind of growth in any group of plants, this delaying of meiosis would have allowed the development of a diploid thallus closely resembling that of the haploid. In the evolution of *Ulva* the emergence of a distinct sporophyte generation was possibly accompanied by the progressive elaboration of the vegetative structure of both generations.

Cladophorales This Order, consisting predominantly of branched filamentous forms, has affinities with both the Ulotrichales and, in respect of the multinucleate cells, with the Siphonales. Nevertheless, the Cladophorales have sufficient distinctive characters of their own to merit treatment as a separate Order.

Cladophora

Of the 160 or so species of this genus, some are marine and others freshwater. The filaments show true branching, buds developing towards the anterior end of the elongated, cylindrical vegetative cells (Fig. 2.26). The internal structure is complex, each cell being multinucleate, and having what has been described as a 'reticulate chloroplast'. It is still uncertain whether this consists of one large chloroplast, or whether it is some kind of aggregation of numerous small chloroplasts. The cell wall comprises three layers, the inner of cellulose, a central layer of hemicellulose, and finally an outer coating, possibly chitinous, which gives the alga its characteristic 'crisp' feeling. Division of the protoplast, which is not regularly related to nuclear division, is accompanied by the formation of transverse septa. These develop from the margin towards the centre, and gradually acquire a complicated lamellate structure at the periphery, which may impart some flexibility to the older filaments. Growth of the filaments is apical in *Cladophora*, but intercalary growth is general elsewhere.

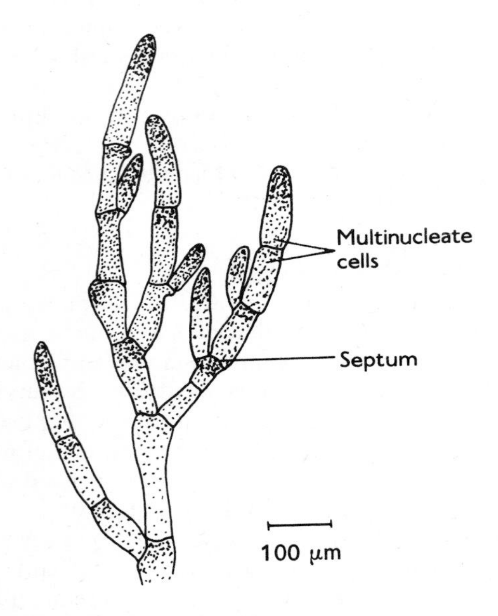

Fig. 2.26 *Cladophora* sp. Upper part of thallus showing characteristic branching and the greater density of the cytoplasm towards the apical ends of the cells.

A very curious feature exhibited by some species of *Cladophora* and certain other genera is the tendency, when subjected to a gentle rolling motion in water, to aggregate into cushions or spheres. These are termed 'aegagropilous' species because of the fancied resemblance of these aggregates to the balls of wool found in goats' stomachs. *Cladophorella*, a plant of damp places, is of interest because its upper cells secrete what appears to be a true cuticle, otherwise unknown in the algae.

REPRODUCTION Both the asexual and sexual reproduction of the Cladophorales resemble that of *Ulothrix*. The zoospores and gametes are produced in non-specialized cells, and copulation is usually isogamous, rarely anisogamous. An isomorphic life cycle has been demonstrated in some species, and may be general in the sexually reproducing forms.

Vegetative reproduction by fragmentation of branches also occurs, and seems to be the sole means of reproduction of the aegagropilous species. In some species survival through unfavourable periods is afforded by the formation of thick-walled resting spores packed with food reserves.

Chaetophorales A distinct advance in the organization of a filamentous thallus is found in the Chaetophorales where the thallus is composed of both prostrate and erect components, and is consequently termed ***heterotrichous***. The prostrate system is typically a flat plate attached to the substratum, from which arises the branched erect system bearing the reproductive organs and often characteristic hairs. Although the basic structure of two different filamentous systems is always detectable, many genera show greater development of one component than of the other, sometimes almost to its exclusion. The maintenance of the dominance of one system has been shown in some species to depend upon environmental factors. In *Stigeoclonium*, for example, magnesium depresses the growth of the erect system, while an excess of nitrogen inhibits that of the prostrate component.

Environmental factors, as well as affecting the quantitative relationships of the two systems, also influence the extent to which the erect system produces hairs and branches. This morphological plasticity, coupled with the ability of many of the Chaetophorales to grow on damp soil (e.g. *Fritschiella*) or as epiphytes, has led to the view that algae of this kind played an important role in the evolution of terrestrial vegetation. Furthermore, the heteromorphic life cycle of our most primitive land plants may have been derived by the two components of an isomorphic heterotrichous form having become differently developed in the two generations.[38] The erect system, for example, possibly became associated with the diploid state of the nucleus, and hence the sporophyte, and the prostrate with the haploid gametophyte. Unfortunately, however, little is yet known about sexual reproduction in the living Chaetophorales, and the nature of their life cycles is correspondingly obscure.

Stigeoclonium

This genus is representative of those members of the Chaetophorales in which both the prostrate and erect components of the thallus are easily recognizable, and each is more or less well developed. The species differ quite widely in external morphology. In general the erect system ends in long, thin, hyaline hairs, and is less branched than the prostrate system

(Fig. 2.27). The latter often forms a pseudo-parenchymatous sheet as a consequence of the close packing of the branches, but detailed investigation of its structure in natural habitats is rendered difficult by the tenacity with which it adheres to the substratum. The vegetative cells contain a peripheral girdle-shaped chloroplast with one or more pyrenoids.

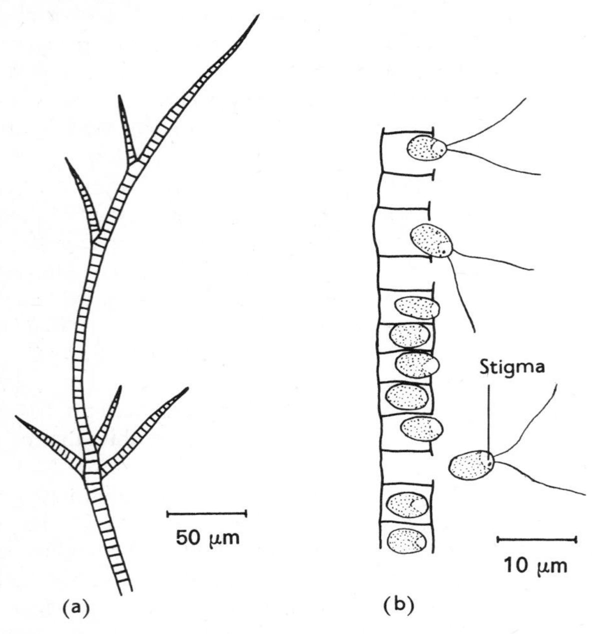

Fig. 2.27 *Stigeoclonium tenue.* (**a**) Terminal portion of erect system. (**b**) Release of gametes. (After West and Fritsch, *A Treatise on the British Freshwater Algae.* Cambridge, 1927)

REPRODUCTION The asexual and sexual reproduction of *Stigeoclonium* resemble that of *Ulothrix*. Production of zoospores or gametes is confined to the erect system, and occurs simultaneously in a large number of cells. The gametes are noticeably smaller than the zoospores, and copulation is isogamous. Sometimes, however, flagellate cells identical in all but behaviour with gametes have been observed to act as zoospores and yield thick-walled resting spores. The site of meiosis in the sexual cycle has not yet been determined. The germination of the various reproductive bodies usually leads first to the development of the prostrate system, and then the erect.

Allied to *Stigeoclonium* is the terrestrial alga *Fritschiella* in which the prostrate system, buried in damp mud, produces nodules of cells which serve as perennating organs.

Draparnaldia and Draparnaldiopsis

In *Draparnaldia* and its close relation *Draparnaldiopsis*, the erect system is dominant, and in the latter genus attains its greatest complexity in the whole of the Chlorophyta. In both genera the erect system shows evidence of division of labour.

The upright axes of *Draparnaldia* consist of large barrel-shaped cells, from which arise highly branched whorls of laterals with much smaller cells (Fig. 2.28). Frequently these laterals are so profusely developed that the axis is quite obscured. When compared with those of the cells of the laterals, the chloroplasts of the axial cells are few, and frequently occupy only a small part of the cell surface.

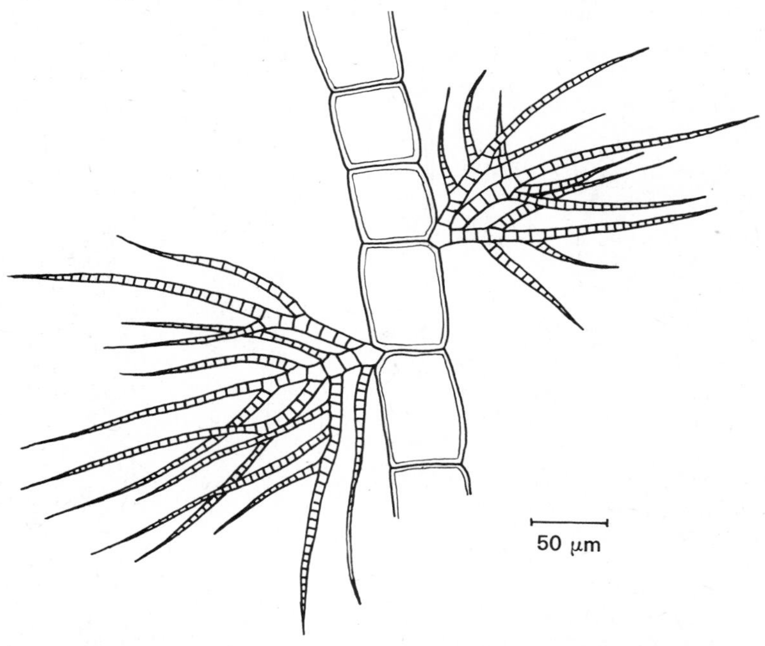

Fig. 2.28 *Draparnaldia glomerata*. Portion of erect system. (After West and Fritsch, *A Treatise on the British Freshwater Algae*. Cambridge, 1927)

Draparnaldiopsis shows further advances in vegetative organization. The main axes consist of alternating long and short cells, and laterals; some of them, of unlimited growth, are produced only from the latter. The shorter laterals show a tendency to behave like rhizoids and to coil around the long axis. In some species this is so marked that the laterals form a sheath around the main axis that is thicker than the axis itself.

In both genera the prostrate system is vestigial, and it is represented by a holdfast, the function of which is assisted by the outgrowth of rhizoids from adjacent cells of the main axis.

Reproduction of the two genera, similar to that of *Stigeoclonium*, is confined to the lateral branches.

Aphanochaete (Fig. 2.29)

This genus contains a number of common epiphytes, and is representative of those members of the Chaetophorales in which the prostrate component is dominant. In vegetative structure it resembles the prostrate system of *Stigeoclonium*, long unicellular hairs being the only traces of the erect system. Sexual reproduction is anisogamous.

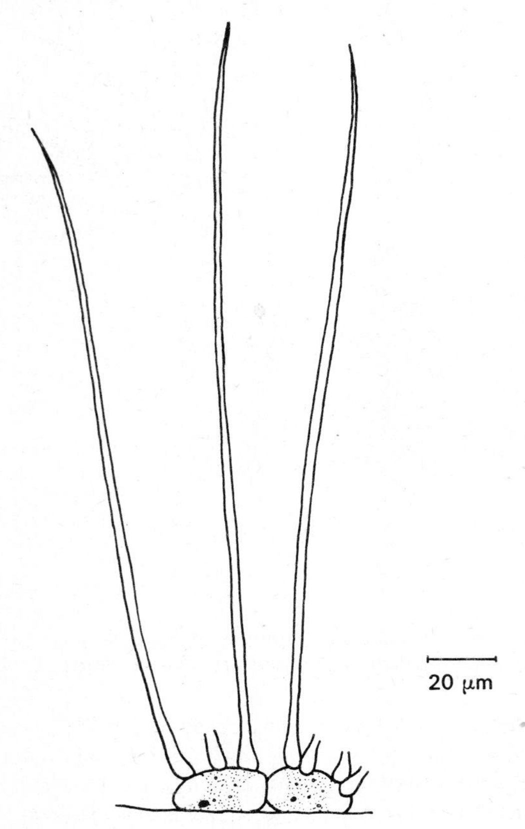

Fig. 2.29 *Aphanochaete polychaete.* (After West and Fritsch, *A Treatise on the British Freshwater Algae*. Cambridge, 1927)

Coleochaete

Many freshwater epiphytes are referable to *Coleochaete* (Fig. 2.30). Here the prostrate system is again generally dominant, although in a few species the habit is more typically heterotrichous. In *C. pulvinata*, for example, the erect system is well developed and forms a hemispherical cushion. Sheathed bristles are a unique and characteristic feature of the thallus of this genus.

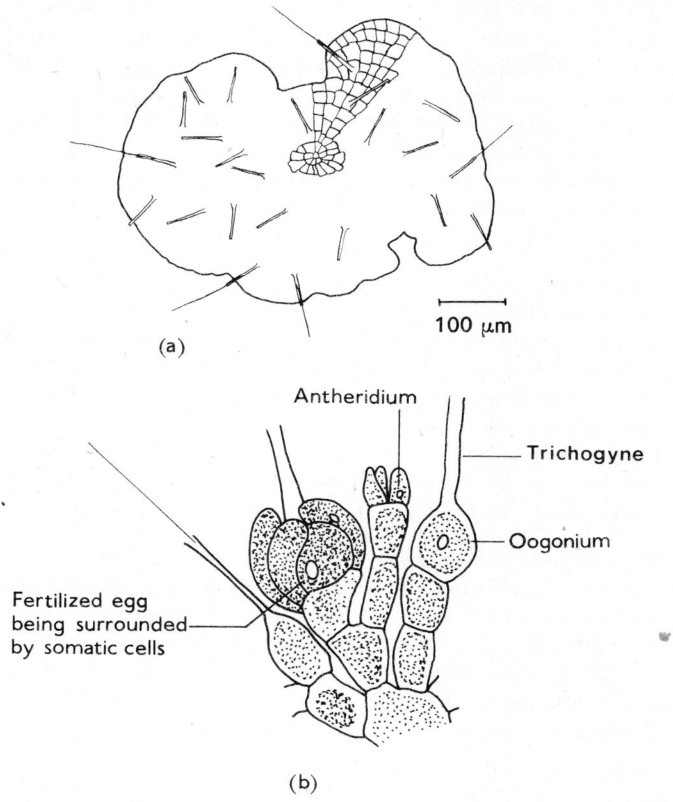

Fig. 2.30 *Coleochaete*. (**a**) *C. sentata*. Surface view. (**b**) *C. pulvinata*. Sexual reproductive structures. (**a** After West and Fritsch, *A Treatise on the British Freshwater Algae*. Cambridge, 1927 ; **b** After Pringsheim, from Fritsch, *The Structure and Reproduction of the Algae*, **I**. Cambridge, 1935)

REPRODUCTION *Coleochaete* is also outstanding in its sexual reproduction. It is alone amongst the Chaetophorales in being oogamous, and the structures associated with the egg are highly specialized. The cells which differentiate into oogonia terminate short branches, although they may ultimately appear lateral because of continued growth from the penultimate vegetative cell. The oogonium develops a long neck, termed a trichogyne (Fig. 2.30b), which opens at the tip when the egg is ready for fertilization. The antheridia arise in the filaments, often just above the oogonium, and each produces a single antherozoid. On liberation, they are attracted, presumably chemotactically, to the receptive spot at the tip of the trichogyne, and fertilization follows.

As a consequence of fertilization, but in a manner still unknown, the vegetative cells adjacent to the oogonium are stimulated into growth and they envelop the zygote in a continuous parenchymatous sheath. This sheath eventually dies, but its cells appear to contribute to a thick membrane which forms around the resting zygote. After a resting period, the zygote germinates to give rise to a plate of 16 or 32 wedge-shaped cells, each of which liberates a zoospore.

Trentepohlia

Adaptation to its terrestrial habitat is the most noteworthy feature of *Trentepohlia*. The thallus, which is differentiated into prostrate and erect components, grows attached to rocks and bark, as well as on soil and humus. It is common in warm, moist situations, and is conspicuous for its orange, rather than green, appearance. The cell walls are thick and often layered, and this feature probably accounts for the ability of the alga to resist desiccation. If growth is inhibited, as in conditions of drought, there is a marked increase in the pigmentation of the cells, and concurrently an accumulation of fat, in which the pigment (haematochrome) is dissolved.

REPRODUCTION Asexual reproduction is by zoospores, and takes place only in damp conditions. Several morphologically distinct types of zoosporangia have been described, and *Trentepohlia* appears to be advanced in respect of this feature. In dry conditions intact sporangia may be distributed by wind, and even purely vegetative reproduction occurs by the dispersal of single cells or groups of cells from the prostrate system. Isogamous sexual reproduction has been observed in some species.

Pleurococcus (*Desmococcus*)

This genus contains what is probably the commonest green alga, *P. vulgaris*, familiar as a friable incrustation on the windward side of walls and tree-trunks. *Pleurococcus*, which consists of small complanate aggregations of cells, must be regarded as a very reduced member of the Chaetophorales, and it is assigned to the Order on the basis of its cytology. It has no known method of reproduction other than simple cell division.

The relationships of the Chaetophorales

The morphological diversity of the Chaetophorales, and the lack of knowledge about their life cycles, make it difficult to assess their relationships with other green algae. Rather isolated specialized types, such as *Draparnaldia* and *Coleochaete*, are a conspicuous feature of the Order. They may indicate that the present Chaetophorales are the diverse relics of a kind of algal organization which became terrestrial, and evolved into forms which are now identified with land plants.

Oedogoniales The Oedogoniales are a well-defined Order, possessing several unique features. Some have attached so much weight to these as to consider that the Order should be removed from the Chlorophyta. The Oedogoniales are not, however, anomalous in such fundamentals as wall structure and pigmentation, and they are probably better regarded as a small group that has diverged from the main line of evolution, the intermediate stages being no longer represented. The Oedogoniales comprise only three genera, *Oedogonium*, *Oedocladium* and *Bulbochaete*. Of these, *Oedogonium* is by far the commonest and best known.

Oedogonium

THE ORGANIZATION OF THE FILAMENT The thallus of *Oedogonium* is an unbranched filament (Fig. 2.31). When young, the filaments are attached

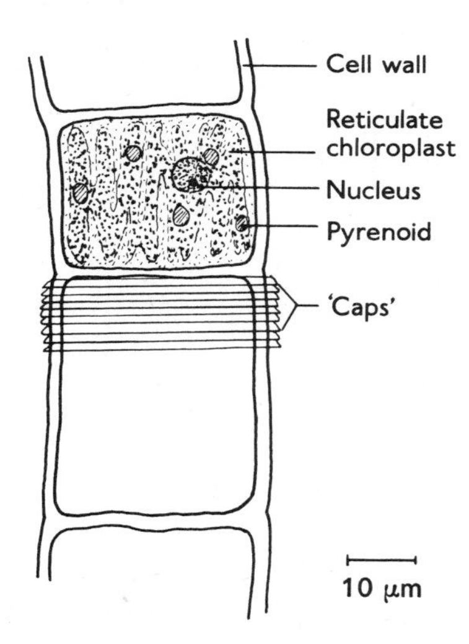

Fig. 2.31 *Oedogonium* sp. Portion of filament.

by a basal holdfast, but unless the water they inhabit is flowing, the mature condition is free-floating. The individual cells have a single, large nucleus, and a reticulate chloroplast, rather reminiscent of the Cladophorales. Pyrenoids are frequent at the interstices of the reticulum. Cell division is normally intercalary, but the way in which the new cell walls are produced is highly peculiar (Fig. 2.32). The first indication that a cell is about to divide is the formation of a ring of wall material towards the upper end of the cell, just below the septum. The nucleus then divides and a septum forms between the daughter nuclei but this septum remains free at the periphery. During the division of the nucleus the ring in the upper part of the cell becomes larger and crescent-shaped in vertical section. Eventually the cell wall breaks transversely at this level, and the ring is drawn out longitudinally to form a cylinder of new wall material. Meanwhile the septum between the nuclei moves up the cell, reaches the bottom of the newly formed cylinder, and then fuses peripherally with the longitudinal walls. The lower cell thus has a wall largely identical with that of the mother cell, while the wall of the upper, except for a conspicuous cap of original wall at the anterior end, is wholly new. The presence of caps, which if the daughter cell goes on dividing may be several in number, at the anterior ends of the cells is a feature diagnostic of the Oedogoniales.

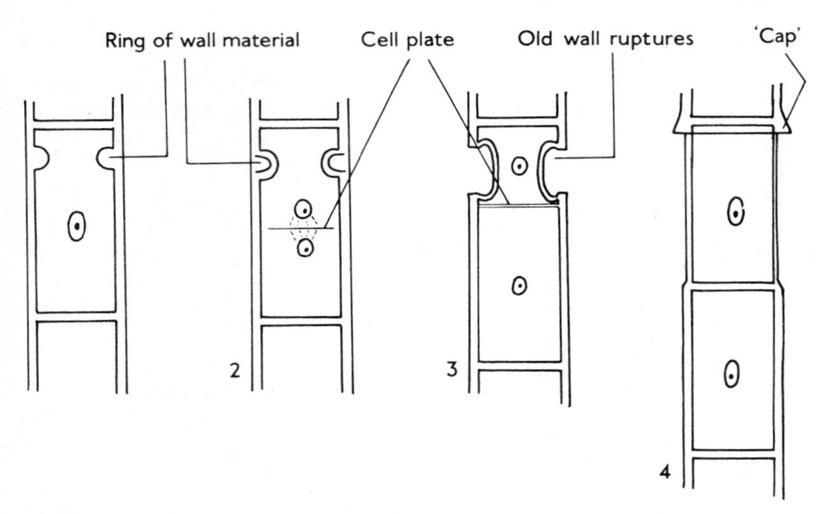

Fig. 2.32 *Oedogonium*. Stages in cell division. Diagrammatic.

REPRODUCTION Both asexual and sexual reproduction occur in *Oedogonium*. Asexual reproduction is by means of zoospores. These arise singly in vegetative cells, and are pear-shaped or almost spherical. Instead of having a small number of flagella arising together, the zoospores of the Oedogoniales are unique in possessing an apical ring of many short flagella (Fig. 2.33a). On liberation, the swarmers are surrounded by a transient mucilage sheath, but this soon disappears as the zoospores become motile. On reaching a substratum, they become affixed at the anterior end, and a new filament is formed.

Both monoecious and dioecious species of *Oedogonium* are known; some of the dioecious species exhibit a difference of morphology between the male and female filaments (dimorphism). In the monoecious, and isomorphic dioecious species, large, almost spherical oogonia are produced from the upper cell of a vegetative division. The lower cell may divide again, and another oogonium develop, but more usually it remains as the supporting cell. A single, dense, oosphere is formed in each oogonium (Fig. 2.33b). The oosphere shrinks away from the cell wall, and develops a colourless receptive spot as it matures. When it is fully mature, a pore appears at the anterior end of the oogonium opposite the receptive spot of the oosphere. Meanwhile, antheridia are produced by division of a vegeta-

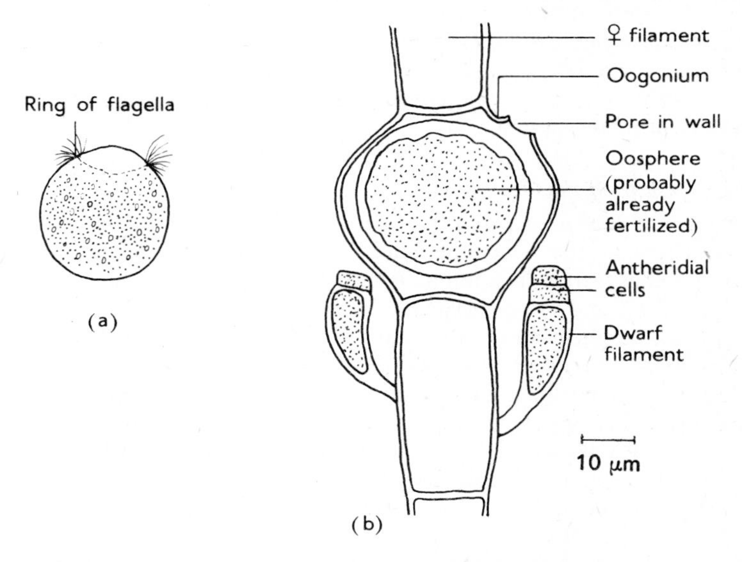

Fig. 2.33 *Oedogonium* sp. (**a**) Zoospore. (**b**) Oogonium with adjacent dwarf males.

tive cell into several disc-shaped portions, each of which normally produces two male gametes. These are similar, both in morphology and method of liberation, to the zoospores. After fertilization, the zygote (oospore) secretes a thick wall and enters a resting stage. Although oospores are sometimes able to germinate immediately, a long resting period seems generally to be necessary. There is some evidence that chilling hastens germination. When growth is eventually resumed, the normal course is for four haploid protoplasts to be extruded, each of which develops a crown of flagella, swims away, and settles to produce a new plant.

In the case of the dimorphic (or nannandrous) species, the female filaments are of the usual size, but the males are much smaller. These dwarf males arise from the female plants by way of a special propagule called an androspore. These are produced in a manner similar to that of the male gametes, except that only one androspore is produced per cell. Resembling zoospores, except for their smaller size, and yellowish colour, the androspores come to settle either on the oogonium or its supporting cell, and give rise to a small filament (Fig. 2.33b). After a few cells have been produced, antheridia are cut off, and reproduction proceeds as in the isomorphic (macrandrous) species.

Bulbochaete and Oedocladium

Bulbochaete resembles *Oedogonium* in essentials, but the filaments are branched and terminate in hairs, similar to those of the Chaetophorales. *Oedocladium* differs from *Oedogonium* in having both branched filaments and a heterotrichous system.

The relationships of the Oedogoniales

The hairs of *Bulbochaete* and the heterotrichous system of *Oedocladium* recall features of the Chaetophorales, and perhaps indicate a distant affinity. Another feature in common is the peripheral girdle-shaped chloroplast in each cell, and the small gamete-like zoospores of some Chaetophorales may indicate the origin of the nannandry of some Oedogoniales.

Conjugales Members of this clearly defined Order are readily recognized by the striking and characteristic symmetry of their cells, by the elaborate chloroplast, and the lack of free-swimming motile forms. Conjugation, which takes the form of fusion of amoeboid gametes, gives the group its name, and is a further diagnostic feature.

The Order includes plants of two different growth forms, the filamentous and the unicellular. The unicellular forms (or desmids) are again divisible into two series, depending upon whether the cell wall is of one piece and devoid of pores (the saccoderm desmids), or of two (or rarely more) pieces and usually porous (the placoderm desmids). The filamentous forms and saccoderm desmids are included in one sub-Order (Euconjugatae), and the placoderm desmids placed in a second (Desmidioideae).

The filamentous forms

The thalli of these forms consist of unbranched filaments, frequently with basal attachment cells in the young state. The cells are often markedly elongate. The nucleus lies somewhere near the centre, and the conspicuous chloroplast is in the form of a flat plate (e.g. *Mougeotia*), a helical band at the periphery of the cell (e.g. *Spirogyra* (Fig. 2.34)), or consists of two stellate portions (e.g. *Zygnema*). Numerous pyrenoids occur in each form of chloroplast.

Growth of the filament is intercalary, all the cells except the holdfast being capable of division. Almost all the filamentous Conjugales are confined to fresh waters.

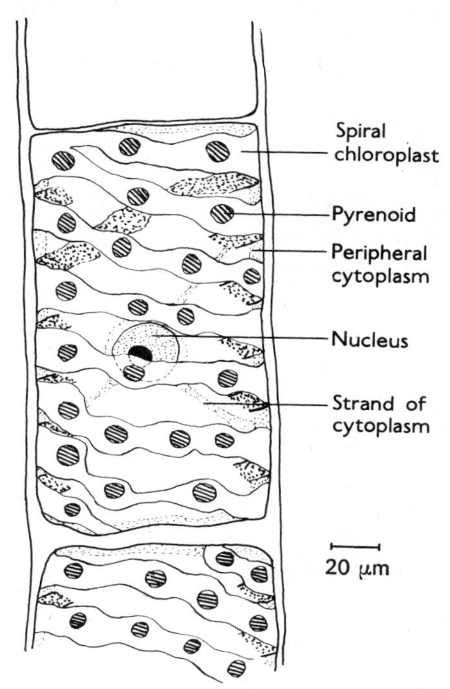

Fig. 2.34 *Spirogyra* sp. Cytology of vegetative cell.

ASEXUAL REPRODUCTION The only method of asexual reproduction is by fragmentation of the filament, and zoospores are never produced. Definite changes occur in the transverse walls preceding fragmentation, but the significance of these changes is not fully understood. In some species the middle lamella of the transverse wall becomes gelatinous, and a turgor difference arises between adjacent cells. This causes one cell to bulge into the other, and the junction between the cells becomes strained to breaking point. In other species a ring or collar of wall material forms on each side of the transverse wall before fracture takes place.

SEXUAL REPRODUCTION Sexual conjugation occurs normally towards the end of the growing season, but factors such as nitrogen deficiency, pH, and altitude have all been found to influence it. Only in *Spirogyra* has conjugation been investigated in detail, but it appears that the process is similar in other genera (Fig. 2.35).

Before conjugation, the two participating filaments must come to lie parallel and close together. The necessary adjusting movements are probably brought about by the localized secretion of mucilage. Since solitary

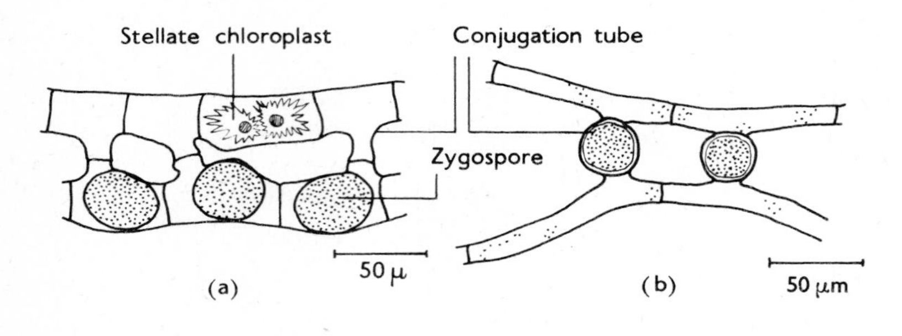

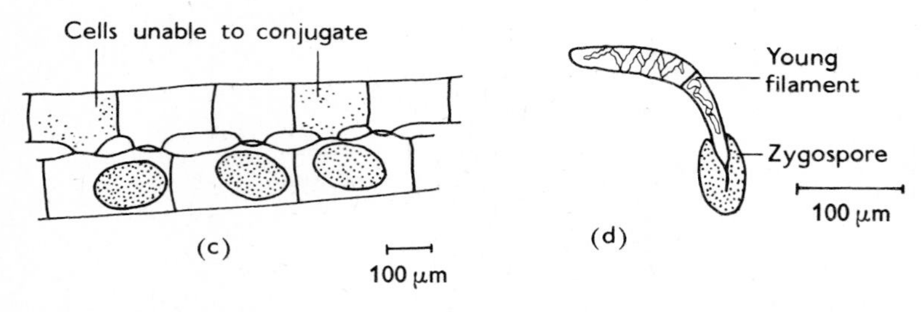

Fig. 2.35 Sexual reproduction in the Conjugales. (**a**) *Zygnema stellinium.* (**b**) *Mougeotia parvula.* (**c**) *Spirogyra setiformis.* (**d**) *Spirogyra velata.* Germination of zygospore. (All after West and Fritsch, *A Treatise on the British Freshwater Algae.* Cambridge, 1927)

filaments do not move at all, the presence of others apparently acts as a stimulus. Once alignment has taken place, further mucilage secretion holds the filaments in position and one cell forms a short protuberance towards the other. On touching the opposite cell this protuberance stimulates the production of a similar structure, and the growth of the two protuberances against each other forces the filaments apart. Eventually growth of the protuberances ceases, the opposed distal walls dissolve and a conjugation tube is formed between the two cells. Meanwhile, complex physiological changes have been taking place inside the protoplasts. There is an increase in starch, and a decrease in permeability and in the osmotic pressure of the cell sap. Before completion of the conjugation tube no sexual differentiation of the two protoplasts is apparent, but as soon as communication is established one protoplast withdraws from its wall, and slowly passes through the tube. Only after the arrival of this male protoplast does the female contract, and a zygote form.

The process of conjugation is subject to a number of variations. In *Spirogyra*, for example, adjacent cells of the same filament may conjugate. In *Zygnema* and *Mougeotia* the zygote is often formed in the conjugation tube. Sexual differentiation of a morphological nature is seen in *Sirogonium*. Here the male gametangium is always associated with two sterile cells, whereas the female has only one.

The zygote becomes a thick-walled zygospore and enters a period of dormancy. The germinating spore gives rise to a single filament. In those species investigated, meiosis has been shown to occur before germination, but three nuclei degenerate.

The saccoderm desmids

These simple unicellular organisms are probably the most primitive of the Conjugales. The cells, which are always circular in transverse section, are ellipsoidal or rod-shaped (Fig. 2.36), and the cell walls lack ornamentation and constrictions. The chloroplasts are plate-like, stellate or in the form of helically wound bands, as in the filamentous Conjugales. Many saccoderm desmids occur in the acidic waters of upland pools and peat bogs.

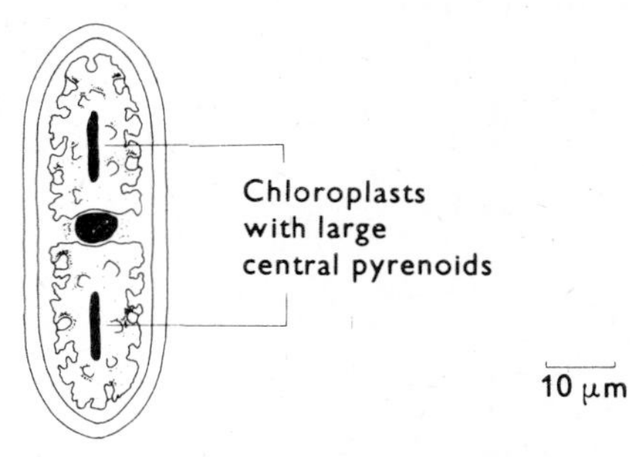

Fig. 2.36 *Cylindrocystis brebissonii.* The two prominent stellate chloroplasts are symmetrically placed in the cell, and the nucleus lies between them.

REPRODUCTION Reproduction occurs principally by binary fission, but conjugation has been observed in most genera. In some forms the conjugation tube is lacking, and fusion of the protoplasts occurs in mucilage. Isogamy is the general rule. Meiosis and germination of the zygote are deferred until after a resting period.

The placoderm desmids

The placoderm desmids, much more numerous than the saccoderm, are an important constituent of phytoplankton, and are commonly thought of as the 'true' desmids. They are unicellular, but the cells have complex

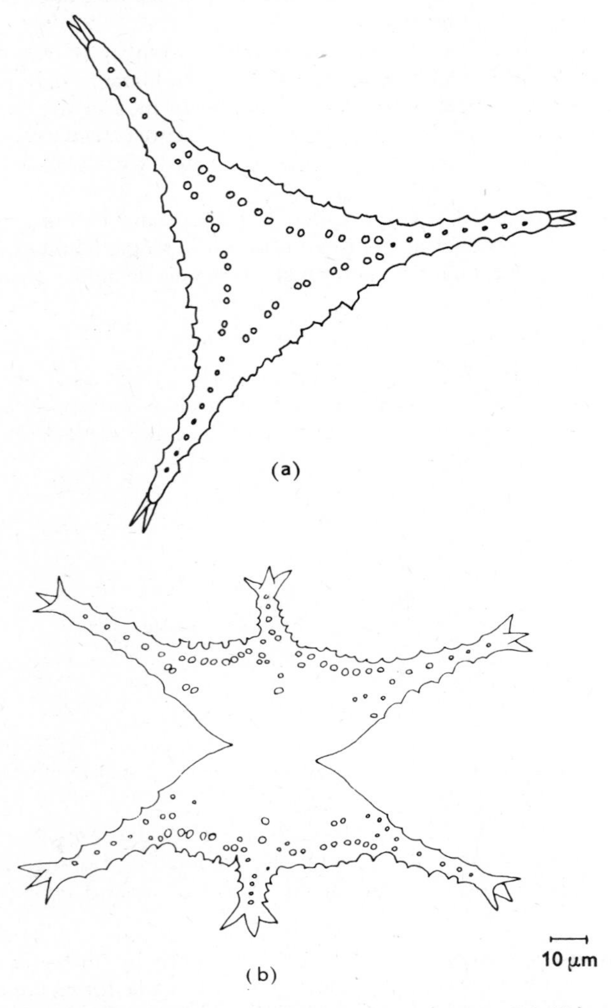

Fig. 2.37 *Staurastrum anatinum*. (**a**) View from above. (**b**) View from side. (After West and Fritsch, *A Treatise on the British Freshwater Algae*. Cambridge, 1927)

shapes and a precise and striking symmetry. Typically, each cell is divided into two halves, one being the mirror-image of the other, connected by a narrow central portion, the isthmus. The cells of some species form filamentous colonies, but more usually they are solitary.

The cell wall is made up of an inner cellulose layer, and an outer layer of variable composition, frequently containing iron compounds or silica. It is principally this outer layer, often patterned with spines, and other protuberances, which provides the specific characteristics, and puts the desmids among the most beautiful of microscopic objects (Figs. 2.37 and 2.38). The cells are frequently surrounded by an investment of mucilage,

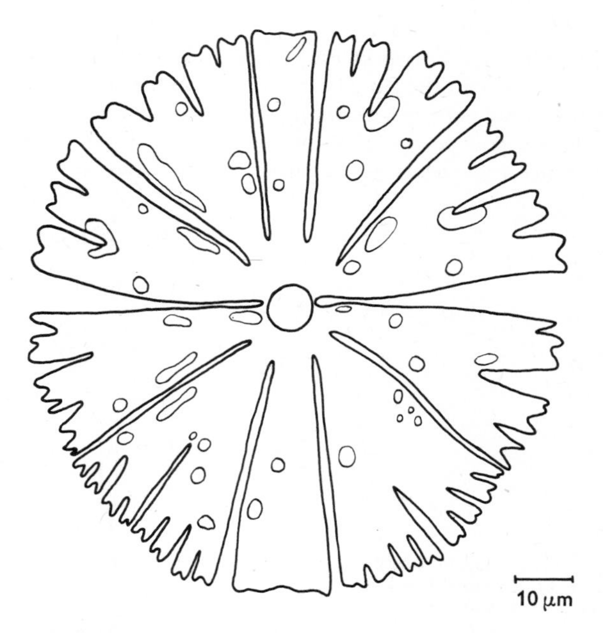

Fig. 2.38 *Micrasterias* sp. from Windermere.

secreted by minute pores in the wall. It is probably by the localized secretion of such mucilage that cells are able to perform slow movements.

Within the cell, the nucleus usually lies in the isthmus, and one or more chloroplasts in each half-cell.

REPRODUCTION Reproduction is principally by fission. This is a process of considerable complexity, not yet fully understood, but in outline is as follows. Following nuclear division, a ring of wall material develops at the centre of the isthmus and grows inwards to form a septum. When complete, the daughter cells separate. Each then regenerates the missing half, and the symmetry is restored. This results in two adult organisms, in each of which half of the thallus is inherited from the previous generation, and the other half newly synthesized.

Conjugating cells become enveloped in mucilage. Protuberances may be formed from one cell to another, but in many species the cells break open at the isthmus and the protoplasts emerge and fuse. Following fusion, a thick-walled zygospore is formed. Meiosis and germination occur after a resting period.

The relationships of the Conjugales

The Conjugales, although having no close affinities with other Chlorophyta, do not differ in the fundamentals of metabolism and pigmentation. Even the amoeboid gametes, a striking feature of the Order, are not unique, since amoeboid gametes are encountered sporadically in other Orders. In *Chlamydomonas eugametos*, for example, in the interval between loss of

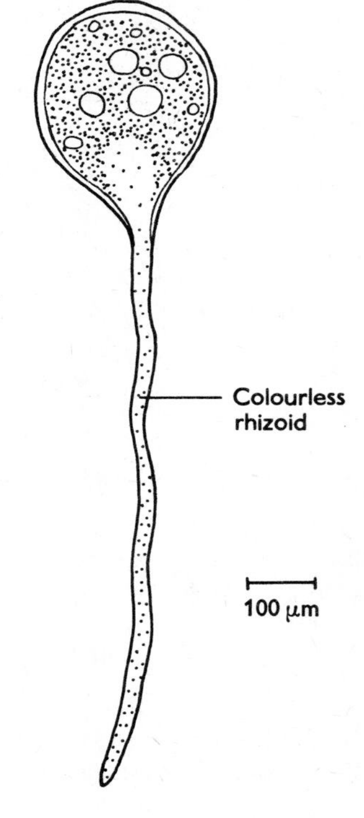

Fig. 2.39 *Protosiphon botryoides*. (After Klebs, from Fritsch, *The Structure and Reproduction of the Algae*, I. Cambridge, 1935)

flagella and fusion, the naked gametes lie adjacent to each other and resemble those of a number of Conjugales.

Siphonales The name of this Order, derived from the Greek word meaning 'tube', refers to the acellular structure of the thallus. The thalli normally contain multinucleate protoplasts, and are consequently termed ***coenocytic***. Dividing walls are absent until the formation of the reproductive organs. The range in form of the thallus is considerable; it may consist of a simple unbranched tube (Fig. 2.39) or a complex mass of interwoven filaments. Those species with complex thalli are invariably marine, and the thalli are often mechanically strengthened by superficial deposits of calcium carbonate.

The Siphonales differ from the rest of the Chlorophyta in containing both α- and β-carotin, instead of β-carotin alone. Additional xanthophylls are also present.

Protosiphon

The simplest vegetative organization found in the Siphonales is represented by *Protosiphon* (Fig. 2.39), often frequent on damp mud and walls. The thallus consists of a single sphere, about 0.3 mm in diameter, anchored to the substratum by a colourless rhizoid which may reach a length of 1 mm. The sphere contains several nuclei, and a reticulate chloroplast with a number of pyrenoids.

REPRODUCTION Reproduction is both asexual and sexual, but the former is little specialized. Young individuals of *Protosiphon* may undergo repeated division, forming a cluster of 4–16 cells, each of which develops as a new plant. Fully developed plants remain capable of budding from the upper portion, the buds soon becoming detached and forming new individuals.

Plants growing in such typical habitats as the edges of lakes or ponds are stimulated into sexual reproduction by submergence. After division of the protoplast, uninucleate biflagellate gametes are liberated which fuse to give a thick-walled zygote. There seems to be neither anisogamy nor physiological heterothallism, since gametes from the same plant may fuse, a situation regarded as characteristic of the most primitive organisms. Unmated gametes may develop into parthenospores, a peculiar stage which is capable of producing either a new individual or more gametes. This prolonged survival of the gamete stage presumably increases the chance of cross-fertilization, and may enjoy some selective advantage.

Bryopsis

Bryopsis, common in warmer seas, resembles *Protosiphon* in its coenocytic structure, but shows a considerable advance in vegetative organization. The thallus possesses a main axis, from which branches arise pinnately (Fig. 2.40). They also produce laterals and some species attain a tripinnate condition. At the insertion of each branch the cell walls are constricted and conspicuously thickened. This may have mechanical significance, because a simple branched tube lacking septa, and with walls of uniform thickness, would become structurally unstable beyond a certain size.

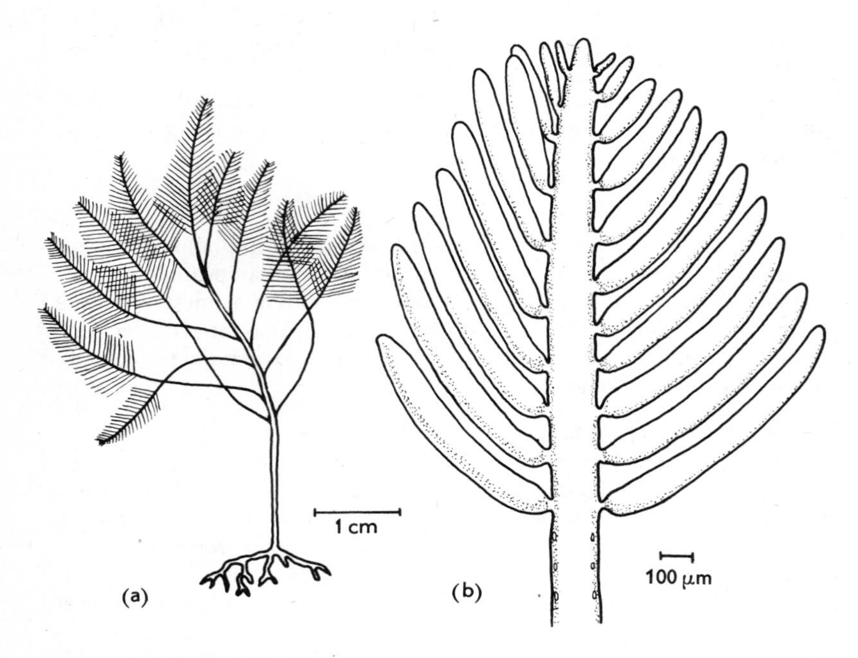

Fig. 2.40 *Bryopsis plumosa*. (**a**) Habit. (**b**) Terminal portion of branch. (After Newton, *A Handbook of the British Seaweeds*. British Museum, 1937)

The frond-like thallus of *Bryopsis* rises more or less vertically, and is anchored by rhizoids produced from a small prostrate filament. The thallus shows a marked polarity, and *Bryopsis* has consequently been much used in the experimental investigation of this phenomenon. In the vegetative condition, the cytoplasm is spread as a thin peripheral layer containing numerous nuclei and chloroplasts.

Vegetative propagation occurs by the detachment of side branches and their establishment as new individuals. No specialized methods of asexual reproduction have been observed.

In sexual reproduction, terminal pinnae, normally the smallest, become transformed into gametangia, having first been cut off from the rest of the plant by a septum. The gametangia are conspicuously opaque as a consequence of the considerable multiplication of the plastids. The gametes are liberated by dissolution of the apex of the gametangium. They are biflagellate, but there is striking anisogamy, and, since each plant usually produces gametes of only one size, a clear approach to dioecy. The larger gametes also contain a distinctly green plastid, whereas the smaller gametes are yellowish.

After fusion, the zygote does not secrete a thick wall, but germinates immediately into a new filament. Since meiosis has been detected in gametogenesis, the thallus is normally diploid.

Caulerpa

Characteristic of this genus is the large, creeping rhizome from which both rhizoids and upright 'fronds' arise (Fig. 2.41). The 'fronds' later show a great variation in shape (Fig. 2.42), different species often being named from their distinctive forms.

Internally, all the species of *Caulerpa* show ingrowths of the cell wall, forming, throughout the thicker portions of the thallus, a web of inter-

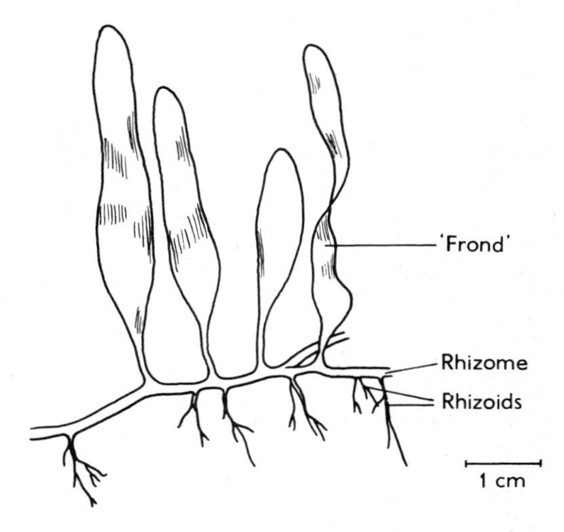

Fig. 2.41 *Caulerpa prolifera*. Portion of plant. (After Taylor, *Marine Algae of the Eastern Tropical and Subtropical Coasts of the Americas*. Michigan, 1960)

connecting strands. This web is of unknown function, but it possibly increases the mechanical stability of the thallus, or enlarges the surface area available to the cytoplasm. The fibrillar polysaccharide of the wall is not cellulose, but β-1:3 xylan, a polymer of a pentose sugar. The walls also contain callose, a β-1:3 glucan, formerly believed confined to the sieve tubes and reproductive structures of higher plants.

REPRODUCTION Reproduction is mainly vegetative by sporadic fragmentation of the thallus. Sexual reproduction probably occasionally occurs, and, although the evidence points to its being basically similar to that of *Bryopsis*, little is known of its details.

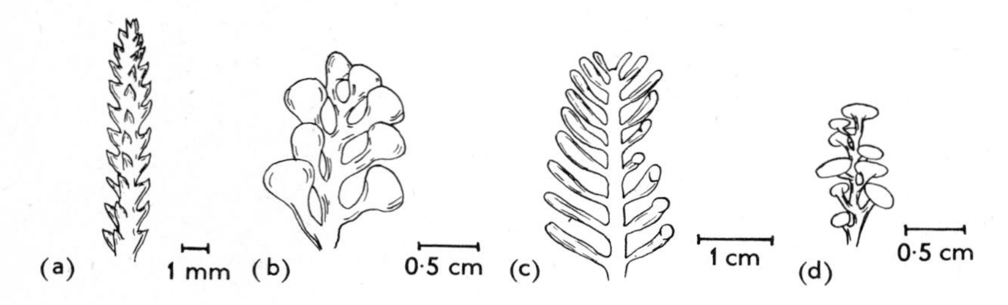

Fig. 2.42 *Caulerpa*. Types of 'fronds'. (**a**) *C. cupressoides*. (**b**) *C. racemosa* var *macrophysa*. (**c**) *C. ashmeadii*. (**d**) *C. peltata*. (All after Taylor, *Marine Algae of the Eastern Tropical and Subtropical Coasts of the Americas*. Michigan, 1960)

Codium

This genus is representative of the most elaborate vegetative organization present in the Siphonales. The thallus is made up uniformly of closely packed interwoven filaments, although the outward morphology varies widely with species. In *C. tomentosum* the thallus is a system of dichotomously branched axes, each about 0·5 cm in diameter, anchored at the base. Other species are flattened, forming a cushion or plate, or spherical. The microscope reveals in all species a central weft of branched filaments, giving rise to a continuous covering of elongated vesicles on the outside (Fig. 2.43).

REPRODUCTION Reproduction again appears to be principally vegetative by fragmentation. Whenever observed, sexual reproduction has been anisogamous. There is evidence, as in *Bryopsis*, for the diploid state of the thallus.

The Siphonales contain a number of other genera of extreme interest. *Halimeda* (Fig. 2.44), for example, is a tropical alga with a complex thallus of variable morphology. Calcium carbonate is deposited in the side-walls of the outer vesicles, and these deposits survive after the death of the plant

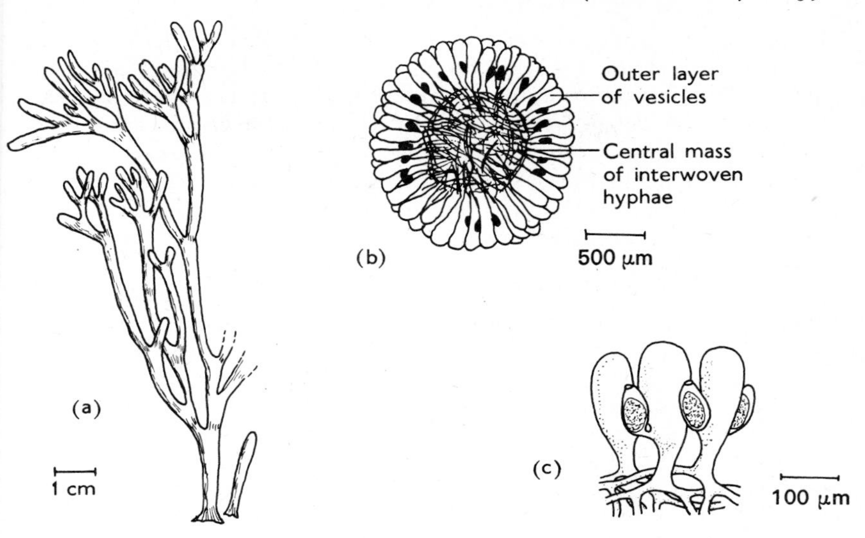

Fig. 2.43 *Codium tomentosum.* (**a**) Habit. (**b**) Transverse section of thallus. (**c**) Gametangia. (All after Newton, *A Handbook of the British Seaweeds.* British Museum, 1937)

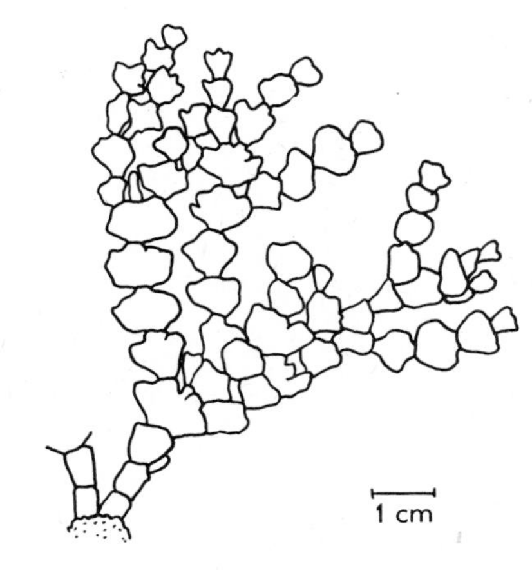

Fig. 2.44 *Halimeda simulans.* Portion of thallus. (After Taylor, *Marine Algae of the Eastern Tropical and Subtropical Coasts of the Americas.* Michigan, 1960)

and accumulate as coral reefs. Fossil records of other lime-secreting genera go back as far as the Upper Silurian (see Table 1.1). *Valonia*, an inhabitant of tropical and sub-tropical seas, consists of a bladder-like cell, occasionally branching, reaching a few cm in diameter. It has been extensively used in experiments on wall structure, permeability, and absorption of electrolytes. *Acetabularia* (Fig. 2.45a) (sometimes placed in a separate Order) has a mushroom-shaped thallus when mature, the central axis producing whorls of deciduous branches during growth. Unlike most Siphonales, *Acetabularia* is a uninucleate cell throughout its vegetative development, the nucleus remaining at the base of the stalk. It has been found that if the stalk is removed from the nucleate portion, the stalk remains capable of some growth and may even begin to form a cap. Nevertheless, the nucleus is clearly essential for continued growth and morphogenesis, and it seems very likely from decapitation and grafting experiments that the nucleus produces a sequence of 'morphogenetic substances' (probably ribonucleic acids) which ascend into the stalk and determine the kind of growth which occurs at its tip. Grafts made between the stalk of one species and the nucleate portion of another have shown the extent to which the cytoplasm of the first species can affect the expression of the genetic information contained in the nucleus of the second.

When vegetative growth is completed the basal nucleus divides into several thousand secondary nuclei. These ascend into the cap, which eventually becomes cleft into uninucleate compartments (Fig. 2.45b). Each of these then becomes a cyst in which meiosis occurs, followed by the liberation of biflagellate gametes (Fig. 2.45c). Although the gametes are morphologically identical, their pairing behaviour indicates sexual differentiation, a situation recalling that already encountered in *Chlamydomonas moewusii* (p. 27).

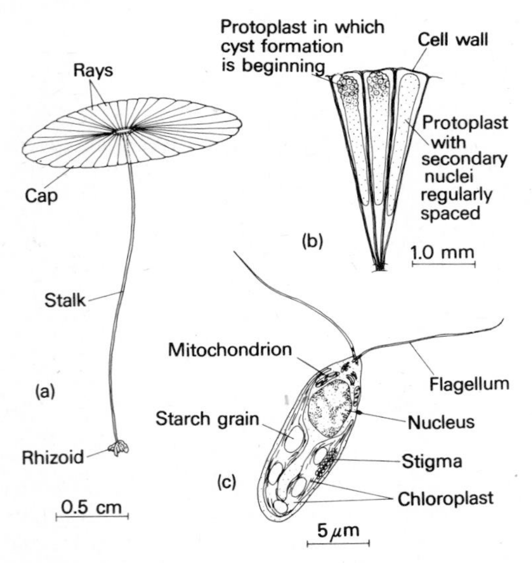

Fig. 2.45 *Acetabularia mediterranea*. (**a**) Cell with mature cap. (**b**) Detail of cap during cyst formation. (**c**) Longitudinal section of gamete.

The relationships of the Siphonales

The relationships of the Siphonales are obscure, but some have seen an origin in a form resembling the present Chlorococcales, an Order in which tendencies towards a coenocytic habit are occasionally evident. An ancestor similar to *Protosiphon* was perhaps derived in this way, and the more complex Siphonales evolved from it. The fossil record (p. 59) shows that the Siphonales are of extreme antiquity.

Charales Although many authors regard the Charales as having Divisional rank, we have included them in the Chlorophyta because they are basically similar in pigmentation, metabolism and the limited nature of their anatomy to the algae. Nevertheless, they have many striking features unrepresented elsewhere in the plant kingdom. There are only six living genera, the remainder being fossil. The living *Chara* and *Nitella* are com-

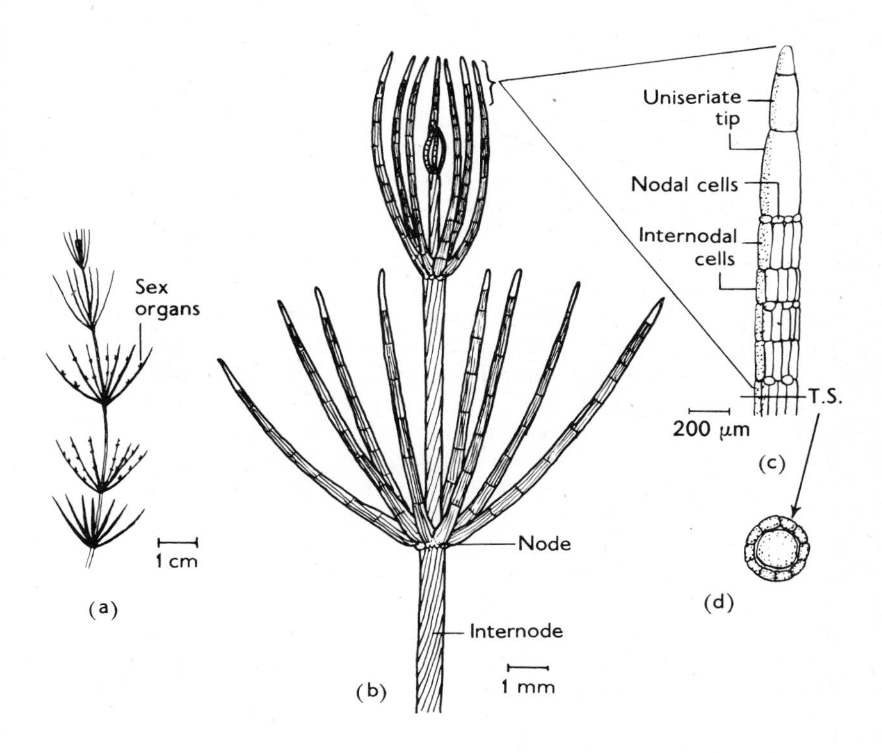

Fig. 2.46 *Chara* sp. (**a**) Habit. (**b**) Apical portion of plant. (**c**) Apex of branch. (**d**) Transverse section of young branch.

mon in base-rich waters. Many species develop an exoskeleton of calcium carbonate. This is clearly a property of some antiquity, since it has enabled the preservation of the fossil forms. The record possibly extends back as far as the Devonian.

THE VEGETATIVE ORGANIZATION The thallus of the Charales always possesses a clear main axis, growing from a dome-shaped apical cell (Fig. 2.46). Segments are cut off at regular intervals and in a single column from the flat base of this cell, and each segment immediately divides again by a transverse wall. The upper cell of this division becomes a nodal initial, and the lower, which does not divide again, an internode cell. The nodal initials remain meristematic and give rise to whorls of branches, mostly of limited growth. The mature plant thus has a morphology reminiscent of *Draparnaldiopsis*, the whorls of branches being separated by the single, much extended, internode cells. In some species, certain cells of the node give rise to upward- and downward-growing filaments which closely apply themselves to the internode cell and consequently provide it with an investment resembling a cortex.

REPRODUCTION Vegetative propagation, the only method of asexual reproduction, is carried out by the separation of groups of cells formed as tuberous outgrowths on subterranean nodes.

Sexual reproduction is oogamous, and highly specialized. The male and female reproductive structures, which consist of antheridia and oogonia surrounded by envelopes of sterile tissue, are sufficiently large to be seen with the naked eye. In both monoecious and dioecious species, the oogonia develop just above the insertions of the lateral appendages, and the antheridia just below (Fig. 2.47). A complicated sequence of divisions gives rise to the mature antheridium (Fig. 2.47) in the centre of which a tangled mass of threads arises from lateral receptacles (manubria) (Fig. 2.48). Each cell of a thread differentiates into an antherozoid, which in structure and shape resembles those produced by some archegoniate plants (Fig. 2.48b). Oogonia, when mature, are surrounded by a spiral of elongated vegetative cells, at the tip of which short columns of rounded cells come together as a corona (Fig. 2.47). Within this sheath is a single egg cell which accumulates food reserves and becomes largely opaque. Antherozoids penetrate slits which appear in the coronal region, and fertilization occurs at a small clear area near the apex of the egg, the so-called 'receptive spot'. After fertilization a cellulose membrane is secreted by the zygote. This, together with the wall of the oogonium and the inner walls of the spiral cells, which become thickened and indurated, serves to enclose the zygote in an almost impervious jacket.

Germination, which occurs after a resting period, is almost certainly accompanied by reduction division, but only one of the tetrad of nuclei so

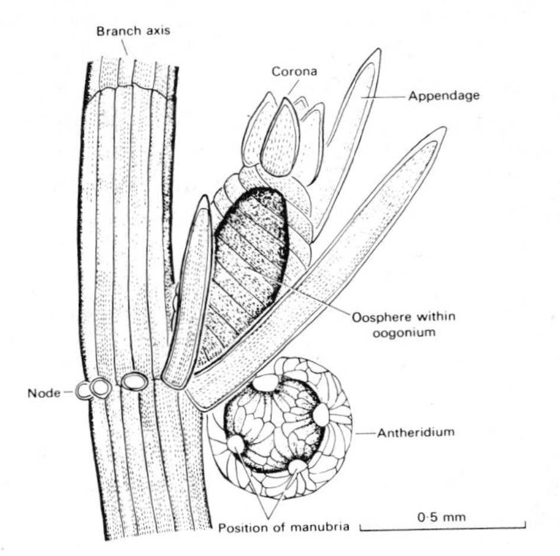

Fig. 2.47 *Chara* sp. Male and female reproductive organs.

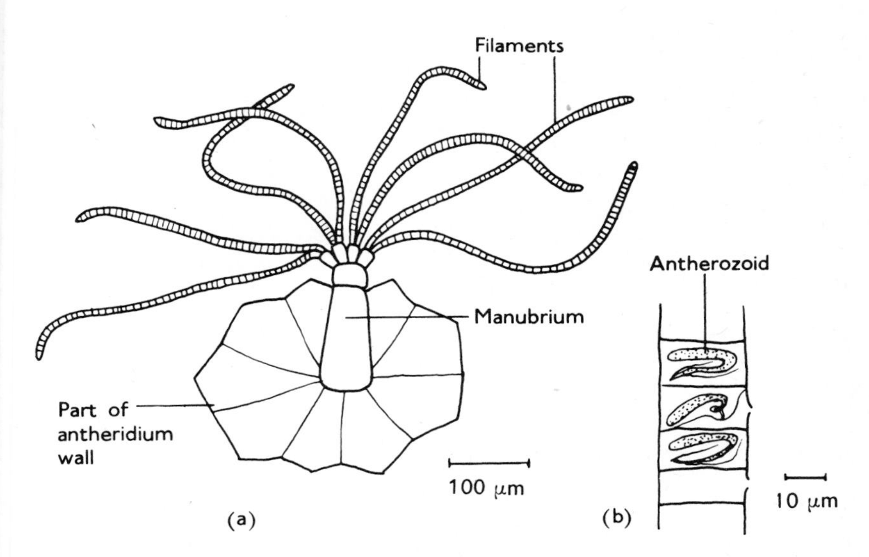

Fig. 2.48 *Chara* sp. Development of antheridium (**a**) and antherozoids (**b**). (After Kerner von Marilaun and Oliver, *The Natural History of Plants*. London, 1902)

produced remains intact; the remainder degenerate. Division of the intact nucleus leads to the production of two cells, which, as the membranes of the zygote break open at the apex, yield from the one rhizoids and from the other an erect green filament. This filament is the ***protonema***, a stage of development peculiar, in the algae, to the Charales. Cell divisions occur in the protonema, and it differentiates into nodes and internodes. The first node gives rise to rhizoids and additional protonema, and the second to a whorl of laterals, one of them developing into a normal main axis, and the others into yet more protonema. The mature plants thus arise as lateral branches of protonemal filaments.

The relationships of the Charales

The relationships of the Charales are largely speculative. The fossils show that they are an extremely ancient type of algal organization, yet the small discoid chloroplasts are more like those of the higher plants than those of the Chlorophyta. Nevertheless, there is no evidence that terrestrial plants came from *Chara*-like ancestors. The Charales are best regarded as a highly specialized, and in certain ecological situations highly successful, Order of aquatic plants, long separated from the main trends of algal evolution.

3

The Algae, II

In this chapter we shall be concerned with those eucaryotic algae which lie outside the Chlorophyta. They show a number of organizational trends resembling those already seen in the Chlorophyta, together with some features not represented in living Chlorophyta, although possibly present at some stage in their evolutionary history.

CHRYSOPHYTA

This algal Division includes three distinct classes, the Xanthophyceae, Chrysophyceae (including the Haptophyceae) and Bacillariophyceae. They are regarded as related because they have a number of outstanding features in common, although all are not necessarily found together in each species. These features are: a cell wall composed of two parts, the presence of flimmer flagella, the deposition of silica in the cell wall, and the accumulation of the food reserve leucosin (a polysaccharide, also called chrysolaminarin).

The Chrysophyta comprise algae of flagellate, coccoid, amoeboid,

Xanthophyceae

Habitat	Aquatic (mainly freshwater).
Pigments	Chlorophylls *a* and *e*, β-carotin, possibly xanthophylls.
Food reserves	Fat, leucosin.
Cell wall components	Cellulose, hemicelluloses, silica. Cell wall usually consisting of two overlapping portions.
Reproduction	Asexual, occasionally sexual.
Growth forms	Flagellate, coccoid, amoeboid, coenocytic.
Flagella	2 unequal, 1 whiplash and 1 flimmer; anterior.

coenocytic and filamentous forms. The highest level of organization attained is the simple filament, suggesting that the rate of evolutionary advance in the Chrysophyta has been slower than in the Chlorophyta, but the course nevertheless to a large extent parallel.

The Xanthophyceae are commonly called the yellow-green algae, their yellowish colour being caused by the large proportion of carotenoids in the plastid. The inequality of the flagella in the motile stages is a striking feature, and responsible for the alternative name for this algal Class, namely the Heterokontae. Although of considerable interest to the biochemist, the Xanthophyceae are but a minor part of the Plant Kingdom, and of little economic or ecological importance.

The flagellate forms

The flagellate habit is represented by a number of species, some of which are able to alter their shapes and methods of locomotion. *Heterochloris*, for example, is pear-shaped, with the two unequal flagella at the narrower anterior end. In certain conditions, however, it may lose this shape and produce thin outgrowths of cytoplasm (***rhizopodia***), the flagella being either retained or lost. Reproduction, which is by binary fission, occurs in the motile phase. Encystment has also been observed. The thickened wall of the cyst is impregnated with silica, and made up of two portions.

The palmelloid and coccoid forms

The palmelloid (colonial) habit is represented by *Chlorosaccus*, a genus resembling *Tetraspora* of the Volvocales (p. 24), but much less common. The pear-shaped cells are enveloped in mucilage, but are usually regularly arranged, and often in groups of four as in *Tetraspora*. Reproduction is

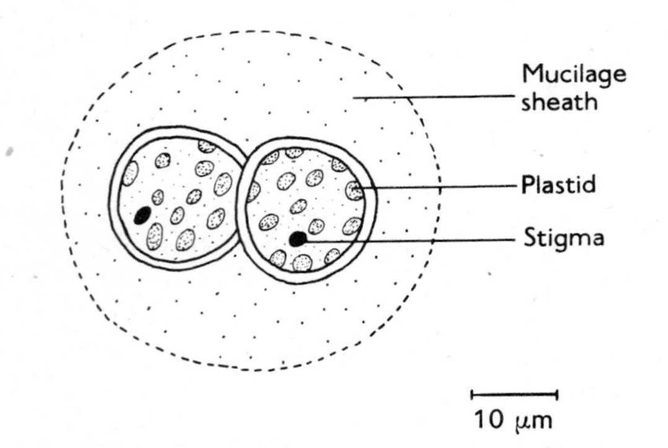

Fig. 3.1 *Chlorobotrys regularis*. Two cells enclosed in mucilage sheath. (After West and Fritsch, *A Treatise on the British Freshwater Algae*. Cambridge, 1927)

solely asexual, individual cells of the colony each giving rise to a single zoospore.

The coccoid habit is exhibited by a number of forms which closely resemble the Chlorococcales of the Chlorophyta, and until recently were classified with them. *Chlorobotrys* (Fig. 3.1), for example, common in bog pools, has a spherical cell resembling that of *Chlorella*. *Meringosphaera*, a constituent of marine plankton, has a cell wall furnished with a number of bristles, which possibly assist flotation.

The filamentous forms

The filamentous habit is not common in the Xanthophyceae. Of the few genera described, only *Tribonema* (Fig. 3.2) is notable, the short threads being occasionally encountered in fresh waters and on damp earth. The

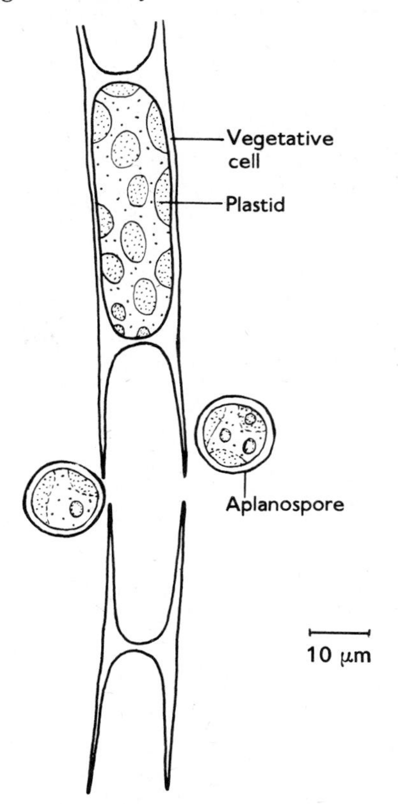

Fig. 3.2 *Tribonema bombicinum*. (After West and Fritsch, *A Treatise on the British Freshwater Algae*. Cambridge, 1927)

cell wall is made up of two overlapping halves. Consequently the filament, when disrupted, breaks up into a number of pieces H-shaped in optical section, closely resembling the situation in the green alga *Microspora* (p. 34). It is the H-shaped piece, rather than the complete cell, which is the basic unit of the filament so far as the wall is concerned, and at cell division, instead of the usual septum being formed, a new H-shape is produced at the centre of the parent cell. Internally, the cell organelles are disposed in much the same way as they are in *Microspora*, except that several lenticular plastids occupy the peripheral cytoplasm, instead of a single band-like plastid. In some species there is evidence of multinucleate cells, perhaps indicating a transition to a coenocytic condition.

In addition to vegetative reproduction by fragmentation of the filament, *Tribonema* reproduces asexually by zoospores, and possibly occasionally sexually. The zoospores are formed singly or in pairs, and are liberated by the separation of the two halves of the cell. They resemble the motile stage of *Heterochloris*. Having swum until a substrate is reached, the zoospores attach themselves and germinate to form a new filament with a basal hold-fast. In place of zoospores, the cells of the filament may give rise to thick-walled aplanospores. As these germinate the protoplast enlarges and causes the wall to fall into two pieces, revealing its structural similarity to that of the cells of the filament. Sexual reproduction, observed very rarely, is isogamous.

The coenocytic forms

The coenocytic habit is found in a few genera, the multinucleate aseptate thalli of which recall those of some members of the Siphonales of the

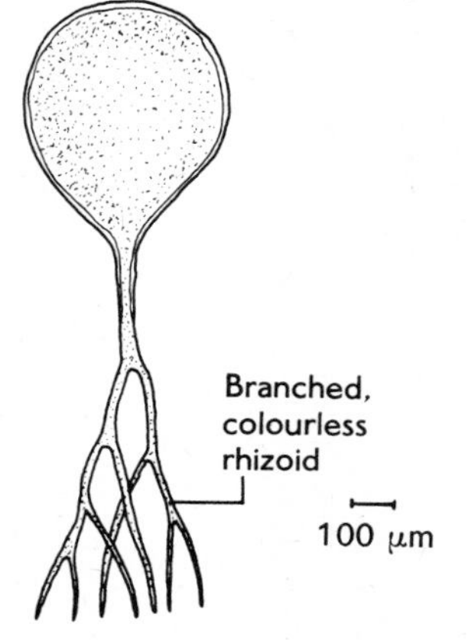

Fig. 3.3 *Botrydium granulatum*. (After West and Fritsch, *A Treatise on the British Freshwater Algae*. Cambridge, 1927)

Chlorophyta. *Botrydium* (Fig. 3.3), for example, is very much like *Protosiphon* (p. 55), while *Vaucheria* was for a long time classified as a green alga.

Botrydium differs from *Protosiphon* in its forked rhizoid, discoid plastids, and its inability to divide vegetatively. Asexual reproduction is brought about by aplanospores in dry weather, or zoospores in wet conditions. Since the zoospores may also fuse in either similar or dissimilar pairs, they are perhaps better regarded as gametes, able to germinate in certain conditions parthenogenetically. If fusion occurs, the resulting zygote divides meiotically to yield 4 or 8 zoospores which escape and produce new individuals.

Vaucheria (Fig. 3.4) is a filamentous form, one species of which is commonly present as a thin green felt over damp soil, often becoming a

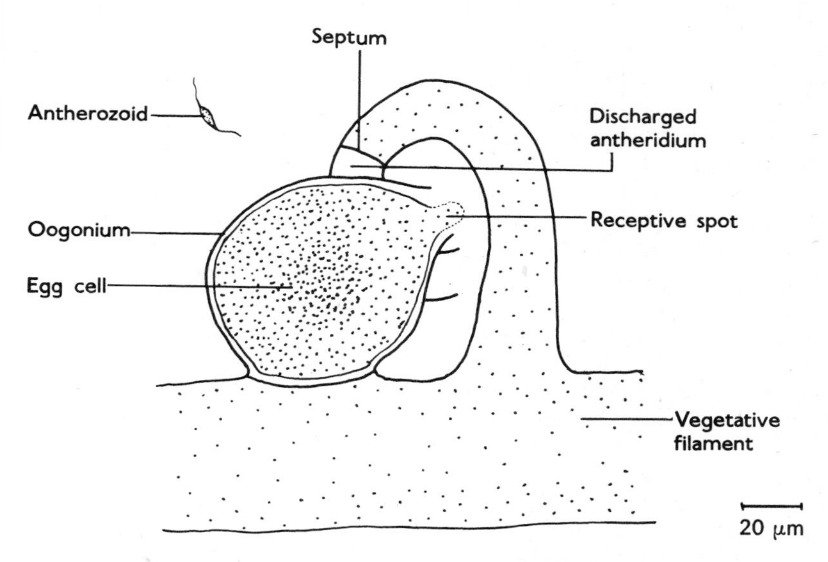

Fig. 3.4 *Vaucheria sessilis.* Reproductive organs. (After West and Fritsch, *A Treatise on the British Freshwater Algae.* Cambridge, 1927)

greenhouse pest. Other species are freshwater or marine. Asexual reproduction is common, and involves the formation of a peculiar zoospore. The apical region of a filament, rich in oil droplets and plastids, is cut off by a transverse septum. The many nuclei within this segment migrate to the periphery of the protoplast, and opposite each nucleus emerges a pair of flagella. The tip of the filament becomes gelatinous, and a multiflagellate zoospore, which can be regarded as a mass of unseparated uninucleate zoospores, escapes. On coming to rest the flagella disappear, a wall is secreted, and the zoospore begins to germinate. Two or three filaments emerge from the spore, one of which usually acts as a holdfast. In some species, particularly the terrestrial, aplanospores are produced in a similar manner. Here, however, flagella are lacking, and the spore is liberated by breakdown of the sporangial wall. Germination may begin before this is complete.

SEXUAL REPRODUCTION IN *VAUCHERIA* The attainment by *Vaucheria* of the filamentous siphonaceous habit has been accompanied by the development of a more highly organized sexual phase than in the simpler members of the Class. Sexual reproduction is first indicated by the formation of antheridia. The tip of a lateral branch, in which many nuclei but few plastids, accumulate, is cut off by a transverse septum. The cytoplasm becomes apportioned between the nuclei, and each uninucleate protoplast then differentiates into a biflagellate gamete. Meanwhile the female organ develops, in many species next to the male on the same filament, or on a closely adjacent branch. The first stage is the appearance of a lateral convexity, made conspicuous by the accumulation of nuclei and plastids within. The convexity becomes spherical, and is eventually cut off by a septum at the level of the parent filament. Meanwhile, in a manner not yet fully understood, the initially multinucleate oogonium comes to contain a single uninucleate egg cell. When the egg is mature a short beak develops asymmetrically at the apex and makes the egg accessible to the male gametes. Only one penetrates the egg cell, and its nucleus comes to lie by that of the egg, not fusing with it until it has swollen to an approximately equal volume. After formation of the zygote, it becomes surrounded by a highly impervious wall and may remain dormant for several months. Germination, which results in the formation of a new filament, is thought to be preceded by meiosis.

Since in most species fusion takes place between gametes from the same filament, sexual reproduction in *Vaucheria* provides for little more than the interpolation of a resting stage in the life cycle. A few species, however, are dioecious, allowing the possibility of genetic recombination.

The limited evolution of the Xanthophyceae

If, as seems probable, the Xanthophyceae are the result of a line of evolution from some ancestral motile form, parallel to that of the Chlorophyta, the problem immediately arises of why the rate of evolution has been so much slower. It is possible that the different pigmentation has been associated with biochemical features restraining variability and evolutionary success. Additionally, the poor development of sexual reproduction would have limited genetic interchange and the range of variation open to selection.

Chrysophyceae (including the Haptophyceae)

Habitat	Aquatic
Pigments	Chlorophyll *a*, β-carotin, xanthophylls (fucoxanthin).
Food reserves	Fat, leucosin.
Cell wall components	Hemicelluloses, silica, calcium carbonate.
Reproduction	Asexual, rarely sexual (isogamous).
Growth forms	Mostly flagellate, some colonial or filamentous.
Flagella	1 flimmer or 1 flimmer and 1 whiplash; anterior. Occasionally a third apical appendage, the haptonema (especially in the Haptophyceae).

Although most members of this Class (referred to as golden-brown algae) remain at the flagellate level of organization, others exist in colonial, palmelloid and even filamentous states, thus showing a close parallel with the Xanthophyceae.

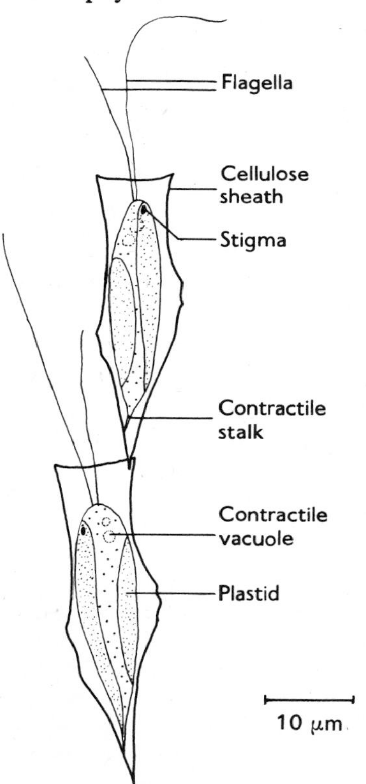

Fig. 3.5 *Dinobryon cylindricum*. (After West and Fritsch, *A Treatise on the British Freshwater Algae*. Cambridge, 1927)

Species of Chrysophyceae are important constituents of freshwater (e.g. *Dinobryon*, Fig. 3.5) and marine plankton, and of the microflora of salt marshes. Amongst the marine forms are the biflagellate Coccolithophoridaceae, remarkable for the plates of calcium carbonate (***coccoliths***) incorporated in their walls. Electron microscopy has revealed that these plates bear remarkable and characteristic sculpturings, and that the chalk deposits of the Cretaceous period consist very largely of the remains of these organisms. The taxonomy of the coccolithophorids is made difficult by similar non-motile forms having dissimilar motile stages. A knowledge of the life cycle is thus essential for accurate classification.[64]

The uniflagellate forms of the Chrysophyceae often show a stump corresponding in position to the whiplash flagellum. Here we possibly have an example of where evolution has led to simplification, the whiplash flagellum eventually becoming no more than a rudiment.

The haptonema, found only in biflagellate forms, is characteristic of the Haptophyceae, but occurs also in other Cryptophyceae. It is a filamentous organ, arising between the two flagella, with no clearly identifiable function. It is often conspicuously long, but in some forms is more often than not coiled up like a proboscis. In transverse section it shows three concentric membranes enclosing a number of fibrils. The outer membrane is continuous with the plasmalemma, and the inner pair represents a tubular extension of the endoplasmic reticulum.

Bacillariophyceae

Habitat	Aquatic and terrestrial.
Pigments	Chlorophylls *a* and *c*; β-carotin, xanthophylls (fucoxanthin).
Food reserves	Fat, leucosin.
Cell wall components	Hemicelluloses, silica.
Reproduction	Asexual and sexual (anisogamous and oogamous).
Growth forms	Unicellular, colonial
Flagella	1 flimmer (in some forms lacking the central pair of fibrils); anterior. (Present only in the male gametes of some members.)

The Bacillariophyceae, commonly known as the diatoms, are a large Class of microscopic algae with intricately sculptured, siliceous walls. The bilateral or radial symmetry of the cells, and the regularity of the delicate markings on their walls (Fig. 3.6), make the diatoms very beautiful microscopic objects, rivalling even the desmids. The diatoms are frequent in fresh water (e.g. *Tabellaria*, Fig. 3.7) and marine (e.g. *Campylodiscus*, Fig. 3.8) phytoplankton, and therefore of economic importance in the manage-

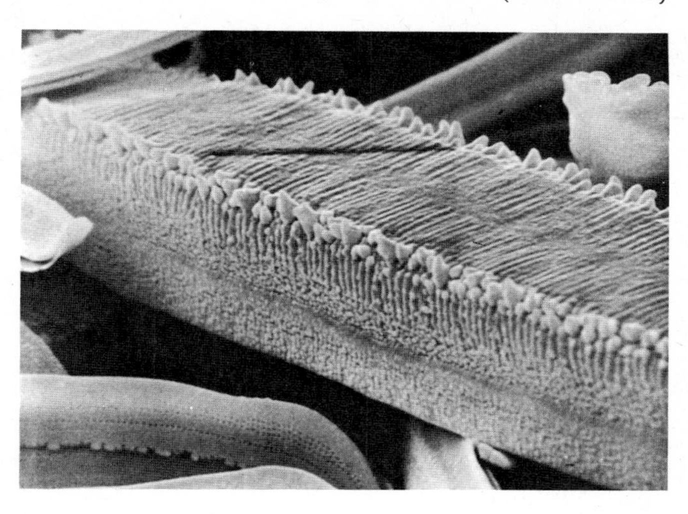

Fig. 3.6 The surface of a diatom (*Diatoma hiemale*) showing the sculpturing of the wall. ×3335.

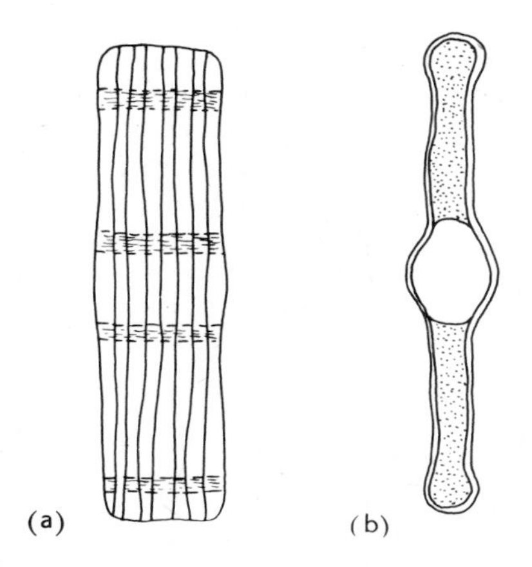

Fig. 3.7 *Tabellaria fenestrata*. (**a**) Girdle view, (**b**) Valve view. (After West and Fritsch, *A Treatise on the British Fresh-water Algae*. Cambridge, 1927)

ment of fisheries. The siliceous walls of the diatoms resist dissolution and decay after the death of the organism, and accumulate as fossils on beds of lakes and seas. Huge deposits of these 'diatomaceous earths' (Kieselguhr) are known from the Tertiary period, and some are mined for use as abrasives, filters and the refractory linings of furnaces.

THE CELL WALL Much study has been given to the cell walls of diatoms, and it is principally upon their characteristics that the classification rests. Although the several thousands of species of diatoms occur in many

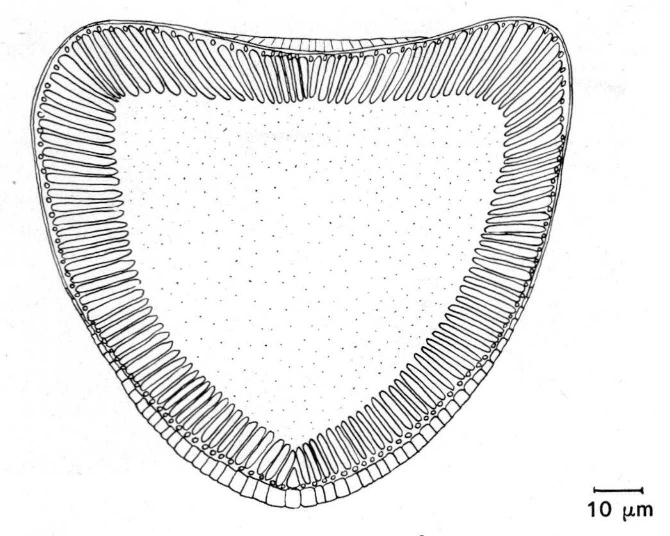

Fig. 3.8 *Campylodiscus eximus*. (After Gregory, 1857, *Trans. R. Soc. Ed.*, 21, Pt. XI)

different shapes, the walls of all consist basically of two parts which overlap like the two halves of a Petri dish (Fig. 3.9). Consequently, the appearance of the cells from the side (girdle view) is different from that from above (valve view). The sculpturing of the wall depends upon inequalities in the thickness of the outer siliceous layer, and in some species pores may be present in the thinner regions.

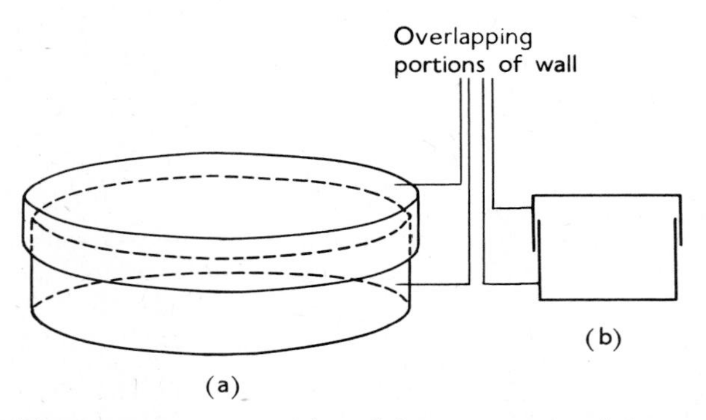

Fig. 3.9 Diagrammatic representation of the arrangement of the two halves of the wall in a diatom. (**a**) 3-dimensional view. (**b**) Transverse section.

CYTOLOGY Within the cell, the cytoplasm forms a thin layer containing one or more plastids, surrounding a large central vacuole. In the radially symmetrical (***centric***) forms, the nucleus is also held in this peripheral layer, but in the bilaterally symmetrical (***pennate***) forms the nucleus is suspended in the centre of the cell by a band of cytoplasm which traverses the vacuole. Many of the pennate diatoms are able to perform small, jerky movements, often returning irregularly to their starting point. Although the mechanism of this movement is not entirely clear, it is believed to result from contact along a longitudinal fissure between the medium and streaming mucilage. This fissure, the raphe, is present in the siliceous wall of all motile pennate diatoms.

CELL DIVISION At cell division the two halves of the wall separate, and one half goes with each daughter cell. Each daughter then immediately secretes another half-wall, thus completing its envelope. Since the mature walls are quite inflexible, and each new half fits into the pre-existing one, in the manner of the bottom half of a Petri dish into the top, the size of the cells inevitably decreases with successive vegetative divisions. Eventually a limit is reached and a regenerative process occurs.

'REGENERATION' AND THE FORMATION OF AUXOSPORES In principle the process of regeneration consists of the protoplast freeing itself from the constraint of the valves, and, in the form of the so-called ***auxospore***, expanding to its initial size. The formation of the normal siliceous walls and sequence of divisions then begins again. There are, however, a number of complexities in the formation of auxospores. In some species they arise after a sexual fusion, isogamous in many pennate diatoms, but anisogamous in the centric. Meiosis has been observed in gametogenesis, and the diatoms are consequently regarded as being diploid in the normal vegetative phase. In very many species (e.g., in some species of the common planktonic *Melosira*) the formation of auxospores appears to be a simple asexual process, but closer investigation of the cytology of diatoms may reveal the not infrequent occurrence of apogamy and parthenogenesis. In certain conditions some species give rise to thick-walled resting spores, but little is known of their germination.

The relationships of the Bacillariophyceae

Although the Bacillariophyceae, as mentioned earlier, share a number of basic biochemical and metabolic features with the Xanthophyceae and Chrysophyceae, particularly with some coccolithophorids, they are clearly highly specialized morphologically. If a common ancestor actually existed, the Bacillariophyceae presumably diverged at an early stage. On the other hand, there are no well-established records of Palaeozoic diatoms, and they do not appear in quantity until the later Mesozic and Tertiary eras.

PYRROPHYTA

Habitat	Principally aquatic.
Pigments	Chlorophylls *a* and *c*, β-carotin, xanthophylls (peridinin).
Food reserves	Starch, fat.
Cell wall components	Cellulose, hemicelluloses.
Reproduction	Asexual, rarely sexual (probably isogamous).
Growth forms	Flagellate, coccoid, filamentous.
Flagella	2, lateral; 1 transverse, the other pointing backwards.

Two Classes, the Desmophyceae and Dinophyceae are included in this Division.

The Desmophyceae are unicellular organisms which are either naked, or possess a wall of two watchglass-like valves. Peculiar elaborations of the wall are found in a number of species confined to tropical seas. Some species emit light and are among the causes of marine luminescence.

The Dinophyceae include the dinoflagellates, which are important constituents of marine phytoplankton. Characteristic of these organisms are the transverse and longitudinal grooves in the cell wall (Fig. 3.10). One

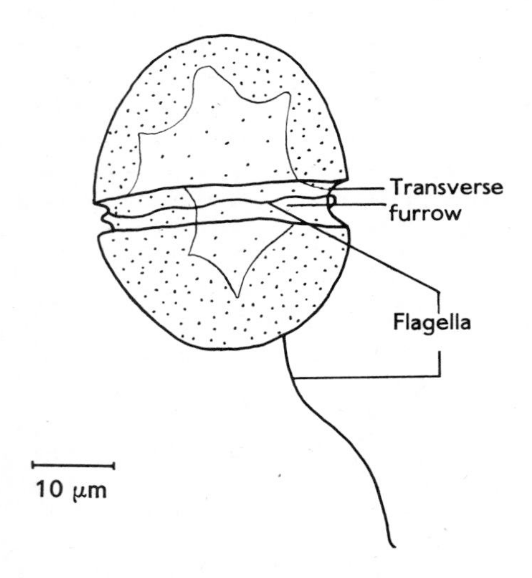

Fig. 3.10 *Glenodinium uliginosum*. The longitudinal groove in the wall is behind. (After West, from West and Fritsch, *A Treatise on the British Freshwater Algae*. Cambridge, 1927)

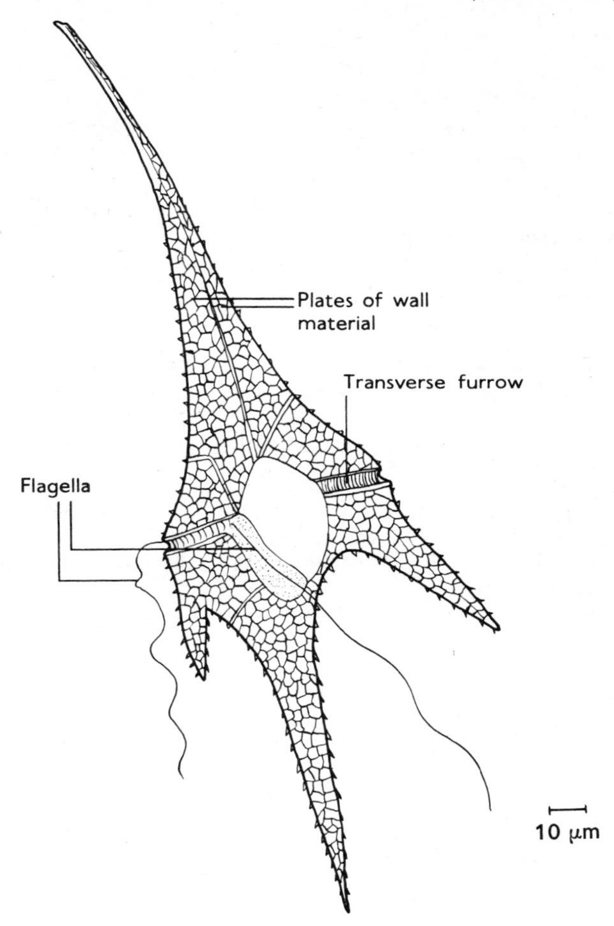

Fig. 3.11 *Ceratium hirundinella*. (After Schilling, from West and Fritsch, *A Treatise on the British Freshwater Algae*. Cambridge, 1927)

flagellum is situated in the transverse groove, where it performs snake-like undulations, while the other, a whiplash flagellum, is inserted in the longitudinal groove and extends posteriorly. The cell wall may be simple, or reinforced by a number of overlapping polygonal plates (e.g. *Ceratium*, Fig. 3.11). The protoplast contains a single large nucleus in which the chromosomes, which are not enveloped in histones as they are in most higher organisms, are visible after appropriate staining as spirally coiled rods. There are one or more plastids, often dark brown in colour as a consequence of a large proportion of the pigment peridinin. A stigma is often present in the motile species and in zoospores.

Unicellular forms predominate in this Class. Some members, however, are coccoid and occur as plankton, and a few filamentous species are included in *Dinothrix* and *Dinoclonium*, the former known only from marine aquaria.

Asexual reproduction in the motile species is commonly by simple vegetative division, but zoospores are produced by the filamentous and coccoid forms. Isogamous sexual reproduction may occur in a few species.

CRYPTOPHYTA

Habitat	Aquatic.
Pigments	Chlorophylls *a* and *c*, α- and ε-carotins, possibly xanthophylls, biliproteins.
Food reserves	Starch.
Cell wall components	Probably cellulose and hemicelluloses.
Reproduction	Asexual, rarely sexual (probably isogamous).
Growth forms	Flagellate, coccoid.
Flagella	2, anterior, slightly unequal, the longer probably flimmer.

This Division includes a few species occasional in freshwater and marine phytoplankton (e.g. *Cryptomonas*, Fig. 3.12). Their organization does not rise above the flagellate or coccoid level. Reproduction is by simple fission, or by zoospores in the coccoid forms.

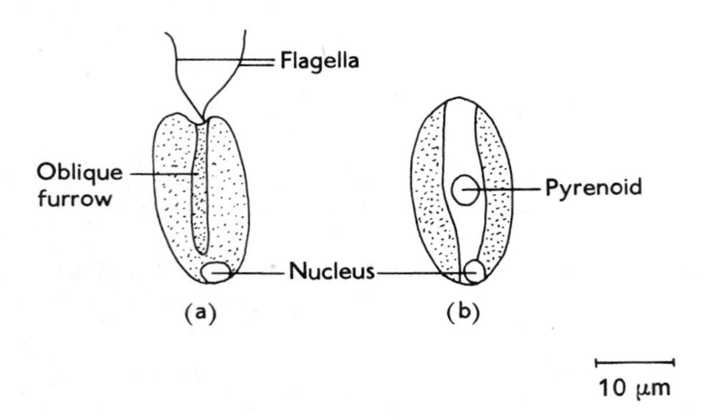

Fig. 3.12 *Cryptomonas anomala*. (**a**) Side view. (**b**) View from above. (After West and Fritsch, *A Treatise on the British Freshwater Algae*. Cambridge, 1927)

EUGLENOPHYTA

Habitat	Freshwater, a few marine.
Pigments	Chlorophylls *a* and *b*, β-carotin, xanthophylls.
Food reserves	Paramylum.
Cell wall components	None.
Reproduction	Asexual, rarely sexual (isogamous).
Growth form	Predominantly flagellate.
Flagella	2, but only 1 (rarely 2) emerging from the gullet, flimmer, anterior.

The Euglenophyta have been assigned to both the Plant and Animal Kingdoms, since the Division contains both green autotrophic and colourless heterotrophic forms. Many of the autotrophic Euglenophyta are also able to thrive in the dark if supplied with suitable metabolites, so they can be regarded as facultative heterotrophs. In some forms the plastids can be eliminated, without damaging the cell, by the use of suitable concentrations of streptomycin, thus creating experimentally wholly heterotrophic races. It seems very likely that the naturally heterotrophic species have evolved by a similar, but spontaneous, loss of plastids from autotrophic forms.

Thus although heterotrophic nutrition, a feature of animals, is today common in the Euglenophyta, the evidence points to the ability to ingest complex food materials being a secondary feature. Heterotrophic nutrition, both facultative and obligate, is in fact known in many other algae. Examples are provided by the Cyanophyta, and by *Polytoma* (Volvocales) and *Prototheca* (Chlorococcales) in the Chlorophyta. The Euglenophyta are nevertheless outstanding in the extent to which they have developed this facility.

Similarly, a cell wall, absent in the Euglenophyta and animal cells generally, is by no means always present in other flagellate unicellular algae. It is lacking, for example, in *Dunaniella* (p. 28), otherwise identical with *Chlamydomonas*, and in *Micromonas* (p. 5). Flagellate zoospores and gametes are also commonly naked.

The mode of nutrition and the presence or absence of a cell wall thus appear to be relatively plastic features at the flagellate level of organization. It should be noted, however, that although there is evidence that photosynthetic activity has been lost in the course of evolution, resulting in heterotrophic forms, there is no indication that it has been spontaneously acquired. Indeed, it is difficult to envisage how a heterotrophic form lacking plastids could create the photosynthetic apparatus *ab initio*.

THE CYTOLOGY OF *EUGLENA* *Euglena* itself is normally autotrophic, the cell containing several discoid or band-like plastids (Fig. 3.13). Some species abound in fresh water rich in organic material, such as seepage

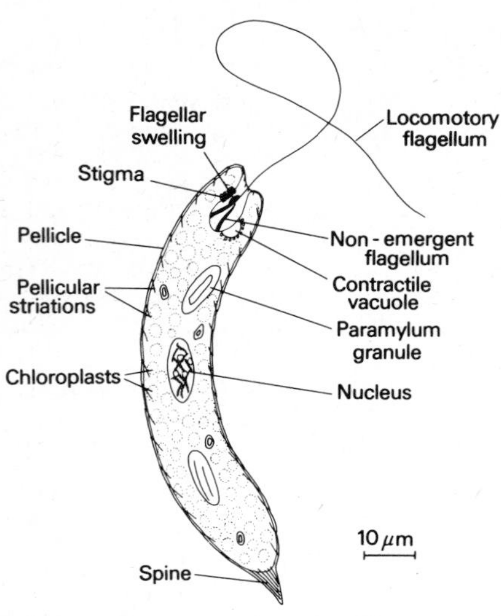

Fig. 3.13 *Euglena spirogyra*. Reconstruction of cell based on photo- and electron micrographs. (From Leedale, Meeuse and Pringsheim, *Arch. Mikrobiol.*, **50**, 68 (1965).)

from dunghills and farmyards, while others occur in damp mud by rivers, salt-marshes and similar places.

The cell is retained in a characteristic spindle-like shape by a rigid outer layer of cytoplasm, termed a ***pellicle***. Electron microscopy has shown that the pellicle contains a band-like component wound in a helix about the cell. The plastids—each of which contains a granule of paramylum, a food reserve allied to starch—and the nucleus lie in the less viscous inner cytoplasm. At the anterior end is a small invagination, the 'gullet', which in related heterotrophic forms may serve for the ingestion of food. A system of vacuoles discharges at intervals into the gullet. There are two flagella (Fig. 3.13), only one of which, clearly flimmer (Fig. 3.14), emerges through the mouth of the gullet. The base of this flagellum bears a thickening, believed to be a photoreceptor, close to the stigma in the adjacent cytoplasm.

Possibly belonging to the heterotrophic Euglenophyta is *Scytomonas*, a common intestinal parasite. The cytology closely resembles that of *Euglena*, and paramylum and fat occur as food reserves.

REPRODUCTION Binary fission is the only common method of reproduction of the Euglenophyta. Following nuclear division cleavage of the whole cell proceeds from the anterior to the posterior end. Sexual reproduction has been reported, but awaits corroboration.

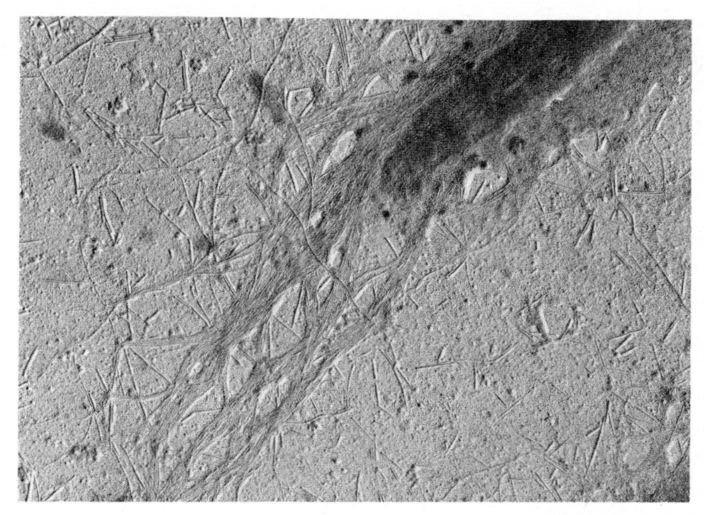

Fig. 3.14 The tip of the flagellum of *Euglena gracilis* showing the 'flimmer' hairs. Shadowed preparation. ×40,000.

Other forms of Euglenophyta

Although predominantly flagellate, a few Euglenophyta are encapsulated or form dendroid colonies. These can be regarded as the result of developmental trends from the flagellate state which parallel those displayed in other algal groups, but which here have reached only a rudimentary level of morphological complexity. The relationships of the Euglenophyta with other algae are obscure, but the striking similarity of the principal pigments of the plastids to those of the Chlorophyta perhaps points to a common origin in the remote past.

PHAEOPHYTA

Habitat	Predominantly marine.
Pigments	Chlorophylls *a* and *c*, β-carotin, xanthophylls (fucoxanthin).
Food reserves	Mannitol, laminarin.
Cell wall components	Cellulose, hemicelluloses.
Reproduction	Asexual and sexual (oogamous).
Growth forms	Filamentous, parenchymatous.
Flagella	2 unequal, 1 whiplash and 1 flimmer (sometimes spiny); anterior or lateral.

Of the many genera of the Phaeophyta (brown algae), only three rare ones are freshwater, the remainder being seaweeds whose macroscopic flattened thalli are familiar inhabitants of the intertidal regions of rocky coastlines. Other genera inhabit the region just beneath the low-tide mark, while a few are found solely in mid-ocean.

The vegetative organization of the Phaeophyta surpasses that of any of the algae so far considered. The simplest thallus, consisting of branching filaments, is heterotrichous, and thus resembles the most complex found in the Chlorophyta. It has consequently been said that morphologically the brown algae begin where the green algae finish.

This remark also applies to the sexual reproduction, for in the Phaeophyceae oogamy is the general rule, and alternation of generations is developed to the point at which the haploid and diploid states begin to diverge morphologically.

The Phaeophyta are divided into nine Orders, but representatives of only the five commoner ones will be considered here. They will, nevertheless, fully illustrate the probable evolutionary trends in the development of the morphology and of the reproductive cycles of the brown algae.

Ectocarpales *Ectocarpus* (Fig. 3.15), a genus common along the

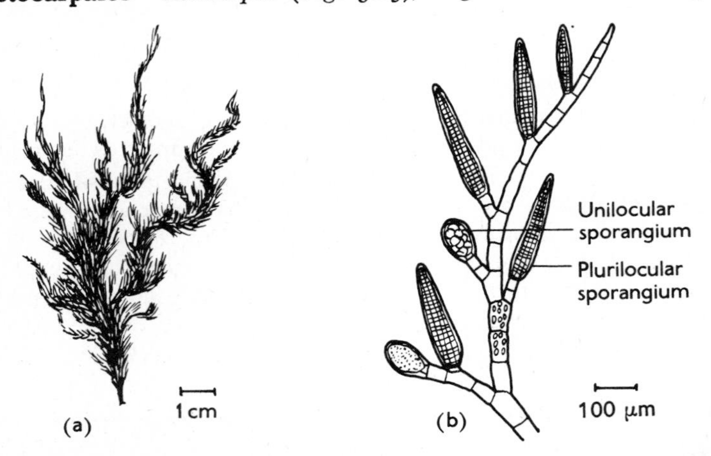

Fig. 3.15 *Ectocarpus confervoides.* **(a)** Habit. **(b)** Sporangia. (After Newton, *A Handbook of the British Seaweeds*. British Museum, 1937)

Atlantic coast of America and in the colder seas of the northern hemisphere, is a small heterotrichous plant, closely resembling in habit the green alga *Stigeoclonium* (p. 39). Cell division in the erect filaments is confined to well-defined intercalary regions. This method of growth, frequently found in the Ectocarpales, is termed ***trichothallic***.

THE LIFE CYCLE: PLURILOCULAR AND UNILOCULAR SPORANGIA *Ectocarpus* exists in two forms, haploid and diploid. In favourable environmental conditions the two generations, except in respect of the reproductive organs, are morphologically identical, but in colder regions the haploid plant may develop little, or even not at all. Reproduction of the diploid plant takes place solely by zoospores, but these are produced in two kinds of sporangia, termed respectively ***plurilocular*** and ***unilocular*** (Fig. 3.15b). The plurilocular sporangia arise from cells which undergo repeated transverse and subsequently longitudinal division, giving rise to spindle-shaped groups of small, more or less cubical compartments, each of which at maturity contains a single biflagellate zoospore. The plurilocular sporangia are commonly either lateral, or terminate short lateral branches, and dehiscence occurs at an apical pore. Dissolution of the partitions allows the zoospores to escape. After a motile phase, the zoospores settle and yield diploid plants identical with their parent.

The cytology of the developing unilocular sporangium is more complex. The nucleus of the initial cell is conspicuously large, and its first divisions are probably meiotic. Although the cytoplasm becomes multinucleate, no walls are produced, and it is only at maturity that the protoplast is cleaved into uninucleate portions. Each portion differentiates into a biflagellate zoospore, typically bean-shaped, and all are eventually released at an apical pore. Although these zoospores behave as those produced in plurilocular sporangia, they yield only haploid plants.

The haploid plants bear only plurilocular sporangia. These produce motile cells which behave either as zoospores, reproducing the haploid generation asexually, or as gametes. Isogamy is the general rule, but anisogamy, almost reaching oogamy, has been reported in some species. The zygote germinates to form a diploid plant.

VEGETATIVE STRUCTURE The manner of reproduction is fairly uniform throughout the Ectocarpales, but there is significant variation in the form of vegetative growth. *Ascocyclus*, for example, has an elaborate prostrate system, and in this respect resembles the green alga *Coleochaete*. The gelatinous, cushion-like thallus of *Leathesia*, although appearing parenchymatous, is formed by the adhesion of filaments. A truly parenchymatous condition is reached in *Punctaria* (Fig. 3.16), where cell division takes place in three dimensions and results in a leaf-like thallus closely resembling that of *Ulva*. The common *Pylaiella* is very similar to *Ectocarpus*, but the sporangia are usually intercalary. Tidal action often causes the axes of *Pylaiella* to roll together, forming a characteristic cable-like growth.

Laminariales This Order, the kelps, includes *Macrocystis* and *Nereocystis*, the largest known algae, together with a number of species harvested commercially as sources of the mucopolysaccharide algin. This

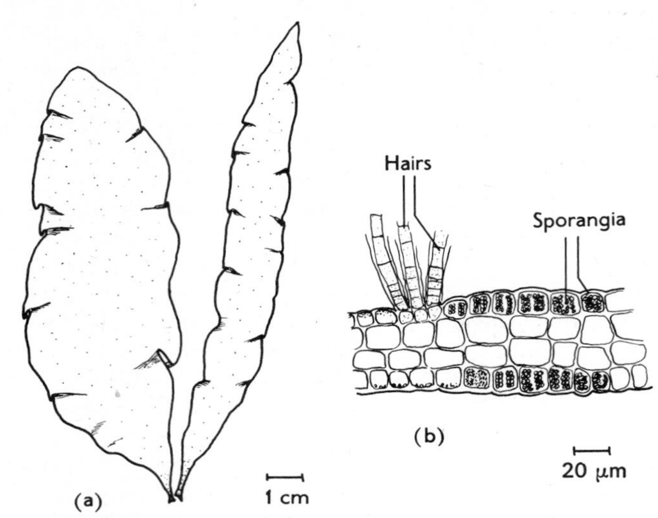

Fig. 3.16 *Punctaria latifolia*. (**a**) Habit. (**b**) Transverse section of thallus with sporangia. (After Newton, *A Handbook of the British Seaweeds*. British Museum, 1937)

yields alginic acid, widely used as an emulsifying agent in the food and paint industries, and in the processing of rubber.

The Laminariales are usually found below low water mark, but a few are regularly exposed at low tide. An example of the latter is the striking *Postelsia* of the Pacific coast of North America, a species with an arboreal habit, suggestive of a miniature coconut palm.

VEGETATIVE STRUCTURE In *Laminaria*, which may be taken as representative of the Order, the large thallus is differentiated into holdfast, stipe and blade (Fig. 3.17). Abrasion by tides and turbulence continually wears away the end of the blade, but the loss is made good by continued growth from a meristematic region at the base. A transverse section of the tough, pliable stipe reveals three anatomically distinct zones. In the outer zone (meristoderm), which is covered by a layer of protective mucilage, some cell division persists indefinitely. Pigmented plastids are also present in the cells. Within is a zone of paler, elongated cells forming a well-defined cortex, towards the interior of which the longitudinal walls become increasingly gelatinous. At the centre is a medulla consisting of intertwined and branching filaments, variously orientated and embedded in a mucilaginous matrix. The medulla and the inner cortex also contain columns of elongated cells broadened at each end, and hence referred to as

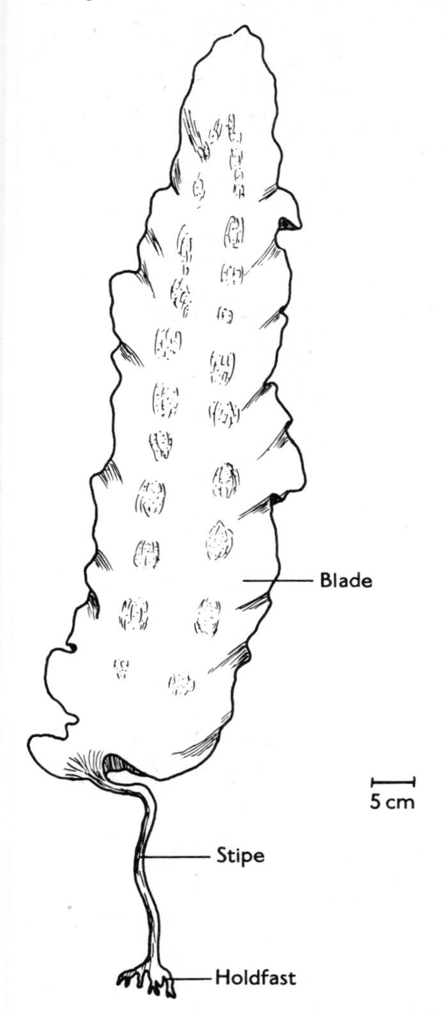

Fig. 3.17 *Laminaria saccharina*. Habit.

'trumpet hyphae'. The transverse walls of these cells are often perforated by groups of pits, recalling the sieve plates of higher plants. The stipes of some other genera of the Laminariales contain columns of cells showing an even closer resemblance to authentic sieve tubes. Like these they contain callose, and there is experimental evidence of similarity of function.

REPRODUCTION Reproductive areas, referred to as ***sori***, develop on the blades at certain times of the year. The sori consist of many unilocular sporangia interspersed with thick, sterile, protective hairs (***paraphyses***) (Fig. 3.18a). Meiosis occurs in the development of the sporangia, and they eventually yield haploid zoospores. These in turn develop into haploid gametophytic plants, much smaller than, and totally different in morphology from, the highly organized sporophytes (Fig. 3.18b). The gametophytes are dioecious, and, since it has been shown in a number of instances that spores giving rise to male and female gametophytes are produced in equal numbers in a sporangium, it appears that sex determination is genotypic.

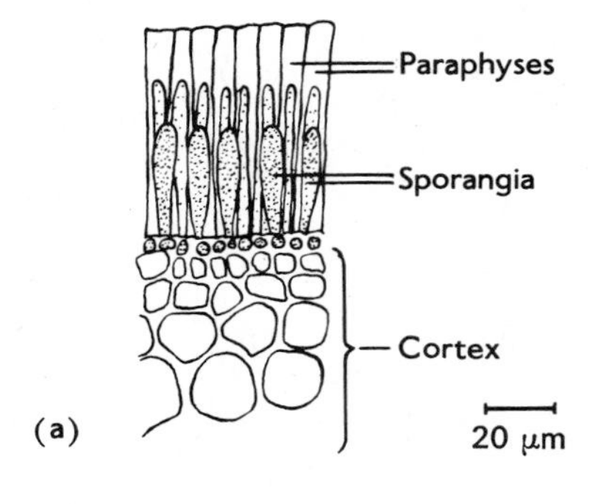

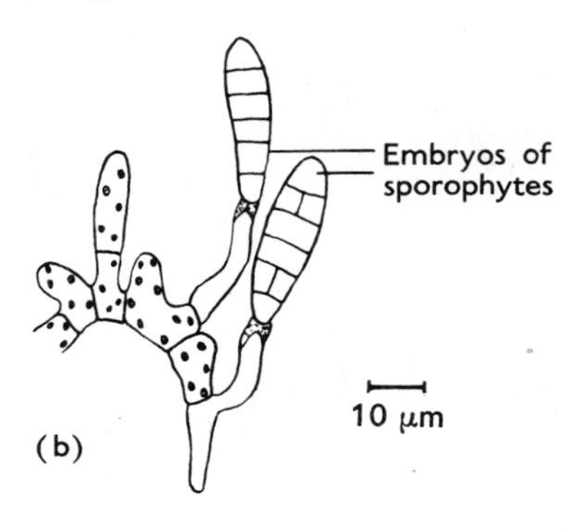

Fig. 3.18 *Laminaria.* (**a**) *L. cloustoni.* Transverse section of fertile region of lamina. (After Newton, *A Handbook of the British Seaweeds.* British Museum, 1937) (**b**) *L. digitata.* Portion of female gametophyte. (After Sauvageau, from Fritsch, *The Structure and Reproduction of the Algae,* **II**. Cambridge, 1945)

Although the gametophytes show a tendency towards heterotrichous growth, any cell seems capable of yielding a gametangium. Oogamy is fully developed, the oogonium producing a single egg which escapes at maturity through a pore at the apex of the cell. The egg, however, does not become free, but remains seated in a cup formed by the thickened margins of the pore. The male plant produces a number of terminal antheridia, from each of which is liberated a single antherozoid with two lateral flagella. Following fertilization, the zygote secretes an external membrane and develops into a new sporophyte without any resting period. The young embryo may remain attached to the female gametophyte for a short period, but it is doubtful whether this represents anything more than purely physical adhesion.

There is no specialized asexual reproduction of either the haploid or diploid generations.

Fucales The Fucales, the various species of which are known as 'wracks', are probably the most familiar of all seaweeds, particularly in the British Isles. The intertidal (littoral) regions of rocky shores often show a number of distinct horizontal bands, each consisting of an almost pure stand of a member of the Fucales. *Fucus* (Fig. 3.19), *Pelvetia* and *Ascophyllum* are genera frequent in these habitats. Desiccation of the thalli during exposure to air is prevented by the secretion of mucilage, and photosynthesis probably continues during low tide. A few Fucales grow in

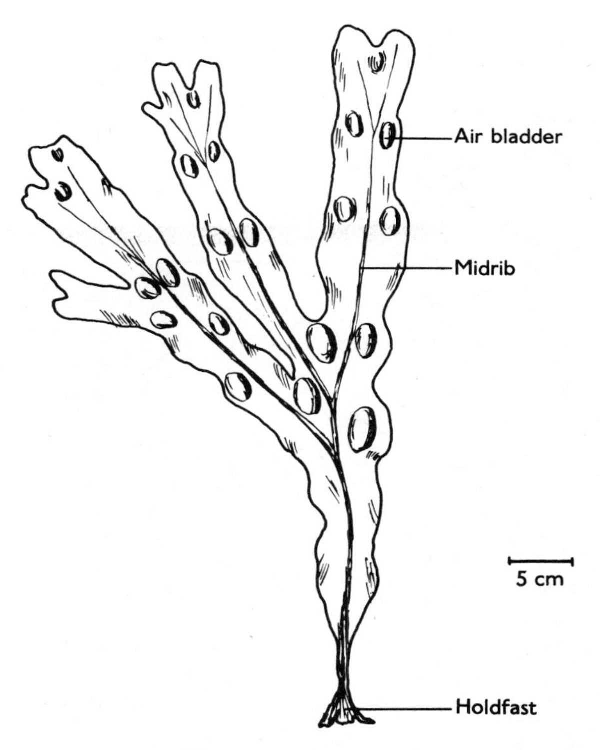

Fig. 3.19 *Fucus vesiculosus.*

deeper waters, and some (e.g. *Sargassum*) are free-floating in the warmer regions of the great oceans. Air bladders (with contents often different from the normal atmosphere) are frequently present in the thalli of the Fucales.

The thallus of *Fucus* is a much smaller structure than that of *Laminaria*, rarely exceeding 20 cm in length, but the differentiation into holdfast and blade is still evident. The flattened thallus is dichotomously branched and grows from apical meristems. A distinct midrib usually lies at the centre of each segment, branching in register with the thallus. The blade is gradually worn away by the action of the sea until, in the region of the holdfast, only the midrib remains, giving the impression of a short stipe (Fig. 3.19).

REPRODUCTION Both monoecious and dioecious species occur. In both, the reproductive structures form at the apices and these in turn become swollen with mucilage.

Microscopic examination reveals flask-shaped invaginations (***conceptacles***) in this swollen region, some of which contain male gametangia and others female, both interspersed with paraphyses (Figs. 3.20 and 3.21). Meiosis occurs during gametogenesis. The gametes are liberated at high tide.

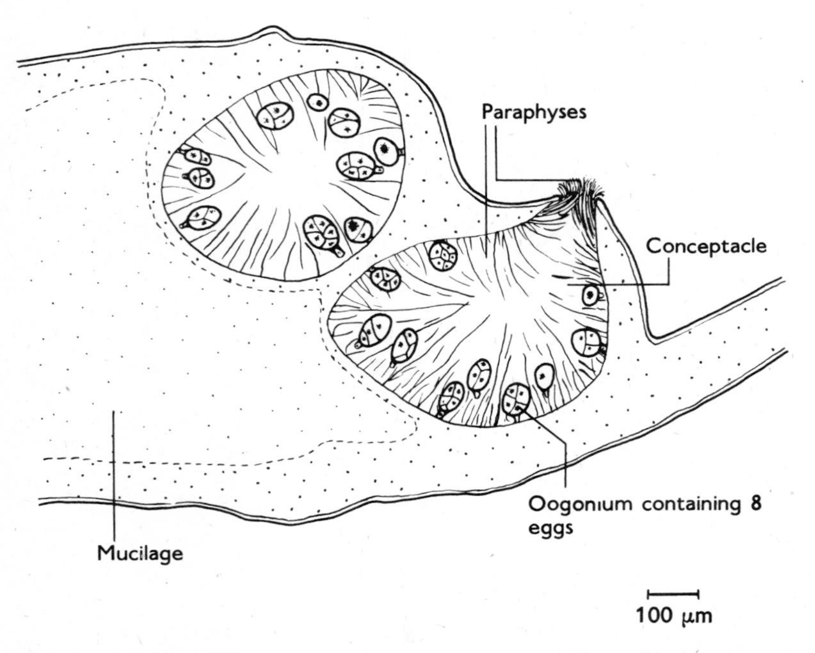

Fig. 3.20 *Fucus vesiculosus*. Transverse section of thallus with female conceptacles.

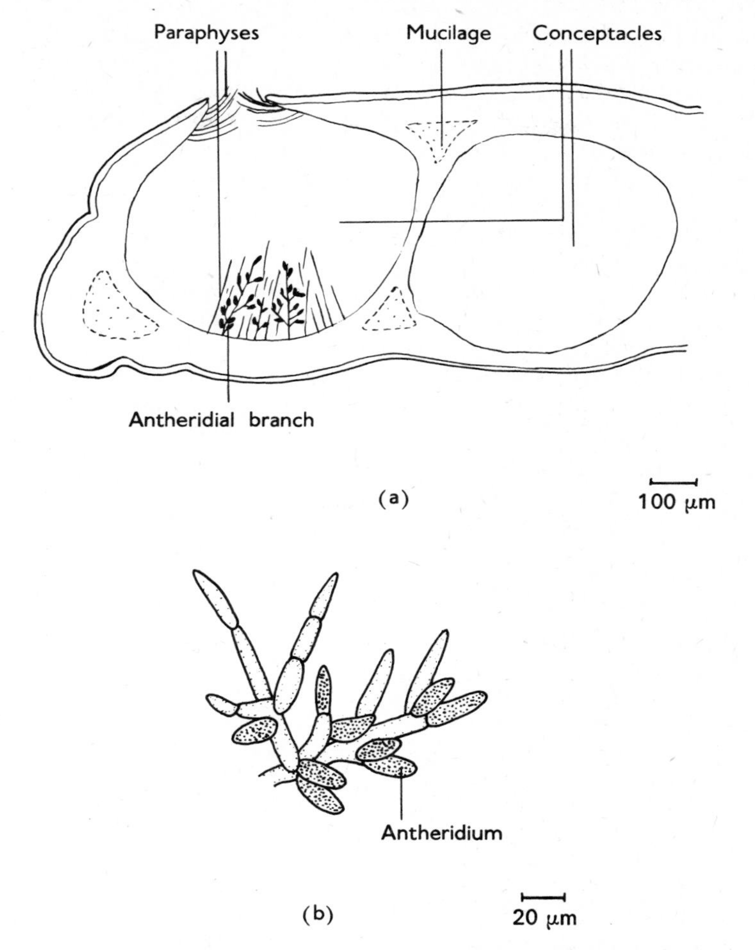

Fig. 3.21 *Fucus vesiculosus*. (**a**) Transverse section of thallus with male conceptacles. (**b**) Portion of antheridial branch. (**b** after Thuret, from Fritsch, *The Structure and Reproduction of the Algae*, **II**. Cambridge, 1945)

As in the Laminariales, oogamy is fully developed, but in *Fucus* the eight eggs produced in an oogonium become quite free and drift passively in the sea. The antherozoids, each of which has two lateral flagella, are attracted to it. Following fertilization, the zygote continues to drift while it secretes a mucilaginous envelope. It eventually settles, becomes anchored by the mucilage, and germinates. The first division of the zygote establishes the polarity of all subsequent growth, since one daughter cell gives rise to the holdfast and the other to the blade. It has been found experimentally that if the zygote is allowed to germinate in an environment that is not uniform, such as in a gradient of light, temperature or hydrogen ion concentration, the dividing wall always forms transversely to the direction of the gradient.[78] The subsequent behaviour of the daughter cells depends upon the nature of the gradient. With a gradient of temperature, for example, the cell on the warmer side yields the holdfast. In nature, of the two cells formed in the first division of the zygote, the holdfast develops from that in contact with the substratum. It seems likely that the orientation of this first division under natural conditions is determined by a combination of the environmental gradients that have been found effective in influencing the polarity in laboratory cultures.

Two different interpretations have been made of sexual reproduction in *Fucus*. One brings it into line with that of the Laminariales, interpreting the gametangia as homologous with unilocular sporangia, the gametophyte being reduced to nothing more than a gamete. The other draws an analogy with sexual reproduction in animals such as Man, where there is no question of an independent haploid generation.

There is no specialized asexual reproduction in the Fucales, but fragments of thalli may regenerate in favourable conditions to yield independent plants. The free-floating species of *Sargassum* reproduce solely in this manner.

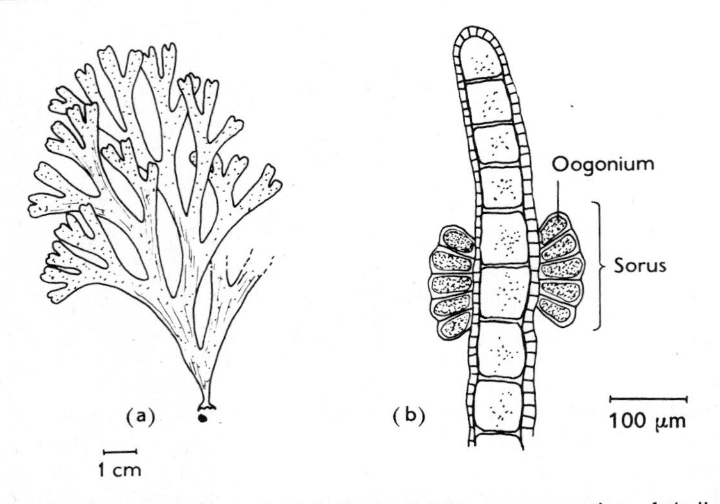

Fig. 3.22 *Dictyota dichotoma*. (**a**) Habit. (**b**) Transverse section of thallus of gametophytic plant showing oogonial sorus. (After Newton, *A Handbook of the British Seaweeds*. British Museum, 1937)

Dictyotales The thallus of *Dictyota*, a widely distributed genus, is flattened and dichotomously branched, but lacks a midrib (Fig. 3.22). It is commonly three cells thick, the two outer layers being assimilatory, while the central, which consists of larger cells, may act as a storage region.

REPRODUCTION Reproduction involves an isomorphic alternation of generations. Unilocular sporangia arise scattered or in groups over the surface of the diploid plant, and each yields, as a consequence of meiosis, four non-flagellate ***tetraspores***, a feature which distinguishes the Dictyotales from all other known algae. The tetraspores germinate immediately and give rise to the gametophyte generation.

In *Dictyota* the gametophytes are dioecious, although in some other genera monoecy prevails. The oogonia and antheridia are produced in groups on the surface of the thallus (Fig. 3.22b). Each oogonium produces a single ovum which, like that of *Fucus*, drifts passively in the water when released. The antherozoids have a single lateral flagellum (flimmer), but electron microscopy has shown that a second is present as a rudiment enclosed in the cytoplasm, recalling the situation in some uniflagellate Chrysophyceae (see p. 72).

Little is known of the germination of the zygotes, but it is probably similar to that of the tetraspores.

There is no specialized asexual reproduction of either generation.

Cutleriales *Cutleria* (Fig. 3.23), representative of this small Order, is a heterotrichous form showing a heteromorphic life cycle, but only one component of the heterotrichous system is developed in each generation.

The sporophyte, which displays only the prostrate component, grows as an incrustation on rocks, and was originally regarded as a distinct genus, *Aglaozonia*. The unilocular sporangia, in which, as usual, meiosis occurs, are superficial, and yield haploid biflagellate zoospores. On settling, these become attached to the substratum by a sucker-like process.

The gametophytes, which arise from the attached zoospores, consist principally of an erect filamentous system. The filaments lie parallel and adhere to each other for much of their length, forming a pseudoparenchymatous thallus. For a short distance at the apex, however, the filaments are free, and here there is distinct trichothallic growth. The gametophyte generation is monoecious, but the male and female gametangia are usually produced on separate filaments (Fig. 3.23c and d). Both the male and female gametes are biflagellate and motile, but the male are minute compared with the female.

The relationships of the Phaeophyta

The Phaeophyta are a circumscribed division of the algae, and little can be said of their wider relationship since no forms are known simpler than

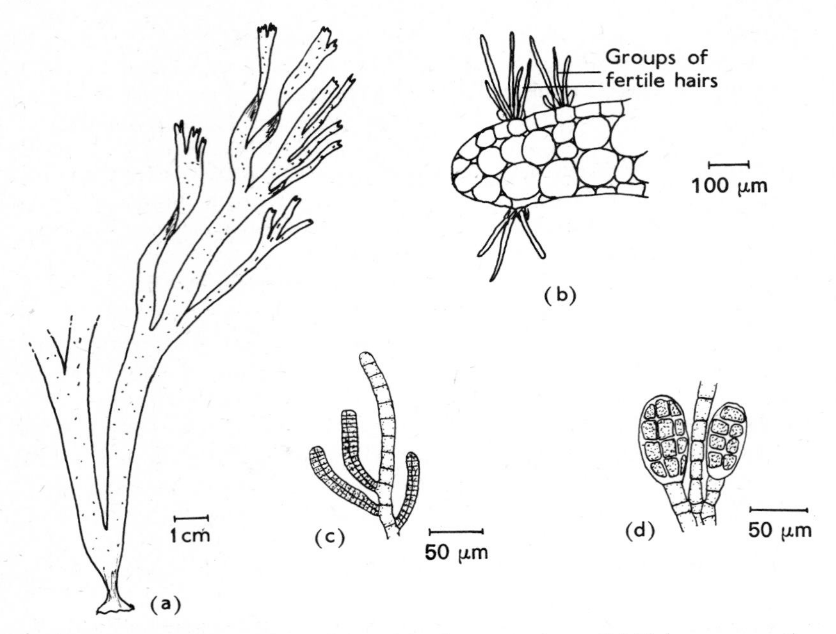

Fig. 3.23 *Cutleria multifida*. (**a**) Habit of gametophyte. (**b**) Transverse section of thallus. (**c**) Portion of antheridial filament. (**d**) Portion of oogonial filament. (After Newton, *A Handbook of the British Seaweeds*. British Museum, 1937)

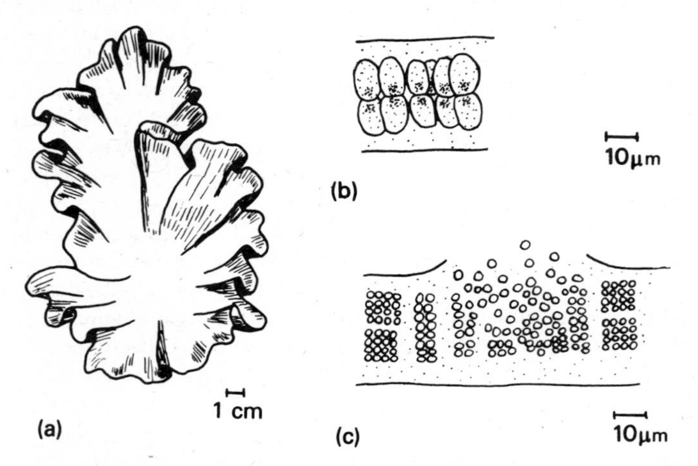

Fig. 3.24 *Porphyra umbilicalis*. (**a**) Habit. (**b**) Transverse section of thallus with monospores. (**c**) Transverse section of thallus producing spermatia. (After Newton, *A Handbook of the British Seaweeds*. British Museum, 1937)

the Ectocarpales. They presumably arose from some primitive flagellate ancestor, and proceeded to exploit a particular kind of pigmentation and metabolism that proved especially satisfactory in marine environments. There is no evidence that the Phaeophyta have ever contained forms becoming adapted, as some of the Chlorophyta, to terrestrial life. The few varieties of the Fucales (e.g. *Fucus vesiculosus* var. *muscoides*), all of limited reproductive capacity, which occur in salt-marshes mixed with halophytic flowering plants appear to be instances of specialization without any far-reaching significance.

RHODOPHYTA

Habitat	Aquatic (mainly marine).
Pigments	Chlorophylls *a* and *d*, α and β-carotin, lutein, biliproteins.
Food reserves	Floridean starch, compounds of sugars and glycerols.
Cell wall components	Cellulose, hemicelluloses, sulphated polysaccharides.
Reproduction	Asexual and sexual (oogamous).
Growth forms	Unicellular, filamentous, pseudoparenchymatous.
Flagella	None.

The Rhodophyta (red algae) are another circumscribed group, usually easily recognized by their bright pink colour caused by the biliproteins phycoerythrin and phycocyanin. The plastids are very simple in structure, and in many respects resemble the protoplasts of blue-green algae,[40] even to the extent (in some species) of possessing phycobilisomes (see p. 17) between the photosynthetic lamellae. Floridean starch, which appears outside and not within plastids, is chemically similar to the amylopectin of higher plant starches.

Although some species frequent rock pools, the majority live in warm seas at depths to which only the shorter blue wavelengths penetrate. These forms are consequently most commonly seen when washed up on beaches. A few species build up a calcareous exoskeleton, and are responsible for the formation of coral reefs. By contrast, the unicellular *Cyanidium*, which, although green in colour, on the basis of the pigments actually present and the structure of its plastids is probably a red alga, thrives in waters issuing from volcanic springs. These are often hot to the touch and may be as acidic as pH 2·0.

The Rhodophyta are the only eucaryotic algae which produce no motile forms. Even sexual reproduction, which is oogamous and complex, depends upon the passive dispersal of the male gamete.

A few red algae are of economic importance. *Porphyra* (Fig. 3.24), for example, is cultivated in the Far East and used as a foodstuff, as is *Chondrus crispus* to a lesser extent in Europe. *Gelidium*, a Pacific genus, is the source of agar.

The Rhodophyta fall into the Bangioideae and the Florideae, of which the former have the simpler sexual reproductive process. There are also cytological differences, one of the most conspicuous of which is the presence of pit connections between the cells in the Florideae. The two classes have probably diverged at an early stage in the evolution of the Division, the Bangioideae representing a more primitive kind of organization.

The Bangioideae

The simplest Bangioideae, mostly epiphytes on other marine algae, are heterotrichous forms. *Porphyra*, however, has a sheet-like thallus, one cell thick, resembling that of *Ulva*. The cells, embedded in a gelatinous matrix, contain a single nucleus and chromatophore, and in the later stages of growth divide at intervals irrespective of their position. Asexual reproduction is by means of ***monospores*** formed simply by the anticlinal division of vegetative cells (Fig. 3.24b).

SEXUAL REPRODUCTION IN THE BANGIOIDEAE In sexual reproduction both the male and female gametangia arise from single vegetative cells. The antheridium is formed by the repeated division of a cell into 64 or 128 portions, each of which yields a male gamete (***spermatium***) (Fig. 3.24c). The oogonium (often in the Rhodophyta termed a ***carpogonium***) is derived by the enlargement of a single cell, but no division occurs and only a single egg is present (Fig. 3.25). The coming together of the spermatium and carpogonium depends entirely upon water currents, but,

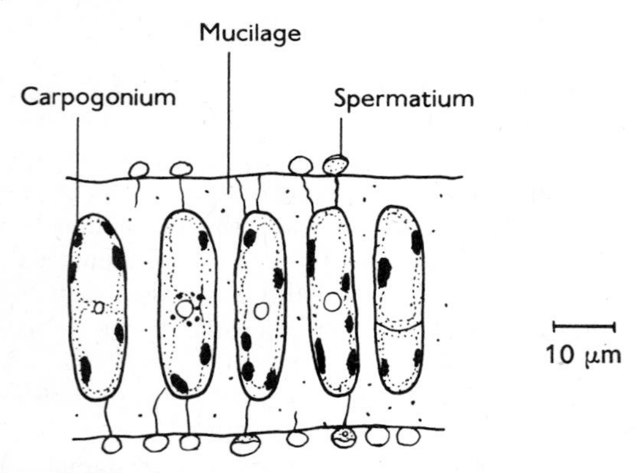

Fig. 3.25 *Porphyra umbilicalis*. Transverse section of thallus producing carpogonia. (After Dangeard, from Fritsch, *The Structure and Reproduction of the Algae*, **II**. Cambridge, 1945)

having approached, a spermatium is then retained by the mucilage surrounding the carpogonium.

Fertilization takes place by the protoplast of the spermatium first putting forth a narrow process which penetrates, possibly by means of localized lysis, the membrane of the carpogonium. The body of the spermatium then passes entirely into the female cell.

After fertilization, the zygote divides meiotically to release several carpospores, each of which becomes amoeboid for a while before secreting a cell wall, and germinating into a new plant. The position of reduction division in the life cycle indicates that the vegetative plant is haploid.

The Florideae

Representative of the simpler Florideae is *Batrachospermum*, one of the few Rhodophyta of fresh water. The vegetative organization, an axis bearing whorls of branches, is very similar to that of the green alga *Draparnaldiopsis* (p. 41). The filaments are enveloped in copious mucilage; this together with the dark pigmentation of the cells causes colonies of the alga *in situ* in ponds and streams superficially to resemble masses of frogs'

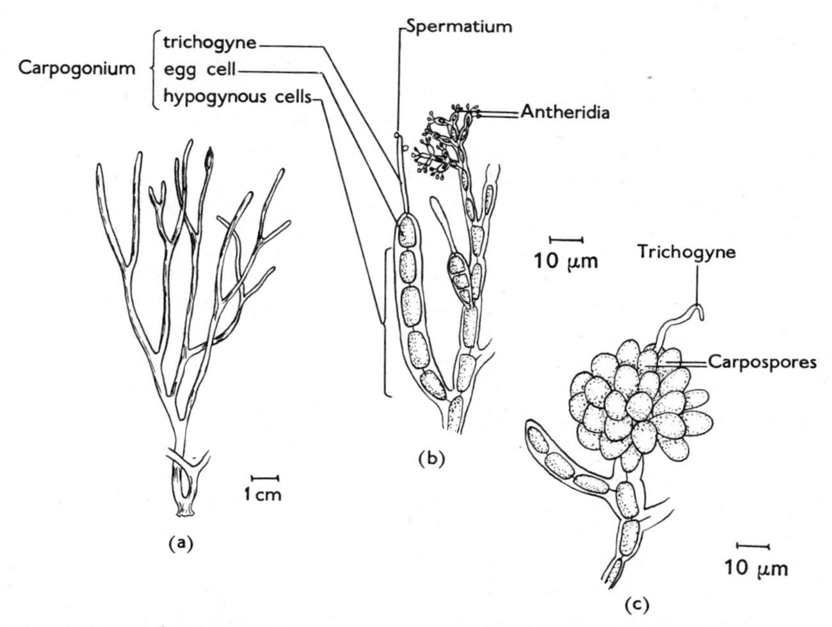

Fig. 3.26 *Nemalion multifidum.* (**a**) Habit. (**b**) Fertile lateral branch. (**c**) Formation of carpospores after fertilization. (After Newton, *A Handbook of the British Seaweeds.* British Museum, 1937)

spawn. *Nemalion* (Fig. 3.26), common in the intertidal zones of North Temperate shores, is basically filamentous, but the central filaments, which lie parallel and adhere to each other, are surrounded by an investment of short laterals enveloped in mucilage. The whole thus assumes a branching pseudoparenchymatous structure.

SEXUAL REPRODUCTION IN THE SIMPLER FLORIDEAE Sexual reproduction, even in these simpler Florideae. displays a number of peculiar features not encountered in other Divisions of the algae. The male and female gametangia, in many species produced on separate plants, arise on specialized side branches (Fig. 3.26b). The antheridia are budded off from mother cells which lie at the tips of short, tufted branches; as many as four or five antheridia may come from one mother cell. Each antheridium liberates a single spermatium through an apical slit. The carpogonium, surmounted by a long, tubular, hair-like process (the trichogyne) terminates a short carpogonial branch. When a spermatium makes contact with a trichogyne above an unfertilized egg, it becomes attached, and the intervening cell walls break down, allowing the contents of the spermatium to pass into the carpogonium. After fertilization, the zygote undergoes meiosis and puts out several short, branched filaments, each bearing a carpospore at its tip (Fig. 3.26c). Germination of the carpospores, similar to that of the carpospores of *Porphyra*, yields the normal gametophyte generation.

Thus, although the life cycle of these simpler Florideae involves alternation between two dissimilar kinds of plants, each is haploid. A further complication is that the carpospores may give rise to a transient juvenile form (the *Chantransia* stage) which is capable of reproducing itself asexually by simple spores. The *Chantransia* stage is eventually displaced by normal gametophytic growth.

VEGETATIVE AND REPRODUCTIVE FEATURES OF THE ADVANCED FLORIDEAE Even greater complexity, organizational and reproductive, is shown by *Polysiphonia*, representative of the more advanced forms of the red algae. The thallus, which consists of a central axis bearing freely branching laterals (Fig. 3.27), shows well-defined apical growth. This results from a dome-shaped apical cell which surmounts a central column of cells discernible throughout the thallus. This central column is surrounded from shortly below the summit by numerous columns of other cells produced by oblique divisions in the apical region. The structure is thus basically pseudoparenchymatous, although in parts, especially at the nodes, it is hardly distinguishable from a truly parenchymatous condition.

The gametophytic plants initiate sexual reproduction much as in *Batrachospermum* and *Nemalion*, although certain cells adjacent to the carpogonium also become distinguished by the density of their contents. These are termed auxiliary cells, and where auxiliary cells and carpogonia are parts of the same branch system they are said to constitute a procarp.

Following fertilization, the zygote again puts forth filamentous growths, but in *Polysiphonia* and its relatives these filaments contain diploid nuclei. They grow towards and fuse with the haploid auxiliary cells. This fusion, however, is restricted to the cytoplasms, and appears to serve as a stimulus, in the region of the fusion, to the production of carpospores. Fertilization and the subsequent events also stimulate vegetative growth at the base of the procarp, so that the developing carpospores become enclosed in a

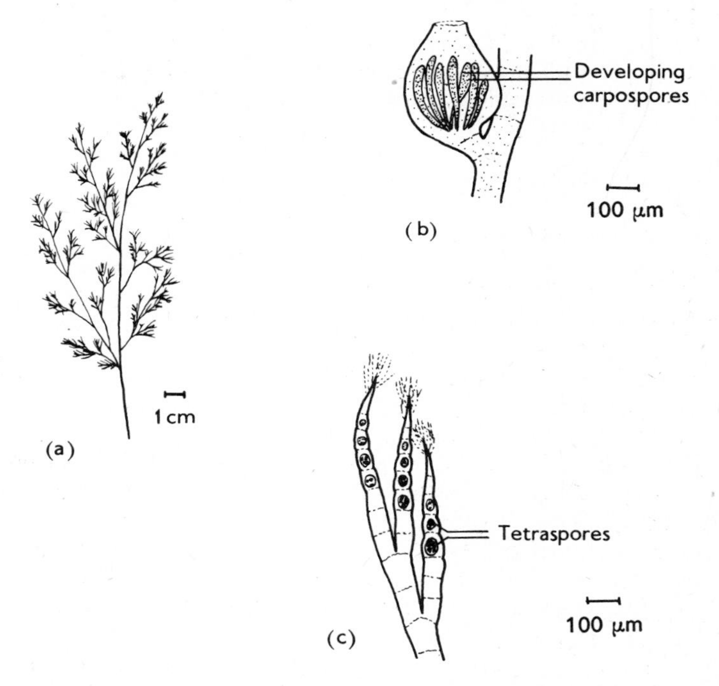

Fig. 3.27 *Polysiphonia nigrescens.* (**a**) Portion of thallus. (**b**) Cystocarp. (**c**) Portion of sporophyte bearing tetrasporangia. (After Newton, *A Handbook of the British Seaweeds.* British Museum, 1937)

filamentous cup with an apical pore. The net result is a fruit body, the ***cystocarp*** (Fig. 3.27b), the pseudoparenchymatous investment of which recalls in both form and formation that of the oogonium of the green alga *Coleochaete* (p. 43).

The carpospores of *Polysiphonia*, which are diploid, germinate directly to give rise to diploid plants morphologically similar to the haploid. They do not, however, bear sex organs, but in their place asexual tetrasporangia. These, which terminate short branches, are initially uninucleate, but in each the nucleus undergoes meiosis and four tetraspores are formed (Fig. 3.27c). The haploid condition is thus re-established, and the tetraspores germinate to form normal gametophytic plants.

THE LIFE CYCLES OF THE FLORIDEAE This brief survey thus provides examples of the two kinds of life-cycle in the Florideae. In the first, shown by *Batrachospermum* and *Nemalion*, there is one diploid phase (the zygote) and two haploid phases (the carposporangial filaments and the normal gametophyte). This kind of life cycle is termed ***haplobiontic***. In the second, represented by *Polysiphonia*, there are two diploid phases (the zygote together with the carposporangial system, and the plants bearing tetraspores) and one haploid phase (the gametophyte). These life cycles are termed ***diplobiontic***. Haplobiontic life cycles are less common in the Florideae than diplobiontic, and they are believed to represent a more primitive condition.

The relationships of the Rhodophyta

As with the Phaeophyta, the assesment of the wider relationships of the Rhodophyta is rendered difficult by the relatively few simple representatives of the division. Although motile forms are generally regarded as absent, some consider that puzzling organisms such as *Glaucocystis*, often interpreted as composite cells incorporating a blue-green algal cell as a symbiont, are in fact representative of a group of flagellates closely allied to the red algae. The similarities, both biochemical and structural, between the plastids of the Rhodophyta and the photosynthetic cells of the Cyanophyta are certainly striking, and possibly indicate the origin of the Rhodophyta in some such source.

There is an undoubted similarity between the reproductive process in *Polysiphonia* and that in Ascomycete fungi, but, if this resemblance is anything more than coincidental, it seems more likely that the Ascomycetes, being heterotrophic, are the derived forms.

The calcareous coral-forming Rhodophyta (e.g. *Lithothamnion*) have a fossil record that extends back to the Cretaceous. All these species belong to the more advanced Florideae.

THE EVOLUTIONARY TRENDS WITHIN THE ALGAE

From the foregoing survey of the principal features and interrelationships of the algae, we can now proceed to a consideration of the evolutionary and morphological trends displayed by the algal kind of organization as a whole.

The aquatic habitat and evolutionary change

A point of general significance in relation to the evolution of the algae arises from their predominantly aquatic habitat. One of the main factors influencing evolution within the Plant Kingdom has undoubtedly been environmental change. Nevertheless, aquatic plants, particularly those that are marine, are to some extent protected from such change, at least in a catastrophic form. The volume of the sea in particular is so vast that changes in such features as salinity and temperature must of necessity be gradual. Algae, therefore, have exploited to an extent greater than that of any other component of the world's vegetation an environment which demands only comparatively slow adaptation to changing conditions. This possibly accounts for the persistence of numerous states of algal organization intermediate between the simplest unicellular and the complex multicellular and of simple isomorphic life cycles (as in *Ulva* and *Ectocarpus*) without spores or zygotes suited to withstand unfavourable periods.

Closer inspection of individual groups reveals, of course, that continuity of structure and pattern is to some extent illusory. The Charales among the Chlorophyta and the Florideae among the Rhodophyta provide examples of the numerous groups of living algae which lack close relatives indicating the paths along which they may have evolved.

The antiquity of the algae

Geological evidence undoubtedly points to the algae being an extremely ancient form of life. Calcareous nodules (***stromatolites***), structurally similar to those produced by some living blue-green algae (e.g. *Lithomyxa*), are known from beds of pre-Cambrian age, possibly over 2,500 million years old.[41] Filamentous remains, containing organic matter and suggestive of blue-green or even green algae, have been found in pre-Cambrian rocks not less than 1,500 million years old.[7] Discoveries of this kind, which are more frequently substantiated than contradicted as palaeobotanists investigate the older rocks, indicate that an algal form of life antedated any other. The fossil record, although admittedly still very fragmentary, also supports the view that simple unicellular and filamentous algae preceded the complex pseudoparenchymatous and parenchymatous forms we have encountered in the three major divisions: Chlorophyta, Phaeophyta and Rhodophyta.

The evolution of the vegetative thallus

If the morphological progression from flagellate unicell, through coccoid, filamentous and pseudoparenchymatous states to large parenchymatous forms such as *Macrocystis* is also an evolutionary progression, we have to face the problem of what has caused and directed this progression. The cause presumably lies in the mutability characteristic of all life, and the direction taken by the progression is no doubt a consequence of natural selection, but we are regrettably almost entirely ignorant of the extent to which the aquatic environment exerts selective pressure. This is particularly so in respect of the marine environment and the planktonic algae. Consequently discussion of algal evolution involves considerable conjecture.

Nevertheless, reasoned speculation is not to be discouraged, and there are undoubtedly many features of the algae of evolutionary significance which merit consideration. Amongst the flagellate forms, for example, it is striking that the development of motile colonies has not proceeded beyond *Volvox*. Presumably, with a diameter beyond about 1 mm the *Volvox* system would become physically unstable, and the co-ordination of the thallus impossible. If the earliest forms were indeed flagellate, the evolution of a sedentary form from a motile would result in the energy that would otherwise be expended in swimming becoming available for growth and reproduction. The enhanced reproductive capacity would result in proportionately greater numbers, and provide the opportunity for the establishment of a line of sedentary organisms.

The tendency for daughter cells to remain united appears to be a basic one, and would account for the occurrence of some kind of colonial and multicellular forms in all the divisions of the algae. In the major divisions we must assume that the advantages of association, possibly again residing in metabolic and reproductive efficiency, led to the elaboration of multicellular thalli in which there gradually appeared division of labour amongst the cells. In a multicellular form such as *Ulothrix*, all the cells, except possibly the basal anchor, divide and liberate reproductive bodies simultaneously. There is no somatic tissue, and the individual is destroyed in reproduction. A form in which the individual persists through several reproductive phases clearly has an advantage in a situation (as might arise with prolonged turbulence) where conditions became temporarily intolerable for the reproductive bodies, but remained tolerable for the parents.

The evolution of complex thalli in which reproduction was confined to special areas or branch systems would thus be favoured. The further opportunity would then arise for various parts of the somatic tissue to become specialized, and assist either in the support or protection of the reproductive structures, as in the Florideae and many other algae, or in the exploitation of particular habitats, as with the air bladders of many Fucales.

The evolution of sexual reproduction

With regard to sexual reproduction, it seems beyond doubt that oogamy has evolved from isogamy. With forms inhabiting moving water, isogamous reproduction must be extremely wasteful, and there are evident advantages if one gamete remains relatively stationary, especially if it secretes chemotactic substances causing the male gamete to accumulate around it. Moreover, a zygote which begins life with a copious food reserve has a better chance of survival than one with little. Increasing size, however, severely limits motility, so again advantages can be envisaged in a situation in which one gamete, the male, remains small and motile, and the other, the female, loses motility and specializes in the laying down of food reserves.

A non-motile zygote may, of course, be disadvantageous if it settles in a situation unfavourable for the plant. This is compensated for in those algae such as *Coleochaete* where the zygote germinates to produce zoospores. Another development, possibly limiting the wastage of zygotes, is shown in *Laminaria* where the zygote germinates while still attached to the gametophyte, foreshadowing a feature of the archegoniate plants.

The life cycles of algae[26]

Sexual reproduction inevitably involves a cyclic alternation between a haploid and a diploid condition. The simplest life cycle found amongst the algae is that in which the diploid condition, generated by the fusion of

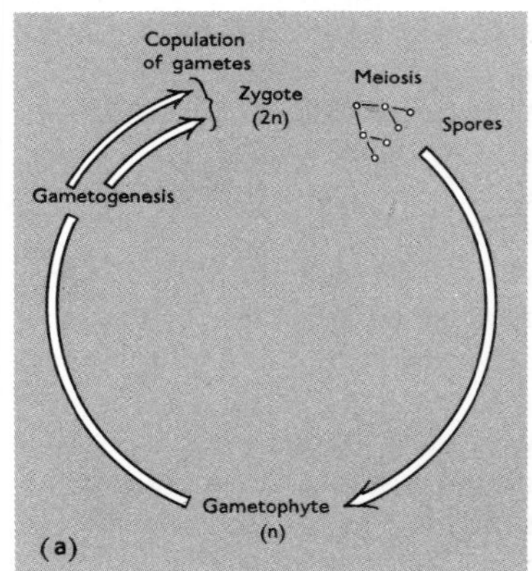

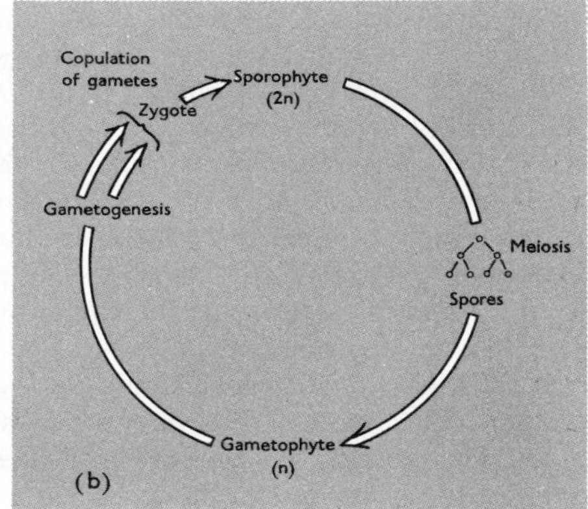

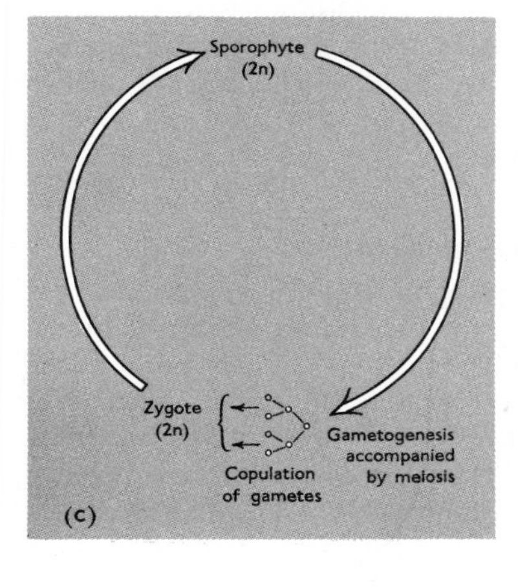

Fig. 3.28 Three kinds of life cycles found in the algae.

morphologically identical gametes, is represented only by the zygote (Fig. 3.28a). Meiosis occurs on germination of the zygote, thereby initiating a new haploid (gametophytic) generation. A cycle of this kind, termed ***haplontic***, is frequently encountered in the simpler algae, and is typical of the filamentous Chlorophyta. Here, however, it may have been retained and developed as an adaptation facilitating survival in unfavourable conditions, since only rarely does meiosis immediately follow syngamy, and the intervening diploid phase is often spent as a thick-walled, resting, zygote.

Closely related to the haplontic cycle, and possibly evolved from it by a delaying of meiosis, is that in which the zygote yields a multicellular, diploid (sporophytic) generation. This eventually produces reproductive bodies, almost always zoospores, by a process involving meiosis. The haploid condition is thus restored and the cycle recommences (Fig. 3.28b). Where, as in *Ulva*, the haploid and diploid plants are morphologically similar, the alternation of generations is isomorphic. Where the generations are morphologically different, as, to take an extreme example, in *Cutleria*, the cycle is heteromorphic (see p. 9). The evidence available does not warrant any general conclusion about whether isomorphic life cycles have evolved from heteromorphic, or the converse.

Superimposed upon these two basic kinds of life cycle are various conditions of anisogamy and oogamy. The peculiar complexities of the isomorphic and heteromorphic life cycles of the floridean Rhodophyta have already been described.

A third kind of life cycle, similar to that of most animals, occurs in the diatoms, Fucales, and in several other isolated instances throughout the algae. The diploid condition predominates, and the haploid is represented only by the gametes, meiosis occurring during gametogenesis (Fig. 3.28c). Again the evidence does not allow any general conclusion about how this kind of cycle, termed ***diplontic***, originated. As with the haplontic, it may in certain instances have selective value. In *Fucus*, for example, as compared with *Laminaria*, it is perhaps an advantage to have the gametophytic phase, possibly vulnerable to the vicissitudes of inter-tidal life, reduced to the unilocular sporangia and gametes.

The importance of the algae in the evolution of plants

Thus in the algae, the simplest of autotrophic organisms, a number of progressions, which can be regarded as representing channels of evolution, can be traced from unicellular to parenchymatous organization, from isogamy to oogamy, and from simple to elaborate life cycles. These all represent steps of fundamental importance in the evolution of plant life, and this fully justifies giving the algae considerable attention in any review of the Plant Kingdom. At its upper limit, the structural and reproductive complexity of an alga approaches that of a land plant. *Laminaria*, for example, possesses a thallus with marked morphological and anatomical differentiation, a strikingly heteromorphic life cycle, and a sporophyte attached in the early stages of its development to the gametophyte. Considered solely in terms of the level of organization, the transition from an advanced alga such as *Laminaria* to an archegoniate or even angiospermous land plant is small when compared with the evolution of that alga from a unicellular flagellate.

4

The Bryophyta
(Mosses and Liverworts)

The mosses and liverworts, although morphologically somewhat dissimilar, are classified together as the Bryophyta.[81] We treat them as comprising a Division of the Plant Kingdom, equivalent to those of the algae, possessing the following characteristics:

BRYOPHYTA

Habitat	Mainly terrestrial.
Plastid pigments	Chlorophylls *a* and *b*, β-carotin, xanthophyll.
Food reserves	Starch, to a lesser extent fats.
Cell wall components	Cellulose, hemicelluloses.
Reproduction	Heteromorphic alternation of generations; gametophyte the conspicuous generation. Sex organs with a jacket of sterile cells. Antherozoid with two whiplash flagella. Embryogeny exoscopic. Sporophyte producing non-motile, cutinized spores. Vegetative propagation of the gametophyte by fragmentation or specialized gemmae.
Growth forms of gametophyte	Thallus flattened, with some internal differentiation, or consisting of a main axis with leafy appendages.

The Bryophyta are the simplest terrestrial plants, although in some parts of the world, such as the bogs of temperate regions and the elfin forests of tropical mountains, they are a dominant part of the vegetation. Some species form dense communities at several metres' depth in antarctic lakes. Vast bogs in the northern hemisphere have been built up largely by the growth of the moss *Sphagnum.* The dead stems and leaves accumulating below the growing surface become consolidated to peat, often many feet in depth.

In some places peat is an important fuel, and in granulated form is widely used in horticulture as a source of humus.

The largest bryophyte, *Dawsonia*, is a tufted moss of swampy places in south-east Asia and Australia. Individual stems of this genus may reach or even exceed a metre in length, but this is quite atypical of the group. Most bryophytes are lowly plants, many of them inconspicuous and not easily seen without a hand lens. The cellular differentiation within the larger bryophytes is greater than in the algae, but lacks the complexity found in vascular plants. Amongst the mosses it reaches its maximum in *Polytrichum*, and amongst the liverworts in *Symphogyna*, in both of which groups of thickened and elongated cells occur in the central region of the stem. They approach in form the tracheids of higher plants, and lignin may even be present, but detailed patterns of thickening are absent.

Although a cuticle has been demonstrated in some bryophytes, in general they are little able to resist desiccation and are consequently principally to be found in damp and humid localities. The exceptions, such as species of the liverwort *Metzgeria* and of the moss *Orthotrichum*, common on tree-trunks, seem to produce sufficient cellular colloids to retain the water necessary for the maintenance of life in even prolonged dryness.

The morphology of the bryophytes is on the whole more complex than that of the algae. In the mosses, for example, the mature thallus regularly takes the form of a stem bearing leaves. On the other hand, the immature gametophyte (***protonema***) of many mosses closely resembles a heterotrichous green alga (see p. 39), and may, as, for example, in the moss *Pogonatum aloides*, persist for many months. Like the algae, the bryophytes produce no roots, although both the mature and immature forms of the thallus bear rhizoids. In a few forms, such as the liverwort *Symphogyna* and the moss *Polytrichum*, the aerial parts of the plant are continuous with fine subterranean creeping axes, superficially resembling the filiform rhizomes of the smaller filmy ferns. So far as is known, only one bryophyte (*Cryptothallus mirabilis*, a thallose liverwort)[84] is subterranean in habit and heterotrophic.

The bryophyte life cycle

Of the two types of plant in the bryophyte life cycle, the haploid gametophyte is the more persistent. The sporophyte (which consists of little more than a capsule, a stalk (***seta***) and a basal foot) grows upon it, and is wholly or partly parasitic and usually of limited life span. All bryophytes rely on free water for the dispersal of the antherozoids and fertilization.

Classification

The Bryophyta fall naturally into three Classes, the Hepaticae (liverworts), the Anthocerotae (hornworts), and the Musci (mosses). The features used in this classification are the nature of the thallus, and (where present) of the leaves, the extent of the development of the juvenile phase of the gametophyte (protonema), and the presence or absence of an opening mechanism in the capsule.

Hepaticae

Despite the diversity of the liverworts, there is little doubt that they form a natural group. There is no very distinct protonemal phase, such as that found in the mosses, and the thallus almost always shows recognizable dorsiventrality. A characteristic feature of many liverworts is the presence of oil bodies in the cells, thought by some to render the tissues unpalatable to grazing insects. The antheridia break open irregularly, instead of by a distinct cap cell, and in the sporophyte generation the capsule matures before the elongation of its stalk, the converse of the situation in the mosses.

Of the seven Orders of Hepaticae, the common Marchantiales, Jungermanniales and Metzgeriales will be considered in some detail, and the small Orders Sphaerocarpales and Calobryales mentioned on account of special features which claim attention.

Marchantiales VEGETATIVE STRUCTURE The Marchantiales are exclusively thalloid. Although some species are simple in appearance, internal

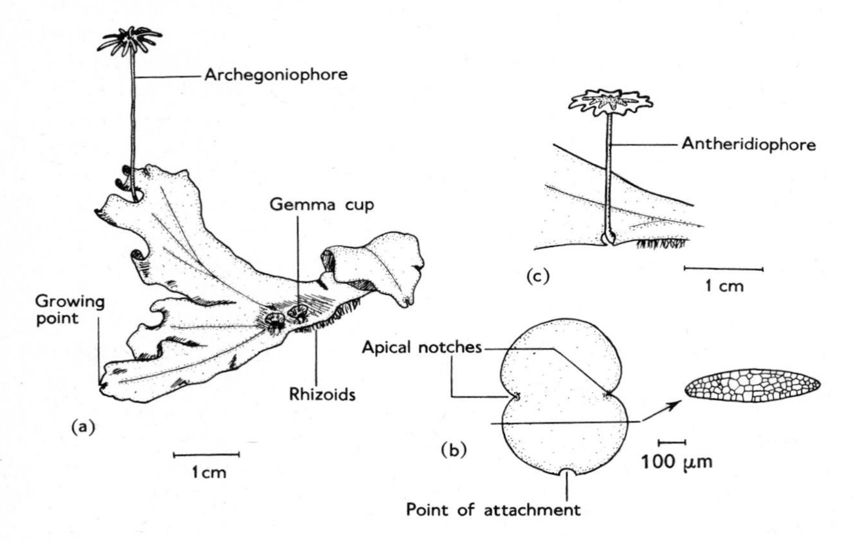

Fig. 4.1 *Marchantia polymorpha.* **(a)** Habit of female plant. **(b)** Structure of gemma. **(c)** Antheridiophore.

organization more complex than that found in any other thalloid liverworts is encountered in this Order.

The thallus of *Marchantia* itself (Fig. 4.1), found growing in damp places and areas of burnt ground, is dichotomously branched, with a thickened central rib and the surface divided into hexagonal areas visible with the naked eye. On examination with a hand-lens, a pore can be observed at the centre of each hexagonal area, which in transverse section is seen to consist of an air chamber containing photosynthetic tissue (Fig. 4.2). The pore, like the stoma of a higher plant, probably allows aeration of the thallus with the minimum dehydration, but is incapable of changing its aperture. Below the chlorophyllous tissue is a compact body of cells largely lacking chloroplasts. The lower side of the thallus bears up to eight rows of scales and unicellular rhizoids, the walls of some of which bear peg-like invaginations ('peg rhizoids').

Amongst other members of the Marchantiales which show a chambered thallus are *Concephalum*, the thallus of which yields a characteristic fragrance when crushed, and *Preissia*, in which the aperture of the pore is capable of some variation in response to atmospheric humidity. The midrib of the thallus of *Preissia* is also distinctive in containing elongated fibrous cells.

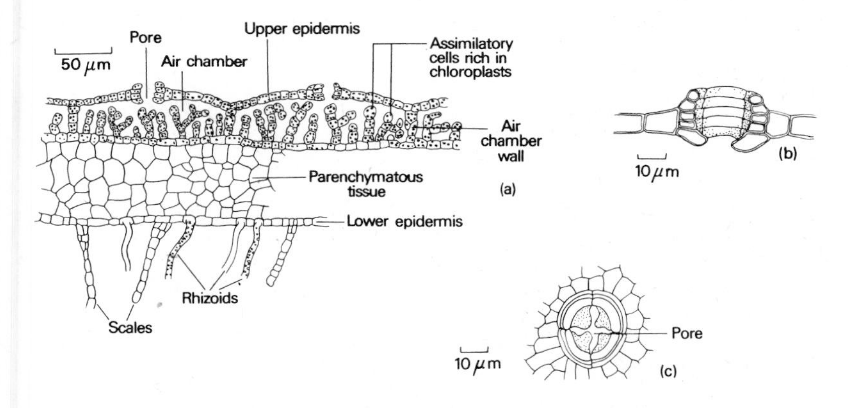

Fig. 4.2 *Marchantia polymorpha.* (**a**) Transverse section of thallus. (*b*) Transverse section of pore. (**c**) Surface view of pore.

Riccia represents the simplest kind of structure found in the Order. The lower part of the thallus is again a compact colourless tissue, but the upper part consists of columns of chlorophyllous cells, separated by narrow air channels. The upper cells of the columns are colourless and fit closely together, leaving no distinguishable pores. In *Riccia fluitans* (Fig. 4.8), a specialized form, the thallus is divided almost entirely into air chambers separated by partitions one cell thick.

SEXUAL REPRODUCTION *Marchantia* is dioecious. The male and female gametes are produced on upright, umbrella-shaped structures termed antheridiophores and archegoniophores respectively (and gametangiophores collectively) (Fig. 4.1a and c). Both structures develop from one half of a dichotomy of the apical cell, and are therefore homologous with a bifurcation of the thallus. Their morphological nature is clearly demonstrated by the rhizoids which grow down grooves in the stalks (Figs. 4.3 and 4.4), and by the characteristic photosynthetic chambers which develop in the caps of the mature gametangiophores. The female organs (archegonia) arise in radial rows on the upper surface of the cap. During the maturation of the archegonia, the cap grows more above than below, with the result that the archegonia become transferred to the lower surface. In the mature

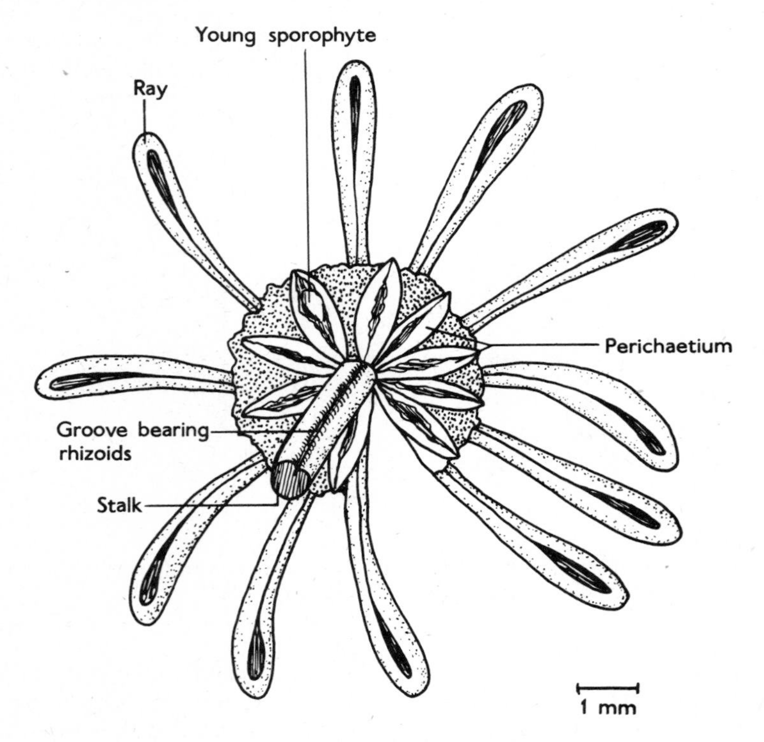

Fig. 4.3 *Marchantia polymorpha*. Archegoniophore seen from below.

archegoniophore (Fig. 4.5) each row of archegonia is separated from its neighbours by a curtain-like outgrowth, termed a ***perichaetium***. In addition, sterile processes emerge radially from the upper surface of the cap between the rows of archegonia, giving the whole its familiar stellate appearance (Fig. 4.3).

The archegonium, as always in the bryophytes, is formed in its upper parts by a single layer of cells, and has a strikingly long neck (Fig. 4.6). The egg lies at the dilated base of the central canal and, when mature, appears to be suspended in fluid. It is surmounted by a ventral canal cell, and a number of neck canal cells. These degenerate at maturity, and their products, when hydrated, give rise to a mucilage through which the antherozoids swim to reach the egg.

The antheridiophore lacks the complexity of the archegoniophore, being merely an elevated cap (Fig. 4.4), with the antheridia on the upper surface. Although superficial in origin, the mature antheridia are sunk in pits, each opening to the exterior by a narrow pore. Each antheridium is borne on a short stalk and bounded by a single layer of sterile jacket cells. When mature, it contains thousands of small, cubical cells (***antherocytes***) in each of which differentiates a biflagellate antherozoid. This, as is usual with the motile male gamates of land plants, consists principally of an elongated nucleus, the cytoplasm being a narrow investment containing a few organelles, probably only mitochondria.

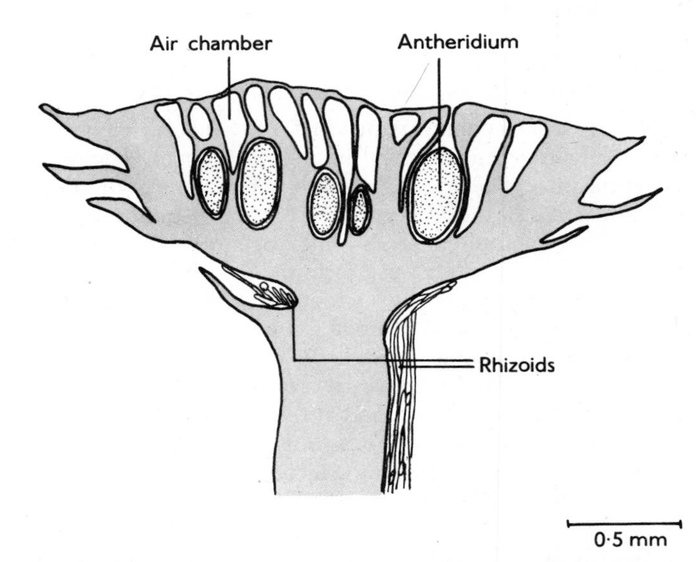

Fig. 4.4 *Marchantia polymorpha*. Vertical section of antheridiophore.

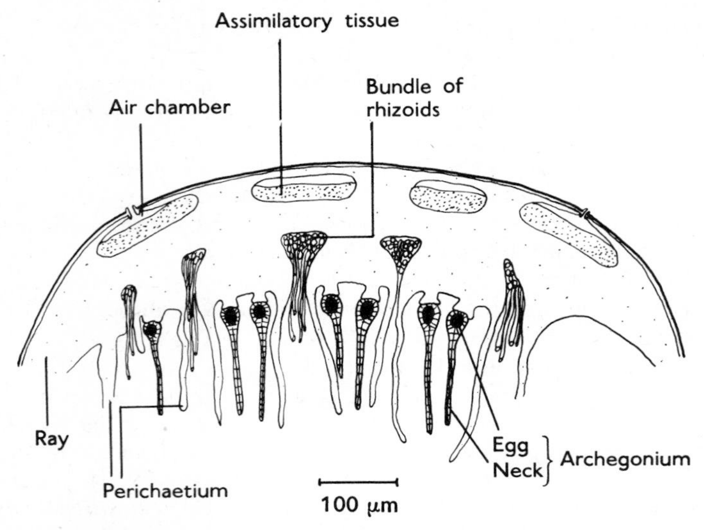

Fig. 4.5 *Marchantia polymorpha*. Tangential vertical section of young archegoniophore.

The mature antheridia open in moist conditions and the antherocytes are discharged. After a short period these in turn break open and release the antherozoids. For fertilization to be possible the male and female plants must be growing together. It seems likely that the gametangia become mature and fertilization occurs, before elongation of the gametangiophores.

DEVELOPMENT OF THE SPOROPHYTE OF *MARCHANTIA* Although detailed observations are few, germination of the zygote probably begins within 48 hours of fertilization. The first division is by a horizontal wall, transverse to the longitudinal axis of the archegonium. Since it is from the outer cell that the apex of the sporophyte arises, embryogenesis is said to be ***exoscopic***. The products of the inner cell form the foot, by which the sporophyte remains anchored in the gametophyte. Continued growth and differentiation, which are dependent upon nutrients drawn from the gametophyte, lead to an embryonic sporophyte consisting of three distinct regions. At the summit is the immature capsule containing the sporogenous cells, below this a short seta, and at the base the foot (Fig. 4.7).

At this stage, the young sporophyte is not only enclosed by the proliferated jacket cells of the archegonium, which form a ***calyptra***, but is also surrounded by a further tubular outgrowth of the gametophyte called a ***pseudoperianth*** or ***perigynium***. Division of the sporogenous tissue (***archesporium***) inside the capsule is mitotic until spore mother cells are produced, when meiosis takes place, giving rise to tetrads of spores. These

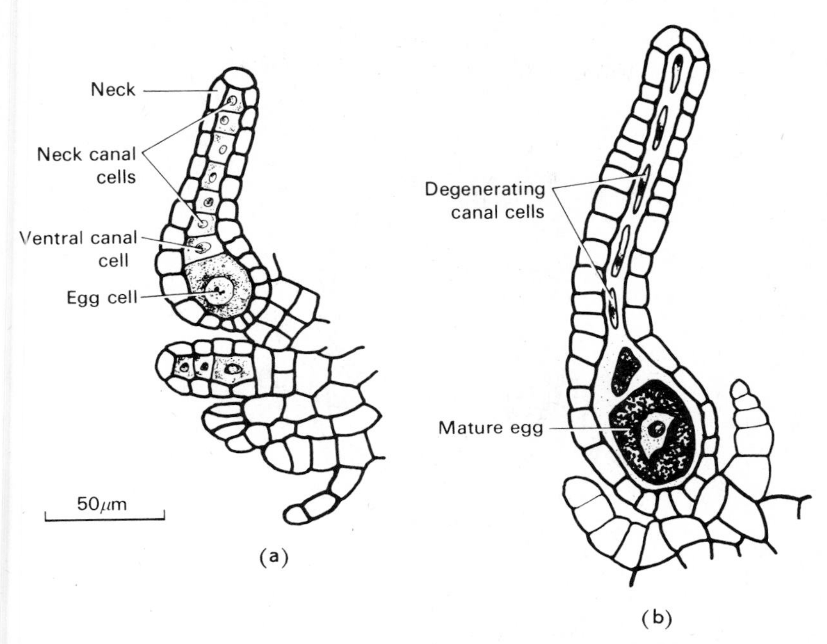

Fig. 4.6 *Marchantia polymorpha*. Archegonia. (**a**) Before breakdown of neck canal cells. (**b**) Mature archegonium.

separate in the capsule, become rounded in outline, and develop thickened walls. Not all the cells inside the capsule become spores; some (referred to as ***elaters*** because of their subsequent behaviour) elongate and lay down spiral thickenings.

Elongation of the cells of the seta eventually causes the calyptra to rupture, and, once exposed to air, the single layer of cells surrounding the capsule soon bursts, so revealing the mass of yellow, haploid, spores. The loosening of this mass and the dispersal of the spores are now assisted by the contortions of the elaters. These contortions are caused by the spiral bands in their walls, presumably consisting of cellulose micelles laid down in particular orientations, shrinking irregularly as they dry. In response to the strains generated in this way the bands attempt to change their curvature, and in turn cause the cell as a whole to make jerky twisting movements.

The spores germinate rapidly on a damp surface, giving rise to short, alga-like, filaments of cells. Division of the apical cell in a number of planes

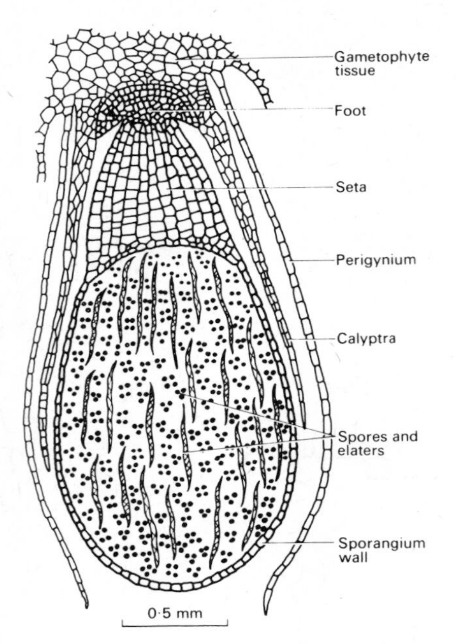

Fig. 4.7 *Marchantia polymorpha*. Longitudinal section of sporophyte rupturing the calyptra. Note the lines of the elaters. (After Parihar, *Bryophyta*. Allahabad).

soon initiates growth leading to the mature form of the gametophyte.

SEXUAL REPRODUCTION IN OTHER MARCHANTIALES No other genus has gametangiophores as elaborate as those of *Marchantia*. In *Conocephalum*, for example, the archegoniophore is a simple cap without emergent rays. In *Riccia* gametangiophores are entirely absent, both archegonia and antheridia merely lying at maturity in pits in the dorsal surface of the thallus. The sporophyte generation is again dependent on the gametophyte for nutrition, but at maturity it consists of only a sac of spores, with no seta or foot. In fact, by the time the spores are mature, no diploid tissue remains, and dispersal must await the decay of the gametophyte.

ASEXUAL REPRODUCTION Proliferation of the vegetative plant often follows from bifurcation of the thallus accompanied by rotting of the rear portion. In this way, an area becomes quite rapidly colonized by many seemingly individual plants. Additional to this, *Marchantia* has a notably elegant means of asexual reproduction. Multicellular bodies, called ***gemmae***, develop inside cup-like growths on the upper surface of the thallus (Fig. 4.1a). Each gemma is slightly biconvex, with two diametrically opposed, marginal notches, each containing a small meristem (Fig. 4.1b). When mature, the gemmae become detached from the short stalk on which they are borne, and are readily dispersed. Experiments have shown that the newly detached gemmae have no innate dorsiventral symmetry. This becomes fixed at germination by gradients of light, temperature and other factors in the immediate environment. Each meristem grows out to form a new thallus and, finally, two individuals result from the rotting away of the central portion.

It has been demonstrated experimentally that the germination of the gemmae while in the cup is inhibited by auxins diffusing basipetally from the apical meristem of the parent thallus.[25]

The evolution of the Marchantiales

Marchantia seems to represent the highest level of organization achieved by a wholly thalloid gametophyte. Are we therefore to regard the simple *Riccia* as a primitive Marchantialean plant, and *Marchantia* as an advanced form? Although this would appear plausible, some striking breeding experiments with *Marchantia* point in the other direction.[17] A number of mutants were raised from species of *Marchantia* in culture and hybridized in various ways, with the result that a whole series of forms was obtained which reproduced features found in other genera of the Order. The thallus of the var. *dumortieroides*, for example, lacks air chambers and resembles that of *Dumortiera*, a genus which, except for this feature, is close to *Marchantia*. Similarly the var. *riccioides* resembles *Riccia* in its narrow branching and the immersion of sex organs in the prostrate thallus. This reservoir of variation in *Marchantia* suggests that its evolutionary antecedents may have yielded the other genera of the Order by a process of simplification. *Riccia* would then be regarded as a reduced form.

On the other hand, perhaps both *Marchantia* and *Riccia* should be regarded as evolved forms. The archegoniophores of *Marchantia*, which elevate the seta-less capsule and facilitate the wide dispersal of the thin-walled spores in air currents, can reasonably be regarded as an advantageous development. In *Riccia*, however, the spores are thick-walled and long-lived. Despite its apparent rudimentary sporophyte, *Riccia* is probably no less well adapted than *Marchantia*, but to a different, Mediterranean-like, environment.

Jungermanniales and Metzgeriales VEGETATIVE STRUCTURE Both thalloid and leafy states are represented in these Orders, the largest among the Hepaticae. Most genera achieve only a small size in temperate regions, but species reaching several centimetres in length are common in the humid Tropics, where they are frequently epiphytic. The thalloid genera are typified by the common *Pellia*, which grows dichotomously as *Marchantia*, but differs in outward appearance. A poorly defined midrib of elongated cells

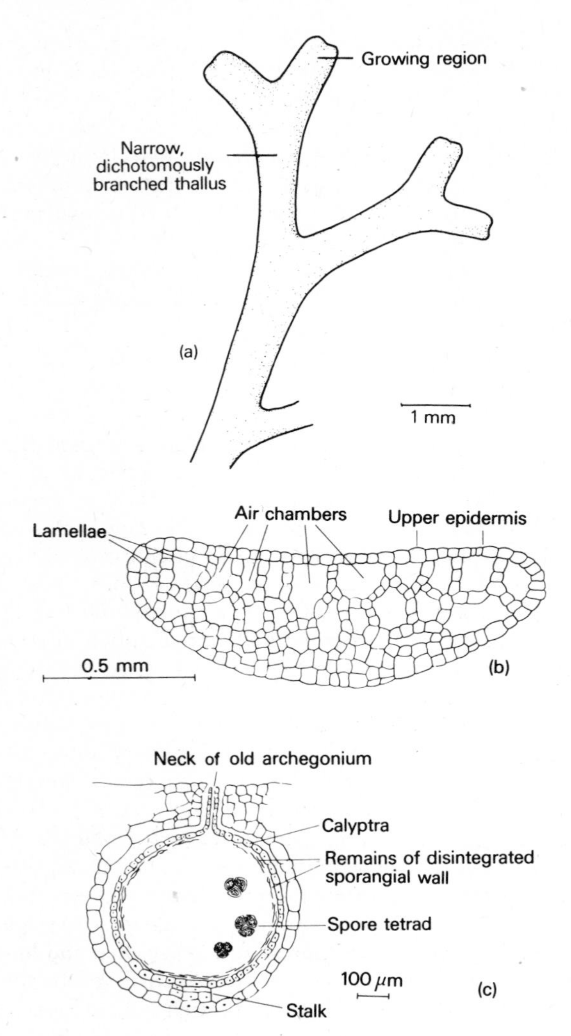

Fig. 4.8 *Riccia fluitans.* (**a**) Habit. (**b**) Transverse section of thallus. (**c**) Transverse section of thallus containing mature sporophyte.

extends to each apical region. Examination under the microscope shows that the thallus is indeed simple, having none of the specialized photosynthetic tissue of the Marchantiales. Some genera have much more distinct

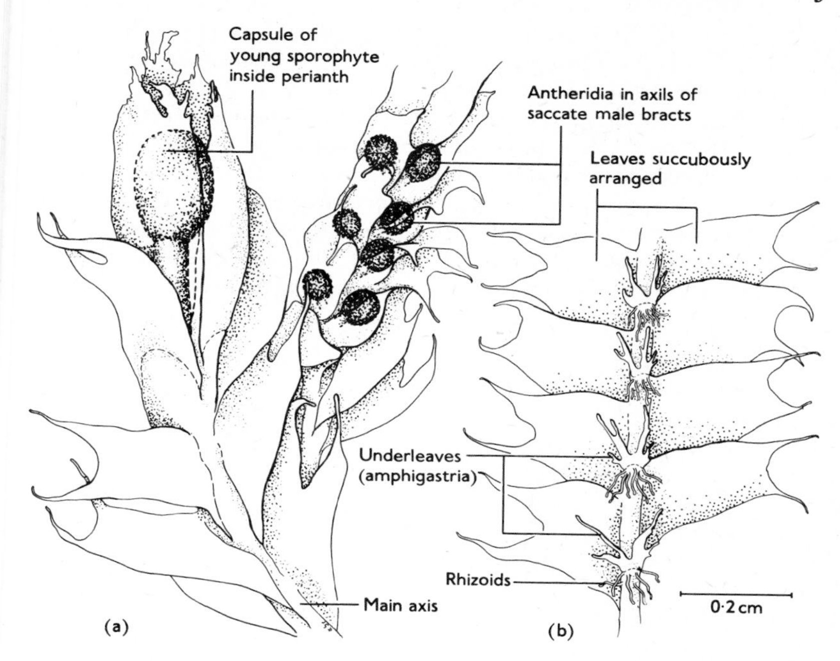

Fig. 4.9 *Lophocolea cuspidata.* (**a**) Fertile shoot seen from above. (**b**) Ventral surface showing amphigastria and the succubous arrangement of the leaves (see text).

midribs (*Pallavicinia*), while others in which the thallus is regularly dissected (*Fossombronia*), begin to resemble the leafy forms. Morphological differentiation in the thalloid forms reaches its peak in some of the tropical representatives. In *Symphogyna*, for example, a filiform underground rhizome gives rise, in a sympodial fashion, to a sequence of erect aerial thalli. These, to which the photosynthesis is confined, and on which the sex organs are borne, may reach a height of 2 cm.

The leafy liverworts also show a wide range of vegetative morphology, but here principally in the form of the leaves. These may be simple, more or less circular, plates of cells, as in the common *Odontoschisma sphagni*, or more often twice or several times lobed. Sometimes the leaves achieve great delicacy. In *Blepharostoma*, for example, the lobes consist only of a single file of cells. In *Frullania* the leaf has two lobes, the lower of which is shaped like a minute helmet, and possibly serves as a water sac.

In most leafy liverworts the stem is inclined or prostrate, and its symmetry clearly dorsiventral. Although the leaves are usually in three ranks, only those on the dorsal side are fully developed. Those of the third, ventral, row (termed ***amphigastria*** or under leaves) remain small, and are often shed a short distance behind the apex.

Classification of the leafy liverworts is based largely on the features of the leaves, including the orientation of their insertions on the stem. When the anterior margins of the leaves lie regularly beneath the posterior of those in front, the arrangement is said to be ***succubous*** (Fig. 4.9) and when the converse ***incubous***. The leaves of liverworts regularly lack nerves of the kind seen in mosses but the lower lobes of the bilobed leaves of *Diplophyllum* possess a conspicuous central row of elongated cells.

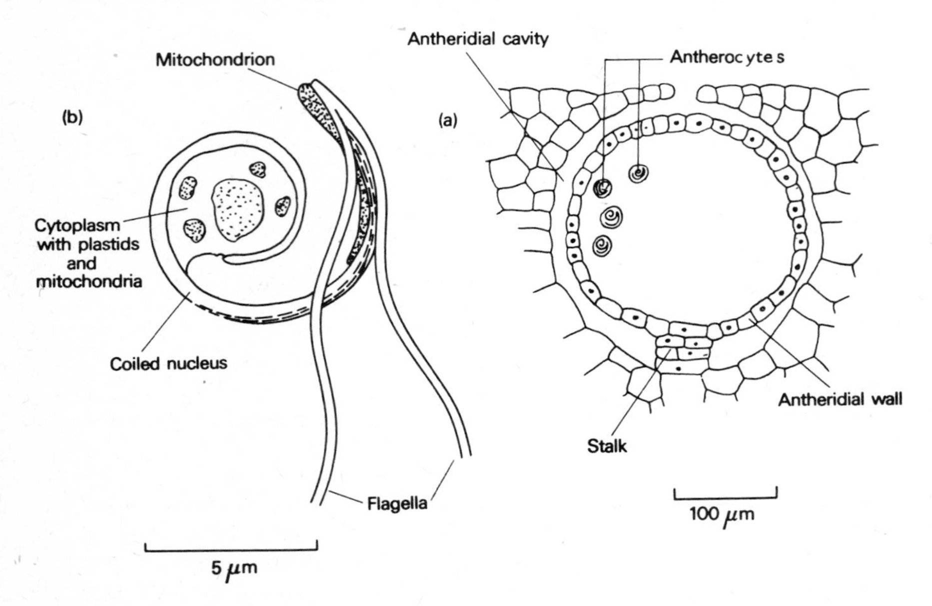

Fig. 4.10 *Pellia.* **(a)** Transverse section through antheridium. **(b)** Mature antherozoid.

SEXUAL REPRODUCTION Reproduction is essentially similar to that described for the Marchantiales, except that specialized gametangiophores are never produced. The antheridia, superficial in origin, usually occur singly, and either come to lie in a cavity in the upper surface of the thallus (e.g. *Pellia*, Fig. 4.10a), or, in leafy forms, lie in the axils of leaves of special branches of limited growth (Fig. 4.9a). The archegonia are usually grouped, and are produced either laterally, as in *Pellia* (Fig. 4.11), or at the tip of the main shoot, as in most leafy liverworts. When they are apical, they, and ultimately the sporophyte, terminate the growth of the main shoot, so that

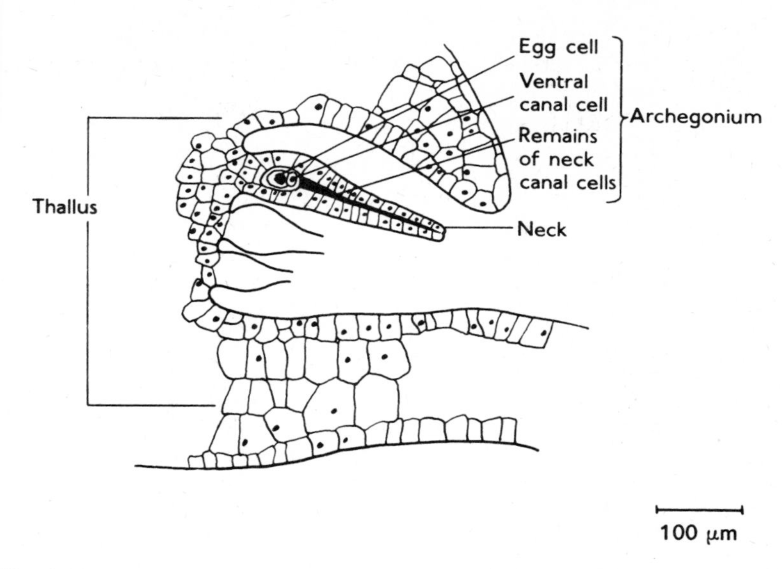

Fig. 4.11 *Pellia*. Vertical section of thallus showing archegonia.

vegetative growth is continued by a lateral, sympodial branching resulting. Both monoecious and dioecious forms occur, sometimes in the same genus. The common *Pellia epiphylla*, for example, is monoecious, but *P. fabbroniana*, common in calcareous districts, is dioecious.

The sporophyte of the Jungermanniales and Metzgeriales (Fig. 4.12) has a higher proportion of sterile tissue than that of the Marchantiales. The seta, for example, may often extend to more than a centimetre, but, as in the Marchantiales, the mature sporophyte lacks photosynthetic tissue. The capsule develops while still enclosed in the calyptra and the ultimate extension of the seta, which involves solely elongation of cells, is extraordinarily rapid. In *Pellia* rates of 1 mm per hour have been recorded in quite normal conditions. The elongating seta of *Pellia* also bends towards the light. Experiments have shown, however, that curvature occurs only in the regions illuminated. There is no transmission of the stimulus, a conspicuous feature of phototropism in coleoptiles and seedlings. This difference is presumably related to the absence in the sporophyte of *Pellia* of any distinct centre of growth.

The mature capsule of the Jungermanniales and Metzgeriales contains a mixture of spores and elaters. In *Pellia* the spores undergo several divisions before being shed (Fig. 4.13), and remains of the spore mother cells are visible between them.[85]

The wall of the mature capsule usually dehisces into four valves which, as a consequence of differential thickenings in the wall, become sharply reflexed. The dispersal of the spores is again aided by elaters (Fig. 4.13).

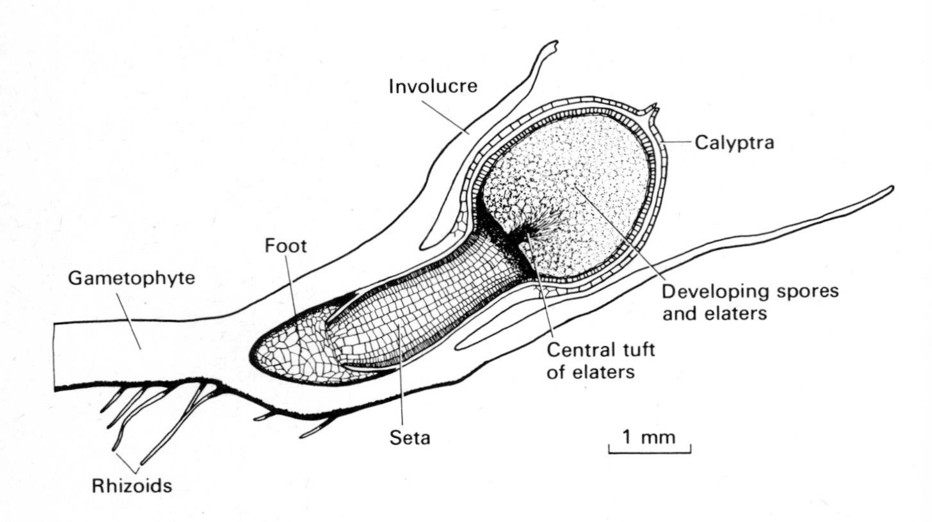

Fig. 4.12 *Pellia*. Longitudinal section of maturing capsule.

In *Pellia* the elaters, some of which remain as a brush attached to the top of the seta (Fig. 4.12), are similar to those of *Marchantia*. In some genera, however, the elaters are 'explosive'. In *Cephalozia bicuspidata*, for example, the elaters are loosely attached at one end to the valves of the capsule. After the capsule opens they begin to dry and in consequence to twist. Suddenly they violently untwist, hurling both the elater and its adhering spores into the air. The sudden expansion of the elater is believed to be caused by the shrinking column of fluid in the drying cell being put under such tension that it eventually spontaneously vaporizes, thus increasing its volume many times.

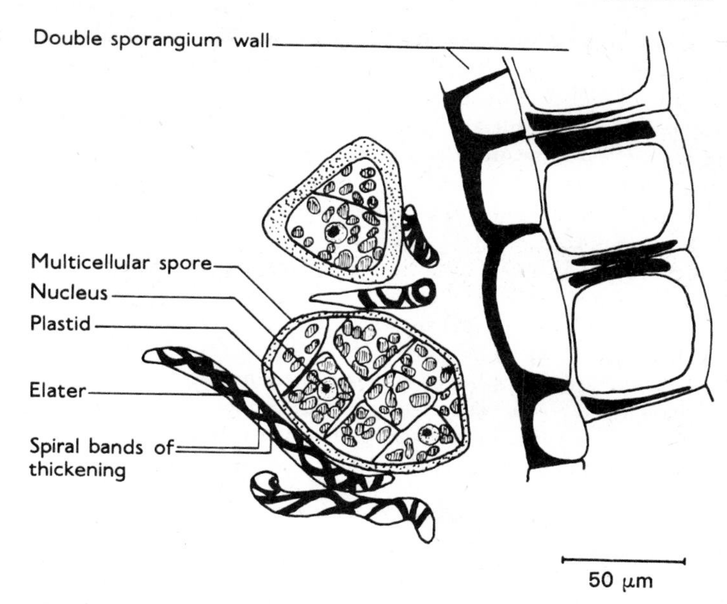

Fig. 4.13 *Pellia.* Portion of capsule showing detail of the wall, multicellular spores and elaters.

ASEXUAL REPRODUCTION Vegetative multiplication of many species takes place by regeneration from fragments of mature plants. Multicellular gemmae are not uncommon and are sometimes conspicuous, as the clusters of reddish, two-celled, gemmae on the margins of the upper leaves of *Sphenolobus exsectiformis*, a plant frequent on rotting wood. In *Blasia pusilla*, a thalloid form, multicellular gemmae are produced in remarkable flask-shaped receptacles on the dorsal side of the thallus.

Sphaerocarpales The three genera of this Order are similar to the simpler Marchantiales, such as *Riccia*, in habit, but the thallus has frilly margins suggestive of leaves. The antheridia and archegonia occur in small clusters, each cluster enclosed in a distinctive involucral sheath. As in *Riccia*, the sporophyte is little more than a capsule.

In some species of *Sphaerocarpus* the spores remain stuck together in their original tetrads in the mature capsule. This feature has been utilized in genetical work since the products of one meiosis can be cultured and examined individually. This allows for 'tetrad analysis', and the direct determination of the amount and kind of recombination between suitable genetic factors. *Sphaerocarpus* also has the distinction of being the first plant in which sex chromosomes were demonstrated,[1] the female gametophyte possessing a large X chromosome, and the male a small Y, in each instance accompanied by seven autosomes. Further, the earliest indication of the nature of the genetic material also came from this plant. It was found that the wavelength of ultraviolet light most effective in inducing mutations in the nuclei of the antherozoids was also that most strongly absorbed by deoxyribonucleic acid.[50a]

Naiadita, a fossil liverwort from uppermost Triassic rocks, is believed to belong to the Sphaerocarpales.[45]

Calobryales This Order, though small, is highly distinctive and the two genera it contains have a similar growth form. The upright stems, which are radially symmetrical and bear three ranks of leaves, rise from a creeping rhizome-like axis lacking rhizoids. The archegonia are effectively terminal. There are no involucral leaves protecting the young sporophyte, but the calyptra is particularly conspicuous. Some bryologists have considered the sexual reproductive structures of the Calobryales to be the most primitive amongst the bryophytes as a whole.

Anthocerotae

This Class contains the single Order, Anthocerotales. Although formerly included with the Hepaticae, the Anthocerotales are now usually placed in a separate Class, mainly on account of their unique sporophyte. *Anthoceros* is representative of its Class.

VEGETATIVE STRUCTURE The gametophyte of *Anthoceros* recalls *Pellia* in external morphology except that there is neither regular dichotomous growth nor a midrib (Fig. 4.14). The vegetative cells are undifferentiated, apart from internal cavities which contain mucilage and occasionally the blue-green alga *Nostoc*, a genus known to fix atmospheric nitrogen. In most species a single chloroplast, containing a complex pyrenoid, occurs in each cell, a situation unknown elsewhere in the Bryophyta or in higher plants—with the exception of some species of *Selaginella* (see p. 156)—but common in the algae. This has led to the suggestion that the Anthocerotae are closer to an algal ancestry than other Bryophyta.

Some species of *Anthoceros* (such as those of the Mediterranean region) regularly form tubers, enabling them to tide over a dry season unfavourable for growth.

SEXUAL REPRODUCTION AND THE FORM OF THE SPOROPHYTE In *Anthoceros*, as in *Marchantia*, the formation of the sex organs has been shown to depend upon photoperiod. In most species gametogenesis is initiated by diminishing day length, so that fertilization occurs during the winter.

The antheridia arise from a cell beneath the surface, one to several antheridia (depending upon species) coming to lie in a closed chamber, the roof remaining intact until the antheridia are mature (Fig. 4.14b). Archegonia arise superficially, but the wall of the archegonium is continuous

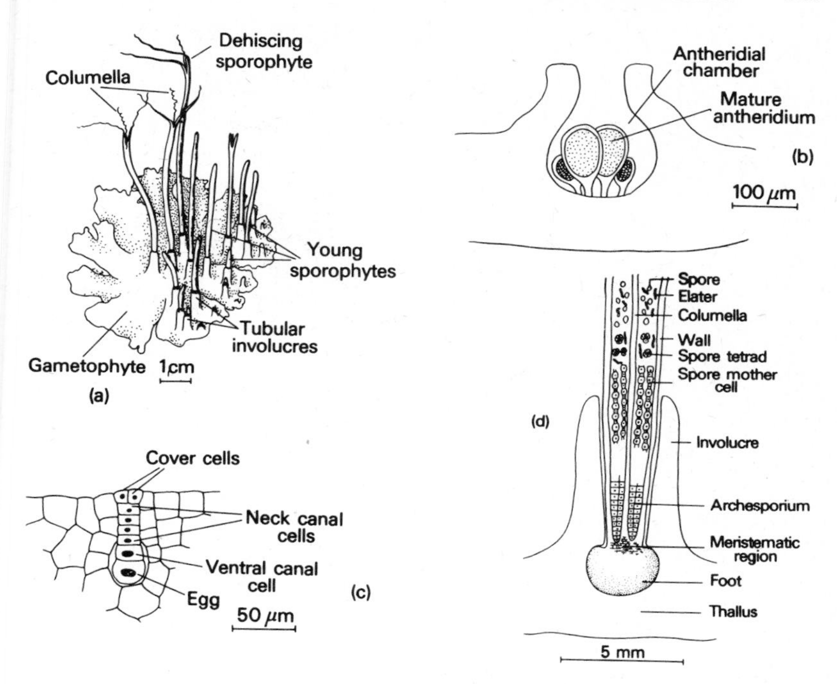

Fig. 4.14 *Anthoceros laevis*. **(a)** Female thallus with sporophyte. **(b)** Section of antheridial chamber. **(c)** Section of nearly mature archegonium. **(d)** Diagrammatic longitudinal section of a sporophyte showing the different regions.

with the thallus, the neck opening at the surface (Fig. 4.14c). This resembles the situation in some vascular archegoniates (see, for example,(Fig. 5.12). Development of the diploid zygote leads to a slender cylindrical sporophyte (Fig. 4.14) with a relatively small proportion of fertile tissue. The sporophyte remains inserted into the gametophyte by a conspicuous lobed foot, and the basal part is surrounded by an upgrowth of the thallus, the involucre.

The development of the archesporium begins as a dome-like layer within the summit of the sporophyte. The sporophyte continues to grow for several weeks from a meristem close to its base. During this growth the archesporium differentiates basipetally as a hollow cylinder, the centre of which is occupied by a sterile columella. When the spores at the top of the capsule are ripe, the capsule dehisces basipetally (Fig. 4.15) along two longitudinal slits, the opening beginning near the tip. As the upper part of the capsule dries, the valves separate completely above, and they begin to twist longitudinally. The consequent contortions of the valves expose the spores and multicellular elaters (which lack distinct spiral thickenings and

Fig. 4.15 *Anthoceros laevis*. Female plants bearing young sporophytes. Approx. × 0·67.

are often referred to as pseudo-elaters) on the central column. The separation of the valves continues downwards as the spores mature, and meanwhile the basal meristem generates new sporophytic tissue at about the same rate. Consequently a single sporophyte continues to yield spores over a considerable period.

Anthoceros and its allies further differ from other liverworts, but resemble the mosses, in possessing photosynthetic tissue in the outer layers of the extending sporophyte. Stomata are also present, as in the capsules of some mosses. The sporophyte is thus not entirely dependent on the gametophyte for nutrition, but, since the sporophyte will still mature even if it is covered with a tinfoil cap, it seems likely that a considerable amount of translocation from the gametophyte normally occurs.

The evolutionary position of the Anthocerotales

The Anthocerotales are remarkable amongst the liverworts in recalling features of both the algae (the presence of the pyrenoid in the chloroplast) and the mosses and higher plants (the presence of stomata and the continued growth of the sporophyte). They are consequently thought by some to stand close to the line of evolution leading from the algae to terrestrial vegetation. The intermediate position of the Anthocerotales extends even to their ultrastructure. The electron microscope confirms that the chloroplasts of *Anthoceros* resembles those of the algae, but the chloroplasts of *Megaceros*, where there are several in each cell, possess irregular grana and are more like those of other archegoniate plants.

Another feature of the Anthocerotales which has excited much attention is the axial form of the sporophyte. Since the sporophyte is often long-lived and may even persist for a time after the death of the parent gametophyte, it may indicate how simple axial plants, such as the psilophytes of the Silurian and early Devonian (p. 147), have evolved. Alternatively the *Anthoceros* condition may be derived, the sporophyte having become reduced and almost deprived of its independence.

Although evidence relating to either possibility is lacking, it seems quite plausible that a growth form similar to that shown by the living Anthocerotales did play some part in the evolution of land plants from algal ancestors.

Musci

The mosses are a class much greater in number and more widely distributed than the liverworts, occurring in almost every habitat supporting life. Apart from being the dominant vegetation in acid bogs, and alpine and Arctic regions, they are familiar features in woodlands and hedgerows, while some species even survive the polluted atmosphere of urban areas, often forming dark green cushions between paving stones and in other damp crevices.

The two morphological phases of the gametophyte are usually more conspicuous than in the liverworts. The protonema, formed when the spore germinates, is typically filamentous. At a certain stage of its development it gives rise to buds, which in turn yield mature leafy plants. The onset of this striking morphological change must depend upon the acquisition by the protonema of some kind of developmental maturity. A naturally-occurring growth factor (bryokinin) is known to stimulate the change in *Funaria*. The change in the form of growth probably follows from the protonema becoming able to manufacture certain additional ribonucleic acids and proteins. Many experiments have demonstrated that the formation of buds by protonemata can be greatly influenced by interference with this aspect of metabolism.[13]

In addition to the development of the gametophyte, other features which distinguish mosses from liverworts are the multicellular rhizoids, the growth of the sporophyte from an apical cell, the complex opening mechanisms of the capsules, and the absence of sterile elaters amongst the spores. The three Orders, Sphagnales, Andreaeales and Bryales, differ principally in the nature of the protonema and the structure of the capsule.

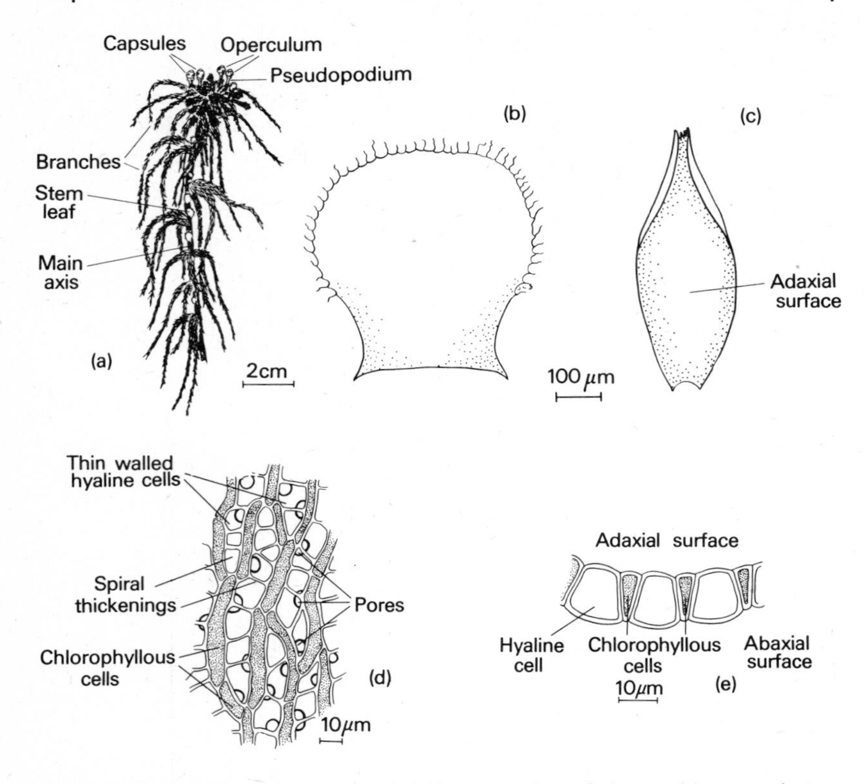

Fig. 4.16 *Sphagnum fimbriatum*. (**a**) Upper portion of shoot with sporophytes. (**b**) Stem leaf. (**c**) Branch leaf. (**d**) Arrangement of cells in leaf, adaxial surface. (**e**) Section of leaf.

Sphagnales The Sphagnales, represented by a single genus, *Sphagnum*, are confined to acid waterlogged habitats. They are the principal component of peat bogs, where they form a more or less continuous spongy layer.

VEGETATIVE STRUCTURE The adult gametophyte comprises an upright main axis from which whorls of branches arise at regular intervals (Fig. 4.16). The leaves, which are closely inserted, have a peculiar structure which is diagnostic of the genus, and also, in its finer details, of the many species (Fig. 4.16c and d). When first formed, the leaves are made up of many diamond-shaped cells. These then cut off narrow daughter cells, but on two sides only. The daughter cells develop chloroplasts, while the mother cell remains colourless, often becomes spirally thickened, and eventually dies. These dead cells also become porose during differentiation, enabling them to act as reservoirs. This peculiar leaf structure accounts for the ability of the *Sphagnum* plant to retain large quantities of water, and consequently for its outstanding bog-building properties.

REPRODUCTION Reproduction of *Sphagnum* is probably principally vegetative, the decay of the older parts eventually causing branches to separate, and thus to become new individuals. Mature plants do, however, produce sex organs in favourable situations, and both monoecious and dioecious species occur. The antheridia, each of which begins its development from a single apical cell, lie in the axils of leaves towards the tips of small upper branches (Fig. 4.17). These antheridial branches are often strongly pigmented and clustered in a conspicuous comal tuft. The female inflorescence consists of a bud-like aggregate of archegonia and bracts borne laterally near the summit of the main stem.

After fertilization the zygote yields a sporophyte (Fig. 4.16a) consisting principally of a capsule, containing a dome-shaped archesporium, and a foot. The seta remains inconspicuous, and the function of elevating the capsule is taken on by the base of the female inflorescence which, as the capsule matures, grows up as a leafless axis, or ***pseudopodium*** (Fig. 4.18). Release of the spores is brought about by air pressure which builds up in the lower half of the capsule as it dries. Eventually this pressure is sufficient to dislodge the clearly differentiated lid (***operculum***) with explosive force, and the spores are effectively dispersed.

The spores germinate to form a filament, but this is rapidly replaced by a small thallose protonema. This in turn gives rise to a bud which develops

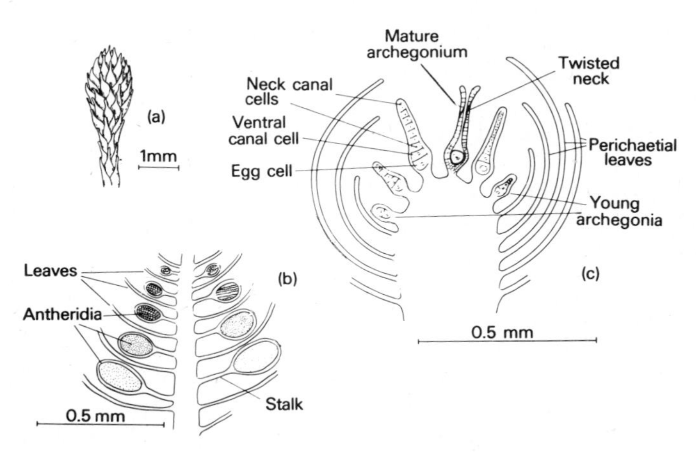

Fig. 4.17 *Sphagnum* sp. (**a**) Antheridial branch. (**b**) Longitudinal section of antheridial branch. (**c**) Longitudinal section of archegonial branch.

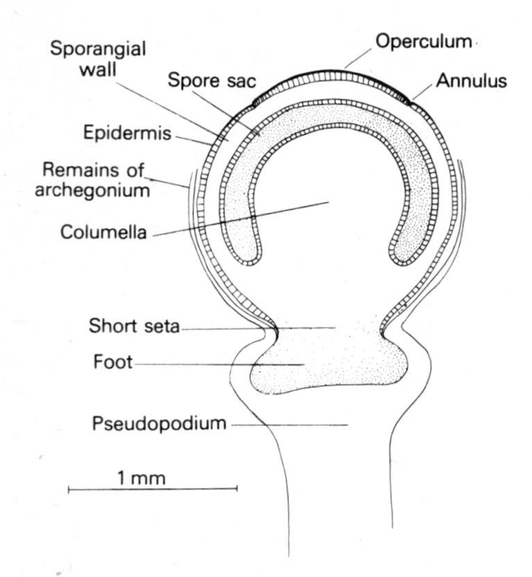

Fig. 4.18 *Sphagnum* sp. Longitudinal section of nearly mature sporophyte.

into the familiar leafy gametophyte, the protonema meanwhile becoming moribund and disappearing.

Andreaeales The Andreaeales are another Order containing only a single genus, distinguished by its peculiar capsule.

The leafy gametophyte of *Andreaea* (Fig. 4.19) rarely exceeds 1 cm in height. It is usually found growing on rock, chiefly in cold, exposed and relatively dry regions. The leaves are olive-brown in colour, composed of rounded cells, and in most species showing no distinct midrib.

Sex organs are formed apically. The sporophyte resembles that of *Sphagnum* in having a domed archesporium (Fig. 4.19c), and in being borne on a pseudopodium at maturity. Dehiscence of the capsule takes place by four longitudinal slits which do not meet at the tip (Fig. 4.19b). The hygroscopic properties of the wall cause the slits to close in damp conditions, and to open again in dry (Fig. 4.19a).

The protonema of *Andreaea* is similar to that of *Sphagnum*.

Bryales The 600 or so genera of the Bryales form a well-defined Order. Although there is a common basic morphology and life cycle within the Order, the variation in size, detailed structure and habitat preferences is considerable. Many mosses are confined to permanently damp situations in woodlands and by springs, but others, e.g. *Tortula ruraliformis*, are able to survive periods of drought in sand dunes and other arid habitats. The

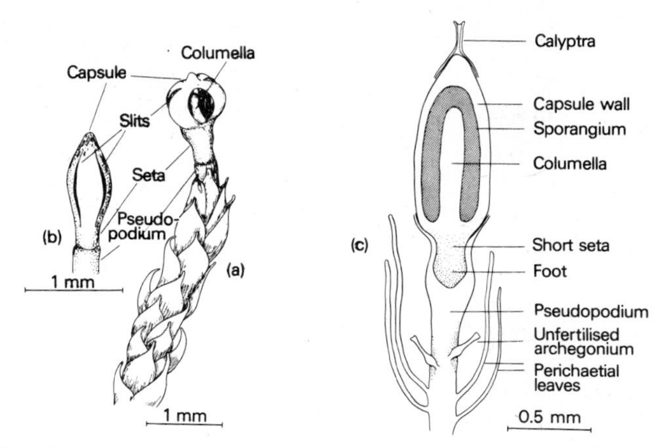

Fig. 4.19 *Andreaea nivalis.* (**a**) Habit of fertile plant showing dehisced capsule in dry condition. (**b**) Dehisced capsule in wet condition. (**c**) Longitudinal section of mature sporophyte.

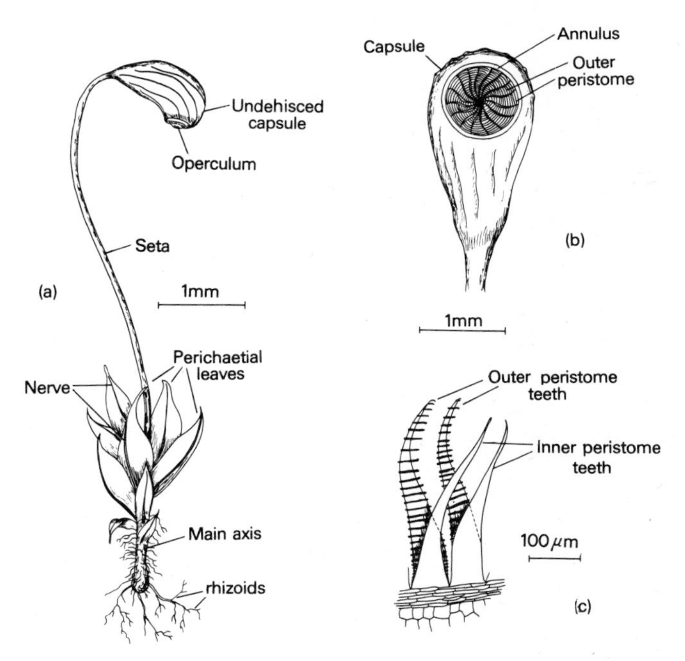

Fig. 4.20 *Funaria hygrometrica.* (**a**) Habit of fertile plant. (**b**) Mature capsule showing intact peristome. (**c**) Portion of peristome viewed from the inside.

cells of these species appear to have acquired the capacity to continue metabolism at a reduced rate while partially dehydrated. At the other extreme are a few sub-aquatic species, such as *Fontinalis antipyretica.*

VEGETATIVE STRUCTURE The most conspicuous form of the moss plant is the adult gametophyte (Figs. 4.20 and 4.25). This consists of a main axis bearing leaves which, although usually spirally inserted, may in some forms come to lie in one plane, giving the shoot a complanate appearance (e.g. *Neckera*, common on banks and rocks). In a few species (e.g. *Fissidens*) the leaves are equitant and arranged in two ranks. The leaves of most mosses consist of a single sheet of cells, although the central region may be thickened and contain a well-defined midrib (often referred to as a 'nerve'), sometimes excurrent in a hyaline point. The most complex leaf is found in *Polytrichum* and its allies. Here a number of parallel longitudinal lamellae grow up from the upper surface (Fig. 4.21), and the chloroplasts occur

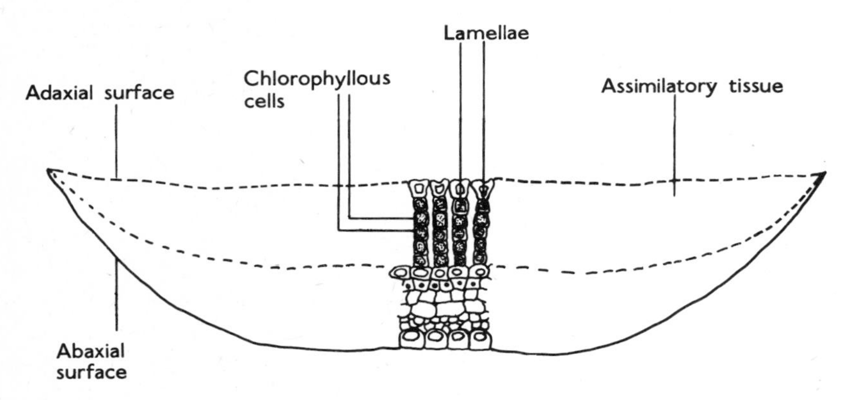

Fig. 4.21 *Polytrichum commune.* **Transverse section of leaf showing the assimilatory lamellae.**

principally in these cells. The shape of the leaf, and the nature and development of the midrib and of the cells at the margin of the leaf are important features in the taxonomy of the mosses.

Anatomically, the mosses offer little that is remarkable, the most complex differentiation being found, as already mentioned, in the stem of *Polytrichum* (Fig. 4.22). Not only is there an approach here to the development of tracheids, but there is also a clear radially symmetrical zonation in structure, recalling that of the axes of some of the smaller ferns. Surrounding a central core of tracheid-like cells (sclereids), containing scattered thin-walled cells (hydroids), is a zone of cells conspicuously large in transverse section, regarded as an approach to phloem. This central region, perhaps a rudimentary vascular tissue, is surrounded by a sheath of parenchymatous cells containing starch, and this in turn by a sclerenchymatous cortex.

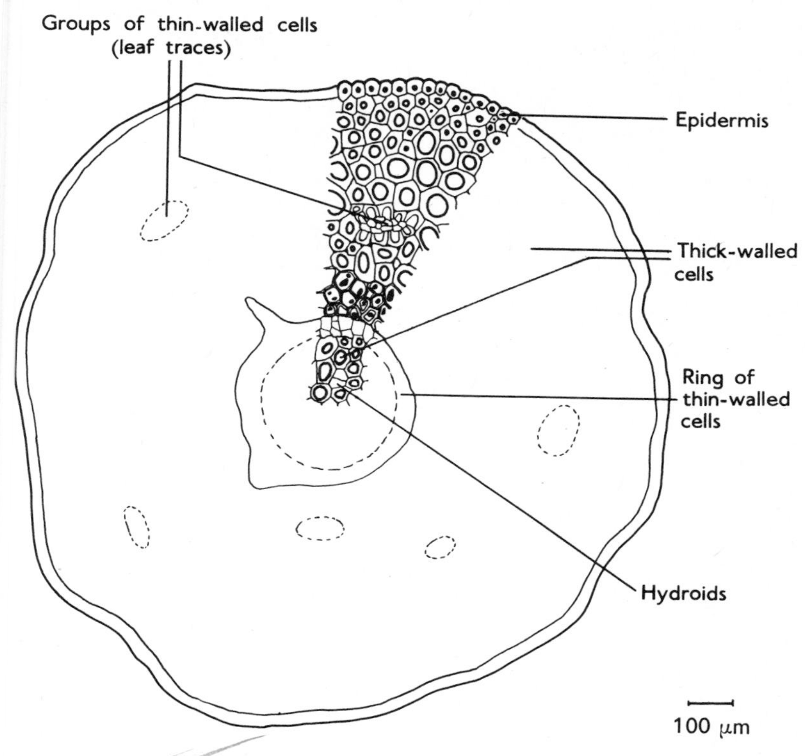

Fig. 4.22 *Polytrichum commune.* Transverse section of stem.

SEXUAL REPRODUCTION In sexual reproduction, the Bryales show every possible arrangement of the archegonia and antheridia. Both monoecious and dioecious species occur, and amongst the monoecious species, the gametangia may be either mixed together in a bud-like inflorescence, or separate. Whatever the arrangement, the antheridia and archegonia are often numerous and interspersed with sterile hairs or paraphyses (Figs. 4.23 and 4.24), recalling the situation in *Fucus* (p. 88). The archegonia usually have long necks, each consisting of several tiers of cells, and the central canal may contain as many as 10 cells. The antheridia are stalked, and one or more cells at the apex usually form a distinct lid at maturity, opening as if on a hinge while the mass of antherocytes is discharged. The cluster of sex organs is usually surrounded by a whorl of closely adpressed leaves, termed a ***perichaetium.***

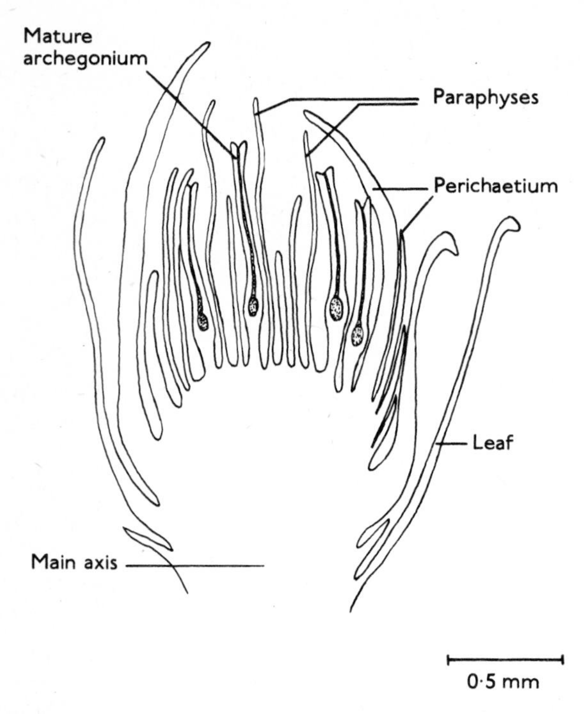

Fig. 4.23 *Mnium* sp. Longitudinal section of female head.

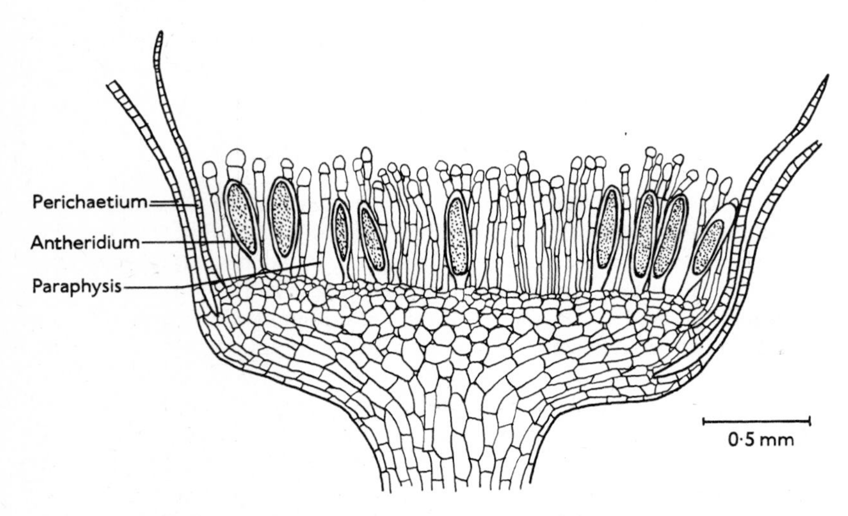

Fig. 4.24. *Mnium hornum*. Longitudinal section of male head.

There is a striking correlation in the Bryales between the position on the plant where the sex organs are produced, and the growth habit. Where the reproductive organs terminate the main axis, and growth is consequently sympodial, the main axis is almost invariably upright. These are the ***acrocarpous*** mosses. In the remainder, where the sex organs are produced laterally (the ***pleurocarpous*** mosses), the main axis is usually creeping (Fig. 4.25). With only a few exceptions, the tufted mosses are acrocarpous.

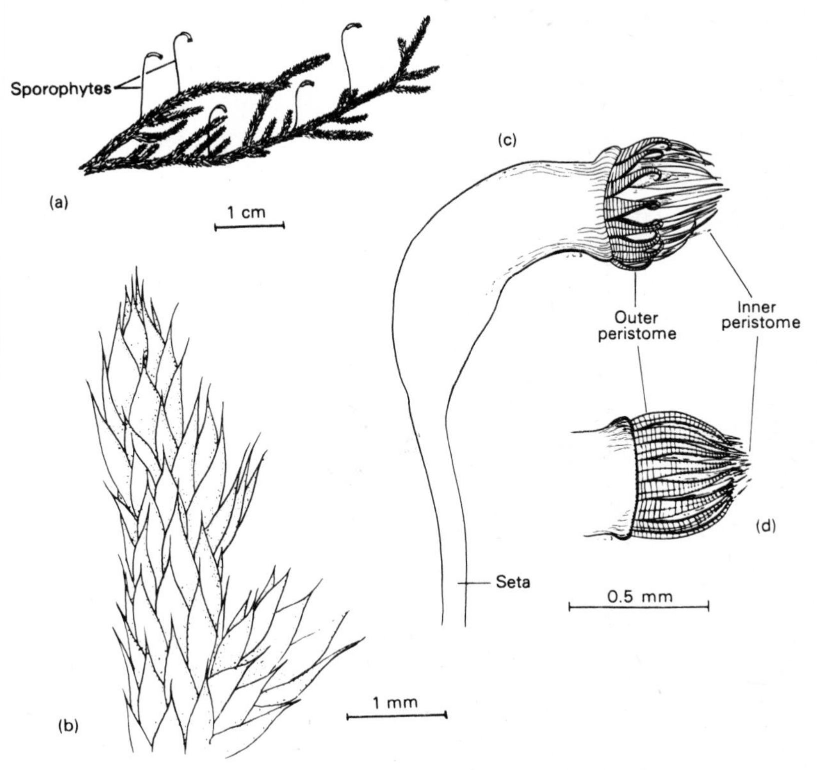

Fig. 4.25 *Hypnum cupressiforme*. (**a**) Fertile shoot system. (**b**) A portion of the shoot showing the closely inserted leaves. (**c**) Capsule, showing peristome in dry state. (**d**) Peristome in wet condition.

The development of the sporophyte begins immediately after fertilization, and a foot, seta and capsule are soon differentiated. The venter of the archegonium is also stimulated into growth by the germination of the zygote, and it expands rapidly to cover the young sporophyte with a cap-like calyptra. Experiments have shown that the presence of the calyptra is essential for the orderly differentiation of the capsule.[12] In many species the calyptra, having become severed from the gametophyte by the extension of the seta, persists as a membrane protecting the ripening capsule. Compared with that of a typical liverwort, the moss sporophyte develops slowly. The seta also begins to extend during the differentiation of the capsule, instead of solely when it is mature.

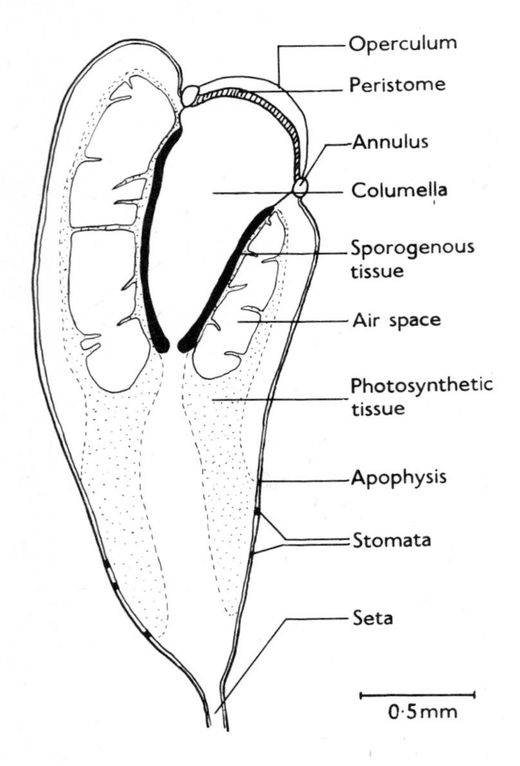

Fig. 4.26 *Funaria hygrometrica*. Median longitudinal section of immature capsule.

The capsule itself (Figs. 4.26 and 4.27) is a complex organ, but its differentiation follows a regular radial pattern, and two concentric regions of tissue can be recognized which follow distinct developmental paths. An inner region, termed the ***endothecium***, gives rise to the archesporium, which in the Bryales is never domed, but is always a cylinder, often with a central sterile columella. Outside the endothecium is the ***amphithecium*** which, in most Bryales, differentiates a ring of remarkable tooth-like structures, the ***peristome***. This remains as a fringe around the mouth of the opened capsule (Figs. 4.20b and 4.25c). Each peristome tooth consists of a number of superimposed segments, and each segment consists of thickenings laid down on the tangential wall common to two adjacent cells. Only these thickenings remain at maturity. Because of the ordered sequence of mitoses in the differentiation of the amphithecium, and the regular spacing of the columns of cells giving rise to the peristome, the number of teeth is constant in any given species and is always a power of two.

Fig. 4.27 *Bryum pallens*. Fruiting condition. Approx. × 2·5.

A certain amount of chlorophyllous tissue occurs in the immature capsule, particularly in the basal region (***apophysis***), where there are also stomata in the epidermis. The sporophyte is thus to some extent autotrophic. In some species (e.g. *Funaria hygrometrica*), air spaces occur between the archesporium and the wall of the capsule. This conspicuous aeration has perhaps been developed in relation to the respiratory demands of the developing archesporium.

Meiotic division of the spore mother cells heralds the last phase in the maturation of the capsule. The columella usually breaks down at this stage, so that the centre of the capsule is occupied solely by spores. The operculum ultimately drops off, exposing the peristome (Fig. 4.20b, c), which now begins to play an important role in the dispersal of the spores. The polysaccharide material forming the peristome teeth is hygroscopic and, since the macromolecular orientations of the thickenings in the two columns of cells giving rise to a tooth differ, tensions are generated in the tooth with changes in its hydration. These are released by sharp twisting and bending movements. The peristome thus forms a very effective scattering mechanism, activated by changes in atmospheric humidity. The exact nature of the movements of the teeth varies with the species. In some mosses, e.g. *Funaria hygrometrica*, the peristome is incurved when wet and recurved when dry, whereas in others, e.g. *Ceratodon*, the behaviour is reversed.

The spores of the Bryales have thin walls, and germinate rapidly on a damp surface. The protonema is usually well developed, resembling a heterotrichous green alga, but distinguishable by the obliquely transverse walls of the prostrate filaments. In spore cultures growing on agar the buds giving rise to the mature plants (Fig. 4.28) frequently arise in a number of concentric zones.

ASEXUAL REPRODUCTION Vegetative propagation undoubtedly plays a large part in the reproduction of the Bryales. Almost any part of the gametophyte—leaf, stem or even rhizoid—is capable of regeneration, either directly or by the production of gemmae (Fig. 4.29), and giving rise to a new individual. Some species are hardly known in the sporophytic condition, but are nevertheless widely distributed. These must be dispersed almost entirely by vegetative means.

The sporophyte is also capable of regeneration under experimental conditions: for example, if segments of the seta are placed on a mineral-agar medium. Almost always, however, growth is of the gametophytic form. In this way diploid gametophytes, and ultimately tetraploid sporophytes, can be obtained. There is some evidence that autodiploid gametophytes exist in nature, and they may have arisen in this way.

THE RELATIONSHIPS OF THE BRYOPHYTA

The origin of the bryophytes

There are so many similarities, particularly in the protonemal phase, between the bryophytes and the algae, that it seems beyond doubt that the mosses and liverworts had their origin in some algal form. Further, it is clear that the Bryophyta share more features with the Chlorophyta than with any other algal group, namely photosynthetic pigments, cell wall components, food reserves and type and number of flagella. Nevertheless, it is also clear that the Bryophyta are considerably more highly organized than the Chlorophyta. This is shown by their terrestrial habit, differentiated thallus, heteromorphic alternation of generations, production of

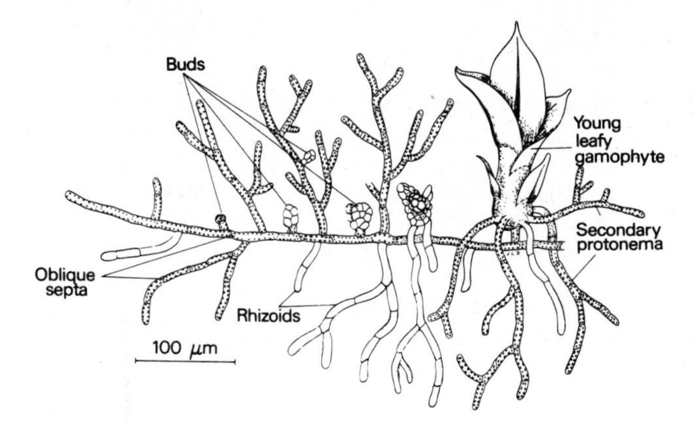

Fig. 4.28 *Funaria hygrometrica*. Development of buds on protonema.

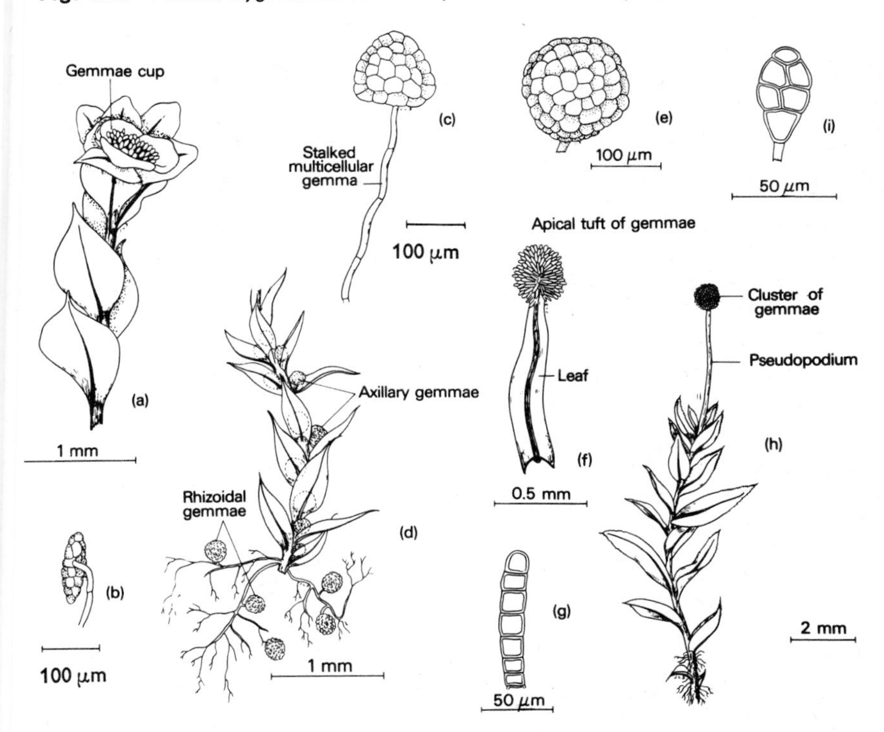

Fig. 4.29 Asexual reproduction in mosses. (**a–c**) *Tetraphis pellucida*. (**a**) Habit of gemmiferous plant. (**b, c**) Mature gemma, front and side views. (**d, e**) *Bryum rubens*. (**d**) Habit. (**e**) Gemma. (**f, g**) *Ulota phyllantha*. (**f**) Young leaf with apical tuft of gemmae. (**g**) Gemma. (**h, i**) *Aulacomnium androgynum*. (**h**) Habit. (**i**) Gemma.

aerial spores, and the enclosed sex organs. Little information is available about how these changes occurred during the evolution of the bryophytes, since not only have the transitional stages left no descendants, but, as yet, no fossil record of them has been discovered. Possible indications of the trends in these vanished intermediates are seen in the Phaeophyta. The sporophyte of *Dictyota* (p. 91), for example, produces non-motile spores, and in several Orders the life cycles are heteromorphic. The transitional organisms probably had a heterotrichous habit, although in some, leading eventually to the thalloid liverworts, it may have been only the prostrate part of the system which developed further.

With the exception of a possible example from the Lower Devonian, *Sporogonites*[2] (Fig. 4.30), the fossil bryophytes first appear in the later Devonian and Carboniferous periods. Despite this early appearance in the fossil record, however, there is no evidence that the bryophytes were ever a major component of the world's vegetation. They have probably remained a minor, but important, component, evolving only slowly, while a succession of larger forms, sporophytic in nature, have dominated the evolution of the autotrophs.

Whatever the ancestral organisms, they must have possessed a sporophytic phase, but the fossil record is wholly silent on how this gave rise to the sporophytes characteristic of bryophytes. As mentioned earlier

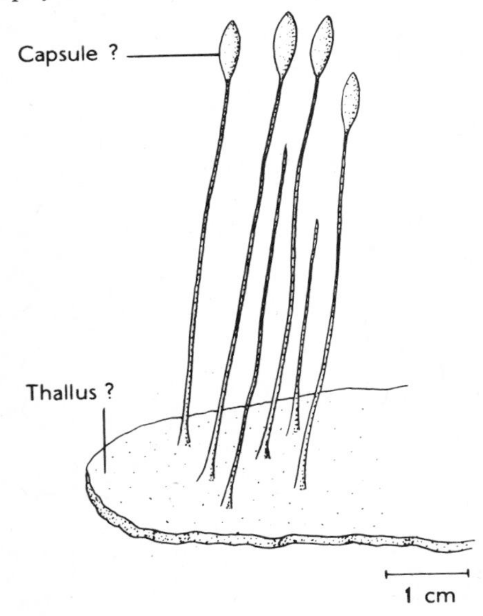

Fig. 4.30 *Sporogonites exuberens.* (From a reconstruction by Andrews, *Palaeobotanist*, **7**, 87 (1960))

(p. 113), comparative morphology and experimental genetics point to those living bryophytes with very reduced sporophytes (e.g. *Riccia*) being advanced forms. Also, if *Sporogonites* be indeed a bryophyte, and its reconstruction correct, a well-developed capsule was already present in the Devonian representatives of the Division.

The morphological and reproductive differences between the mosses and liverworts appear to extend as far back as the Carboniferous, since the general classification of the fossil bryophytes from these ancient rocks is readily apparent. This strengthens the view that mosses and liverworts have been independent evolutionary lines from a very early period, and it is even possible that they had independent origins from transitional archegoniate forms. At the other extreme, neither the mosses nor liverworts have any obvious relationships with the simplest vascular plants, either living or fossil. We must therefore regard them as a group isolated from the main line of evolution of the land plants, exploiting principally a somewhat circumscribed ecological niche.

Evolutionary relationships within the bryophytes

The evolutionary relationships of the mosses and liverworts themselves are hardly less obscure. The liverworts, for example, show a whole series of forms from the creeping thalloid (e.g. *Pellia*), to thalloid with two rows of ventral scales (e.g. *Blasia*), thalloid in which the margin of the thallus is so deeply crenulate that the lobes resemble leaves (e.g. *Fossombronia*), and ultimately leafy forms with upright stems and radial symmetry (e.g. *Haplomitrium* of the Calobryales). There has been much argument about whether this series represents a phylogenetic advance, or whether the first liverworts resembled *Haplomitrium*, the other forms being derived. The view that the radially symmetrical leafy form is primitive is strengthened by the surprisingly complex features present in many of the thalloid forms (e.g. in *Marchantia* and *Anthoceros*). Also a thalloid form such as *Pellia* is so outstandingly simple that it stands under the suspicion of being reduced and specialized.

Similar arguments apply to the mosses, although here the relative uniformity of the group makes comparative morphology even less informative. Few would regard those species whose capsules possess only rudimentary peristomes, or even lack them altogether, as anything other than reduced. The moss capsule with its clearly differentiated operculum and peristome must therefore have been an early and distinguishing feature of the Class We have no direct evidence of how the peristome evolved, but it is possible that the 'pepper-pot' mechanism present at the mouth of the capsule of *Polytrichum* (Fig. 4.31) and the bristle-like peristomes found in a few other genera, indicate steps in a developmental pathway that culminated in the typical peristome of the Bryales (p. 133).

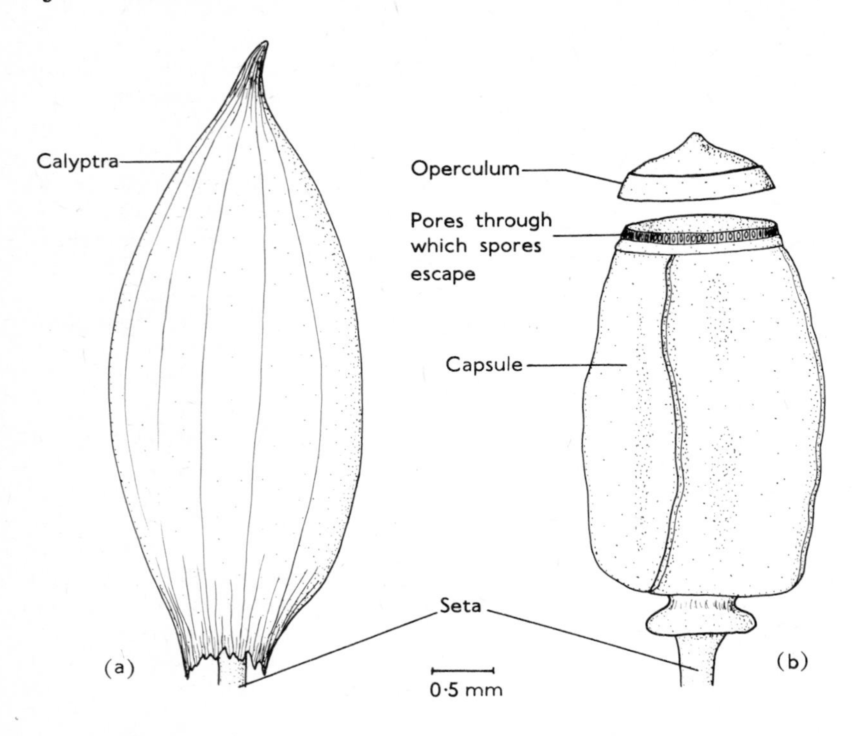

Fig. 4.31 *Polytrichum juniperinum*. Capsule. (**a**) Before removal of calyptra. (**b**) Operculum removed showing 'pepper-pot' mechanism.

Although we must regard the basic morphological features of the bryophytes as having arisen very early in the evolution of land plants, evolution of a more superficial nature has no doubt continued in the Division, probably being influenced by evolution of vegetation as a whole. Many bryophytes, for example, are epiphytes, and a great extension of the epiphytic habitat occurred with the rise of the angiospermous forests. A renewed burst of evolution, consisting principally of diversification of established morphological forms, may well have occurred at that time, enabling the bryophytes to take advantage of the new surfaces becoming available for colonization.

5

The Tracheophyta, I

(Psilopsida, Lycopsida, Sphenopsida)

Under the system of classification we have adopted, all the vascular plants (i.e. those possessing the lignified conducting tissue, xylem) are placed in a single Division, the Tracheophyta. Discussion of the many and diverse Orders which are combined in the Tracheophyta, and their interrelationships, will occupy our remaining chapters.

The general characteristics of the Tracheophyta may be defined as follows:

TRACHEOPHYTA

Habitat	Predominantly terrestrial or epiphytic.
Plastid pigments	Chlorophylls *a* and *b*, carotenoids (principally β-carotin), xanthophylls (usually principally lutein).
Food reserves	Starch; to a lesser extent fats, inulin and other polysaccharides. Proteins.
Cell wall components	Cellulose, hemicelluloses, lignin.
Reproduction	Heteromorphic alternation of generations; sporophyte the conspicuous generation. Sex organs with or without a jacket of sterile cells. Male gametes in some flagellate. Embryogeny various. Spores cutinized. Specialized vegetative reproduction of the sporophyte infrequent.
Growth forms	Predominantly axial.

The Tracheophyta can be conveniently sub-divided into the Psilopsida[3,73] (psilophytes), Lycopsida (lycopods), Sphenopsida (horsetails) and Pteropsida (ferns, gymnosperms and angiosperms).

The angiosperms, with about 200,000 living species, are by far the largest component of living vegetation, but they are nevertheless the most recent of the Tracheophyta to have been evolved, and their dominance is comparatively recent. The remainder of Tracheophyta with the exception of a few gymnosperms, are all archegoniate, and they have a relatively rich fossil record, extending back in some instances well into the Palaeozoic. Taken as a whole, the archegoniate Tracheophyta show a progressive ability to exploit terrestrial habitats. Reduction and modification of the male gametophyte eventually eliminate the necessity for motility on the part of the male gametes and for the presence of fluid at fertilization. These reproductive changes are accompanied by increasing protection of the plant body from unfavourable climatic conditions, thus widely extending the range of environments available to plant life.

Although the simpler and, on the basis of the fossil evidence, more primitive vascular plants are archegoniate, they are, nevertheless, very different from the mosses and liverworts we have just considered. In contradistinction to the situation in the Bryophyta, in the archegoniate Tracheophyta it is the sporophyte which is the conspicuous generation, being longer-lived and possessing considerably greater anatomical complexity than the gametophyte. There are no living plants clearly intermediate between bryophytes and tracheophytes, and the fossil record as yet gives no evidence that any such plants existed. It appears from the present evidence that the common ancestors were remote, perhaps existing only at the time of the first steps in the colonization of the land. If such an ancestor possessed little or no lignified tissue, it would, of course, stand small chance of preservation as a macrofossil.

A feature common to all living Tracheophyta is the presence of cells associated with the sporogenous tissue which break down during maturation of the sporangium. The materials so liberated are utilized in the further development of the spore mother cells and in the formation of the spores. These specialized cells, which often lie in one or more concentric layers around the sporogenous cells, are said to form a ***tapetum***.

We shall begin our consideration of the Tracheophyta with the Psilopsida, since the evidence of palaeobotany and comparative morphology points to their being amongst the most primitive of vascular plants.

PSILOPSIDA

Sporophyte consisting of more or less dichotomously branching axes, often with small leaf-like appendages. Roots absent, the subterranean axes bearing rhizoids. Vascular tissue consisting of tracheids and ill-defined phloem. Sporangia terminal, homosporous. Gametophyte (in living forms) subterranean,

(a)

Fig. 5.1 (**a**) *Psilotum nudum*. Part of the fertile region. The trilocular synangia are subtended by small forked bracts. ×6.6. (**b**) *Tmesipteris tannensis*. Part of the fertile region. The bilocular synangia are attached at the forks of conspicuous bifid bracts. ×5.

(b)

sometimes with vascular tissue, resembling portions of the sporophyte rhizome. Antherozoids flagellate. Embryogeny exoscopic (see p. 110).

Only two genera of living psilopsids are known, *Psilotum* (Fig. 5.1a) and *Tmesipteris* (Fig. 5.1b). The former is pan-tropical and not uncommon, but the latter, confined to Australasia and Polynesia, is rather rare. A number of Palaeozoic fossil forms are known from various parts of the world,

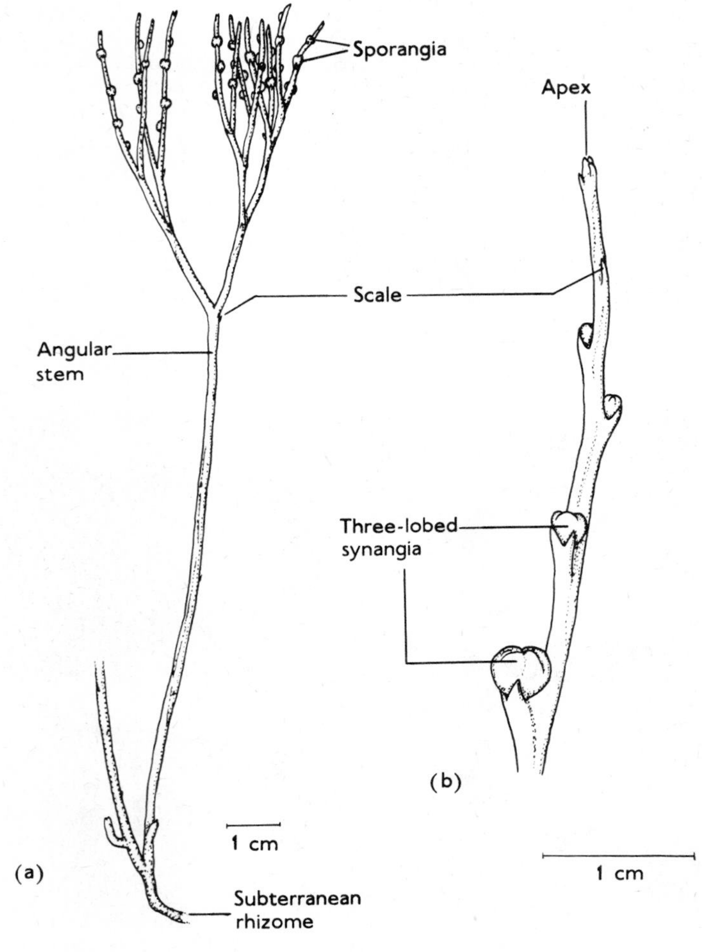

Fig. 5.2 *Psilotum nudum.* (**a**) Habit. (**b**) Fertile shoot.

principally from rocks of Devonian age. The living forms are included in the Psilotales, and the fossil in the Psilophytales.

Psilotales The sporophyte of *Psilotum* (Fig. 5.2), which may be either terrestrial or epiphytic, consists of upright (or, in one epiphytic species, pendulous), dichotomously branching axes arising from a horizontal system of similarly branching rhizomes. The rhizomes bear rhizoids, and contain an endophytic fungus, probably in symbiotic association (***mycorrhiza***). Small scales, which are at first green, but soon become scarious, occur at irregular intervals on the stem. They resemble the leaves of a bryophyte in having no vascular strand, and are perhaps better regarded as appendages than as leaves. *Psilotum*, like all vascular plants, possesses stomata, but here they are in the epidermis of the stem.

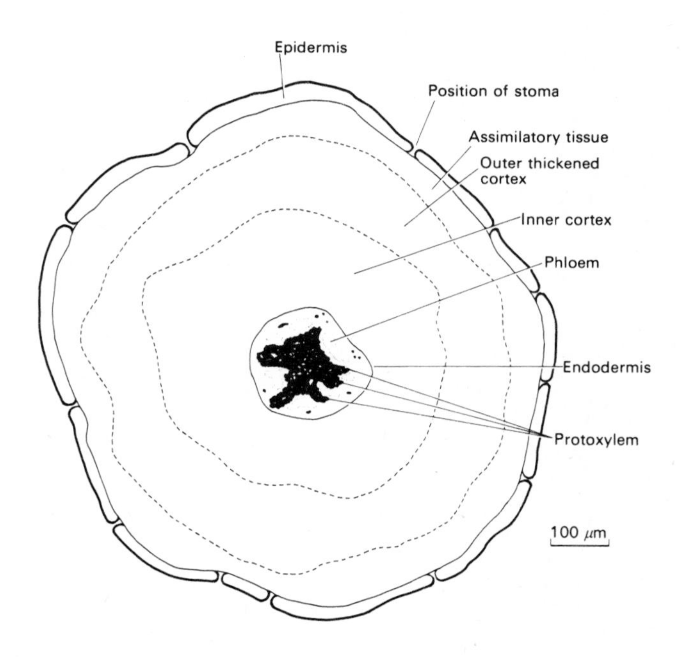

Fig. 5.3 *Psilotum nudum*. Transverse section of lower part of aerial branch.

THE ANATOMY OF THE AXES The stem contains a simple stele (Fig. 5.3), frequently enclosing in the upper regions a central parenchymatous medulla. The xylem, consisting solely of tracheids, is often stellate in transverse section, the arms standing opposite poorly defined ribs at the exterior of the stem. Phloem surrounds the xylem, but, apart from the lateral sieve areas, the sieve cells are little different from elongated parenchyma cells. An endodermis, separated from the phloem by a narrow zone of pericyclic parenchyma, marks the boundary of the stele. Phlobaphene, a condensation product of tannin, is often deposited in considerable quantities in the cells of the inner cortex.

The branches of both the aerial and terrestrial systems of the sporophyte grow indefinitely from single or small groups of apical cells. There is no evidence that the dichotomy of the axes follows median longitudinal

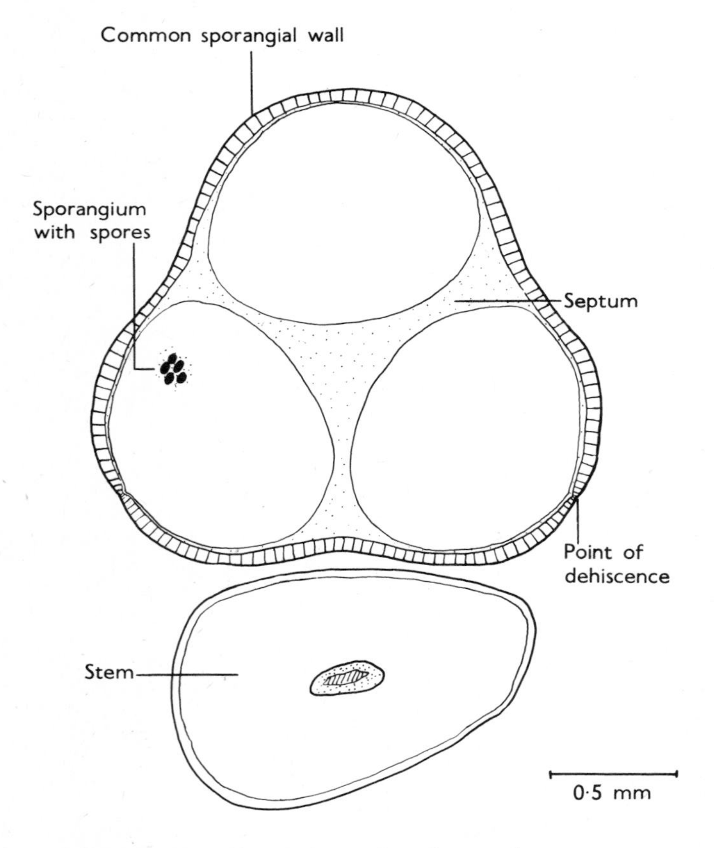

Fig. 5.4 *Psilotum nudum*. Transverse section of synangium.

division of a single apical cell, as in certain algae (e.g. the brown alga, *Dictyota dichotoma* (p. 91)).

REPRODUCTION Spores are produced in the upper region of the sporophyte (Fig. 5.2). The spore-bearing organs, which are somewhat distant from each other, are three-lobed, and each is subtended by a bifid appendage (Fig. 5.4). The lobes correspond to three internal chambers, separated by septa, and each filled with spores. It is still not clear whether this spore-bearing organ is to be interpreted as a trilocular sporangium, or as a ***synangium*** formed by the fusion of three sporangia. Three primordia

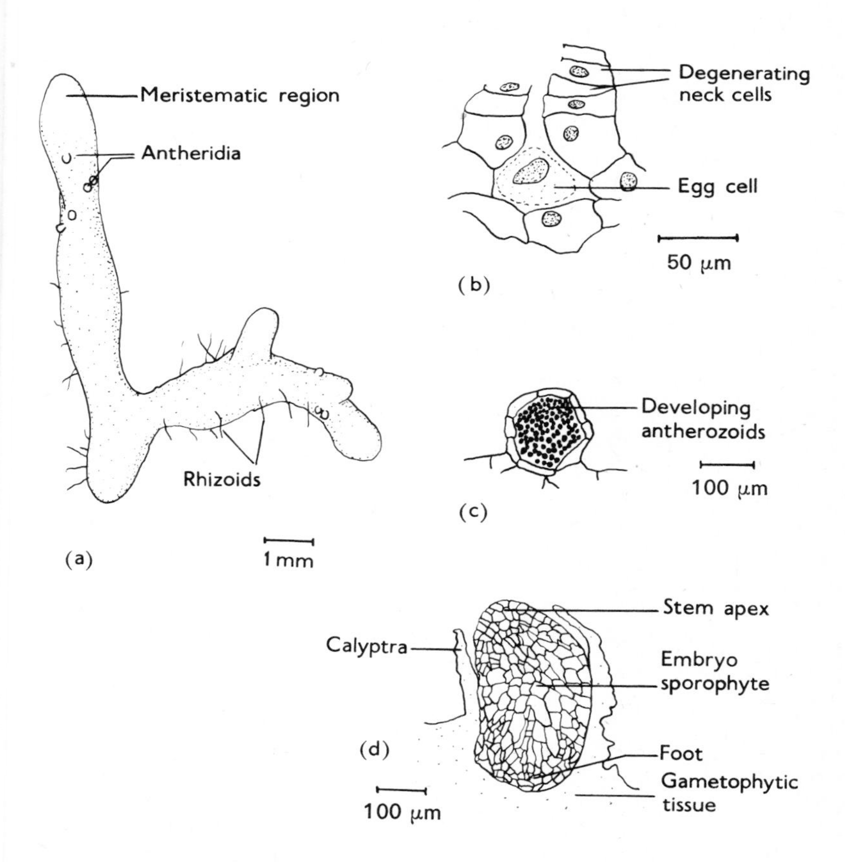

Fig. 5.5 *Psilotum nudum.* (**a**) Gametophyte. (**b**) Mature archegonium. (**c**) Antheridium. (**d**) Young sporophyte. (All after Bierhorst, *Am. J. Bot.*. **40**, 649 (1953); **41**, 274 (1954))

are, however, visible early in the ontogeny of the organ, perhaps indicating its synangial nature. Also, since a distinct vascular strand extends into the base of the synangium, it is usually regarded as terminating a lateral axis, rather than arising in association with a sporophyll.

The synangium has a massive wall, some five cells thick at maturity. During ontogeny, groups of archesporial cells disintegrate forming a tapetum, but these cells are not arranged in well-defined layers. The mature spores have cutinized walls, which, although somewhat irregular, lack distinct ornamentation. They are bilaterally symmetrical.

The gametophyte of *Psilotum* is a subterranean axial structure,[11] dichotomously branching, and resembling short lengths of the sporophytic rhizome. The similarity extends to the anatomy, the finer axes being wholly parenchymatous, and the broader containing a central vascular strand. The peripheral cells, like those of the rhizome, are inhabited by an endophytic fungus. Both the gametophyte and the sporophytic rhizome produce globular multicellular gemmae, a means of vegetative propagation.

Antheridia and archegonia arise from superficial cells in the region of the growing points of the gametophyte (Fig. 5.5). The antheridia, depending upon their size, liberate up to 250 antherozoids. The archegonium has four tiers of neck cells, but at maturity all but the lower one or two tiers degenerate. Fertilization, which depends upon the presence of a film of water, is brought about by spirally coiled, multiflagellate antherozoids.

The first division of the zygote is in a plane transverse to the longitudinal axis of the archegonium, exactly as in the bryophytes. The outer (or epibasal) cell yields the apex of the embryo, and the inner (or hypobasal) the foot. The Psilotales provide one of the few examples of such exoscopic embryogeny amongst the archegoniate Tracheophyta.

Tmesipteris

Tmesipteris is frequently an epiphyte with trailing stems (Fig. 5.6). The gross morphology is similar to that of *Psilotum*, but branching is much rarer. The appendages are larger and more leaf-like, remaining green and possessing stomata. They also frequently have a vascular strand, but the insertion of the appendages is peculiar, being longitudinal instead of transverse, so that they appear more as flange-like outgrowths of the axis than as normal foliage leaves. Spore-bearing organs occur in the upper parts of some of the shoots, each subtended by a bifid appendage. These organs are again regarded as synangia terminating very short lateral branches, but in *Tmesipteris* each consists of only two fused sporangia.

The reproduction of *Tmesipteris* is very similar to that of *Psilotum*. The foot of the young embryo of *Tmesipteris* is lobed, and the whole bears a striking resemblance to the young sporophyte of *Anthoceros*. It is doubtful,

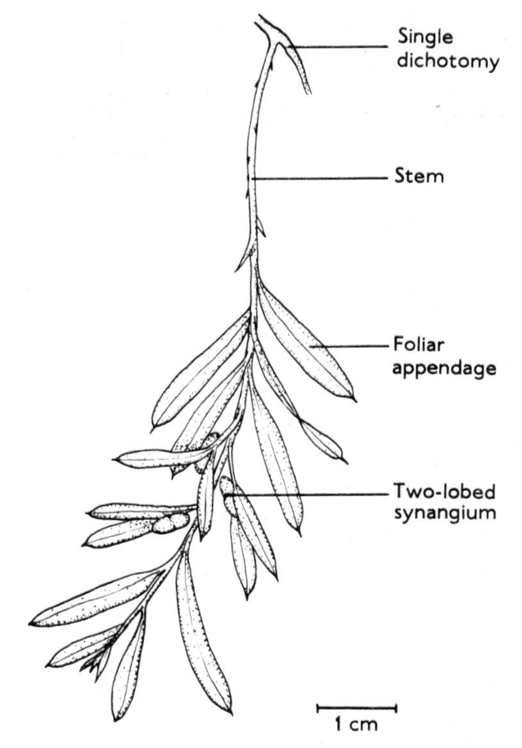

Fig. 5.6 *Tmesipteris tannensis.* Habit.

however, whether this bears any phylogenetic significance, as some have claimed.

Psilophytales The Psilophytales take their name from *Psilophyton*, a Devonian fossil first discovered in sandstones in eastern Canada in 1859 and subsequently reported from rocks of the same age in many parts of the world. Until recently these remains were believed to be of plants resembling *Psilotum* in general habit, but with the lower stems spiny and the sporangia terminating upper branchlets. This reconstruction is now, however, in considerable doubt, and much more representative of the Psilophytales are a number of plants from the Rhynie Chert, a siliceous rock, probably of Lower Devonian age, in Scotland. Those plants, possibly growing in a bog and petrified *in situ* by seepage from volcanic ash, are so beautifully preserved that quite detailed anatomical investigation is possible.

The Rhynie plants again had a general resemblance to *Psilotum*. *Rhynia major*, for example, consisted of aerial axes arising from a horizontal rhizome. The aerial axes, the thicker of which reached a diameter of 6 mm, branched dichotomously and ascended to about 50 cm. Tracheids were present throughout the plant and stomata in the epidermis of the aerial branches. Some branchlets terminated in pear-shaped sporangia, about 12 mm long, containing cutinized spores.

Fig. 5.7 *Rhynia gwynne-vaughani.* (After Delevoryas, *Morphology and Evolution of Fossil Plants.* Holt, Rinehart and Winston, New York, 1962, from a reconstruction in the Chicago Natural History Museum)

Other Rhynie plants (Fig. 5.7) are similar to *Rhynia major*, but differ in morphological and anatomical detail. In *Horneophyton*, for example, the rhizome was tuberous, and the archesporium of the capsule domed, as in some bryophytes. Since the spores of *Rhynia* and other genera can often be seen to lie in tetrads, the sporophytes were probably diploid, as in *Psilotum*, but the nature of the gametophytes remains unknown. Some authorities have identified portions of subterranean axes with bud-like protuberances as gametophytes of *Rhynia* bearing embryos,[63] but it has still to be demonstrated that these axes, which certainly recall gametophytes of *Psilotum*, actually bore antheridia and archegonia.

The evolution of the Psilopsida

The Psilophytales do not show any relationship to any known alga, nor do they resemble (with the possible exception of the sporophyte of *Anthoceros* (p. 121)) any of the Bryophyta. They may have arisen independently from some algal stock, but it is clear that, as present in the Devonian, they are already far removed from any transitional form. This is also true of *Cooksonia*, which closely resembles *Rhynia* and appears to be indisputably Upper Silurian in age. The differentiation of tracheids in the stem, of stomata in the epidermis and the cutinization of the spores, all characteristics of land plants, are all well-established features of the earliest Psilophytales. Nevertheless, simpler, more alga-like forms may yet be discovered in even earlier rocks. Cutinized spores with triradiate markings are already found in the Silurian, the period preceding the Devonian (in which the Psilophytales became widespread), but the plants which bore them are still unknown.

The predominantly axial morphology is a striking feature of both the Psilophytales and Psilotales, but there may be no direct phylogenetic continuity, since *Psilotum*-like remains are not identified with certainty from intervening geological periods. There is, however, a remarkable similarity, both in sporophyte and gametophyte, between *Psilotum* and certain New Caledonian ferns, a feature discussed in more detail later (p. 228).

The chromosome numbers of the Psilotales appear to be generally high, ranging from 52 to 210 in the gametophytic state, and they may thus provide a complex polyploid series, This is not, however, an unambiguous indication of the antiquity of the plants, as some have thought.

LYCOPSIDA

Sporophyte consisting of more or less dichotomously branching axes, but differentiated into root and shoot. Shoot bearing microphylls, each containing a single vein, but leaf trace leaving no gap in the stele. Vascular tissue consisting of tracheids and phloem. Sporangia in or near axils of microphylls, homo- or heterosporous. Gametophyte (in living forms) terrestrial or subterranean. Antherozoids flagellate. Embryogeny endoscopic (see p. 155).

Only five living genera are included in the Lycopsida, which, although more numerous than the Psilopsida, are again a minor component of contemporary vegetation. The fossil record, however, shows that the lycopsids were abundant in the Carboniferous period, many then being represented by quite substantial trees. Only herbaceous forms have persisted until the present day. Five Orders of the Lycopsida are recognized, three containing the living genera and their fossil relatives, and the other two only extinct forms.

Lycopodiales Members of this Order, which includes the living *Lycopodium* (club moss) and *Phylloglossum*, are distinguished by their eligulate leaves, and homospory.

Lycopodium (Fig. 5.8), with about 200 species, is distributed throughout the world from Arctic to tropical regions, with a related range in the growth

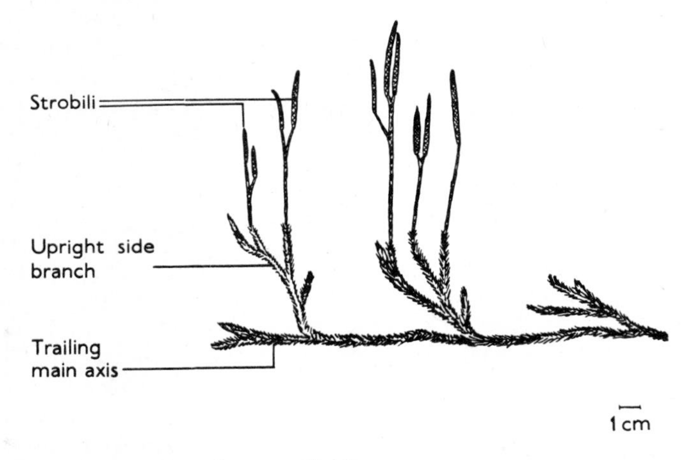

Fig. 5.8 *Lycopodium clavatum*. Habit.

form of the sporophyte. The colder climates favour species with short, erect stems, or creeping stems giving rise to short upright side branches, while those in the Tropics have much laxer growth and are often epiphytes.

The stem is surrounded by ***microphylls***, a kind of leaf which is typically small, simple in outline, and with a single median vascular strand. Some species of *Lycopodium*, especially the epiphytic, are heterophyllous, the lateral rows of leaves being expanded in the plane of the shoot system and the upper and lower rows adpressed and smaller. The fertile leaves (***sporophylls***) may be similar to the sterile, as in *L. selago*, and the fertile regions not clearly set off from the sterile along the axis. More usually, however, the sporophylls differ from the sterile leaves in size, shape and the extent of the chlorophyllous tissue, and are often grouped together, as in *L. clavatum*, in distinct cones (***strobili***) of determinate growth.

Roots arise endogenously, emerging from the underside of the stem in the prostrate species, or from near the base of the stem in upright species. In some of the latter, initiation of the roots occurs near the shoot apex, but instead of emerging there the roots grow down inside the cortex, and break out only when they reach the level of the substratum.

GENERAL FEATURES OF THE ANATOMY The stem, which grows from a group of initial cells (Fig. 5.9), contains a central stele, the anatomy of which shows considerable variation with species. In its simplest form the stele consists of a core of tracheids, more or less stellate in section, with phloem lying between the arms. In other species the xylem and phloem form parallel bands (Fig. 5.10), or the phloem and xylem may be intermingled, anastomosing strands of sieve cells being scattered amongst the tracheids. In every instance differentiation of the xylem begins at the exterior and then proceeds centripetally, leaving no undifferentiated tissue at the centre. The protoxylem is thus exarch, and the stele a simple protostele. The vascular tissue, which is entirely primary, is usually surrounded by a narrow zone of parenchyma, and this in turn by an endodermis, the cells of which have a distinct Casparian strip.

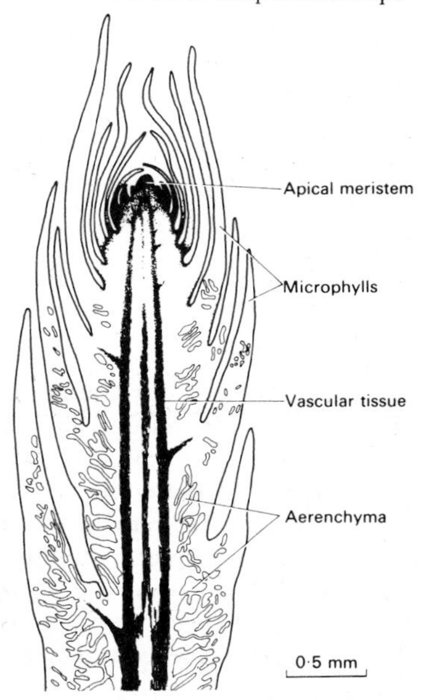

Fig. 5.9 *Lycopodium alpinum*. Longitudinal section of apical region of vegetative shoot.

The leaves, the chief site of photosynthesis, are structurally simple. There are abundant stomata on both surfaces, and internally numerous intercellular spaces. There is, however, no clearly differentiated mesophyll.

REPRODUCTION Sporangia, borne singly in the axil of a sporophyll or close to its insertion (Fig. 5.11), develop from a group of initial cells, and

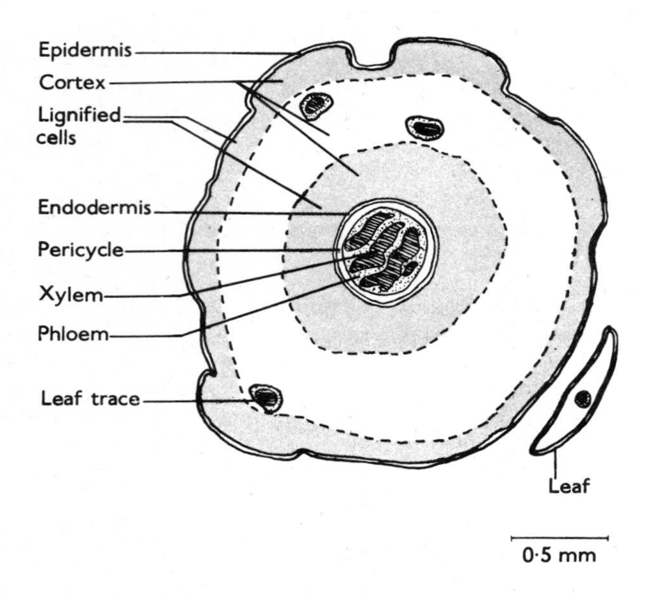

Fig. 5.10 *Lycopodium clavatum*. Transverse section of stem.

are hence termed ***eusporangiate***. Continued cell division within the primordium leads to a central mass of spore mother cells surrounded by a wall several cells thick. The inner layers of the wall function as a tapetum, breaking down to provide food materials which nourish the maturing sporocytes. After meiosis the spores remain in tetrads while each, except for the narrow areas of contact at the centre of the tetrad, becomes enclosed in a thick, cutinized wall. The regions of contact remain relatively unthickened, and form a conspicuous triradiate scar at the apex of each separated spore. The sporangia dehisce transversely along a line of thin-walled cells (***stomium***) and the minute spores (each about 50 μm in diameter) are distributed by wind.

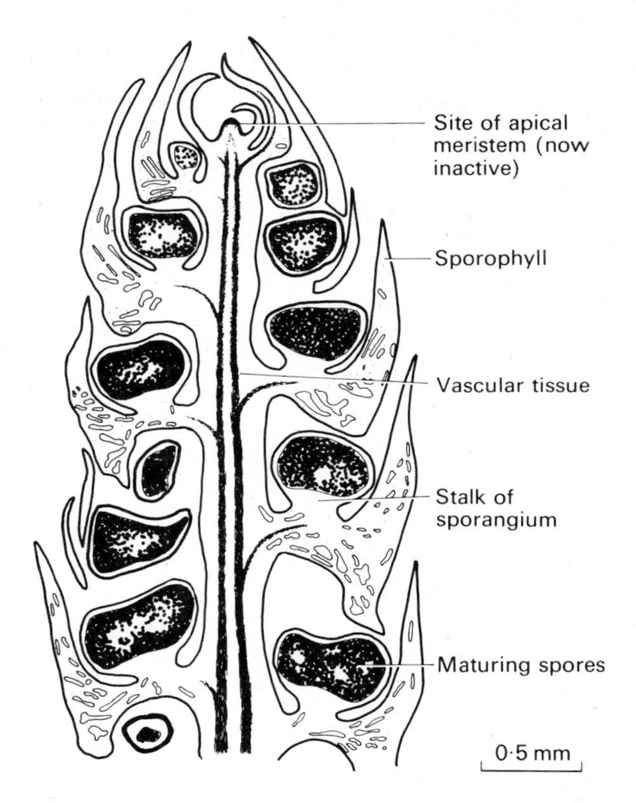

Fig. 5.11 *Lycopodium alpinum*. Longitudinal section of upper region of young strobilus.

In some species of *Lycopodium* the exterior of the spore has quite elaborate reticulate ornamentation, while in others it is almost smooth. The rough spores are not easily wetted and their germination may be delayed for several years, probably until weathering and attrition have rendered the coat permeable. The process may be simulated in the laboratory by immersing such spores (e.g. of *Lycopodium selago*) in concentrated sulphuric acid.[35] Following this treatment the spores germinate freely in pure culture. In natural conditions initially unwettable spores may be washed deep into the soil before germination occurs, and this is reflected in the nature of the gametophyte. For example, in *L. clavatum*, the spores of which have a pronounced reticulate relief, the gametophyte is a subterranean saucer-shaped structure, growing saprophytically and persisting for several seasons. The sex organs are produced on a cushion in the central region

(Fig. 5.12). In the tropical *L. cernuum*, where the spores are smooth, germination and development occur rapidly, producing lobed, cup-shaped gametophytes which possess chlorophyll and last for little more than a single season. In *L. selago*, in which the spores are somewhat intermediate in the development of the wall, there is a corresponding ambivalence in the habitat of the gametophyte. In all instances the gametophyte of this species is a small carrot-shaped body, growing saprophytically and producing sex organs on the upper cushion. It may, however, be either buried, or at the

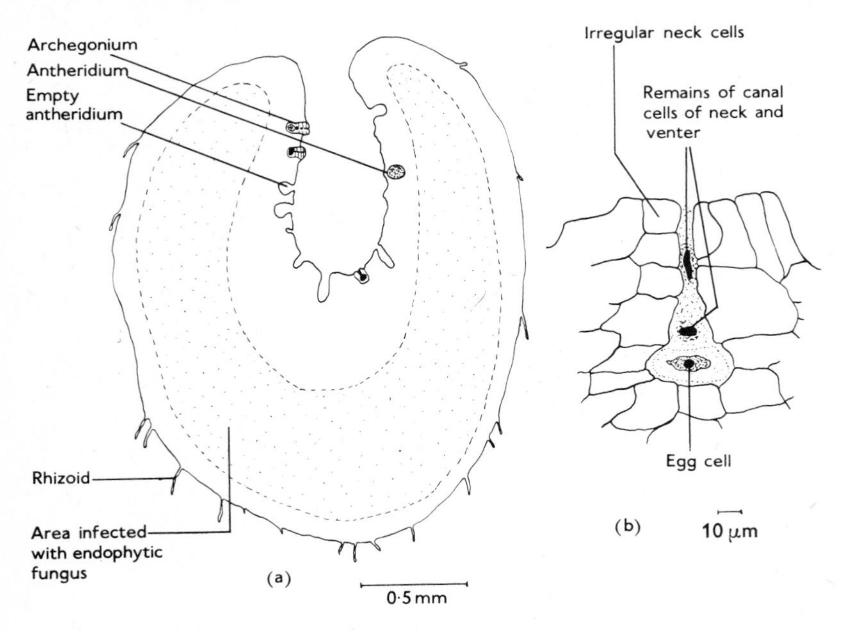

Fig. 5.12 *Lycopodium volubile*. (**a**) Vertical section of subterranean gametophyte. (**b**) Mature archegonium.

surface, and if the latter the upper part becomes chlorophyllous. There is thus not only a general relationship between the habitat of the gametophyte and the features of the spore from which it comes, but also between the habitat and the form and duration of the gametophytic plant.

The sex organs of *Lycopodium* are similar to those of *Psilotum*, except that in many species the archegonia have more conspicuous necks, and both the male and female gametangia are enclosed to a greater extent by vegetative tissue. The antherozoids are biflagellate, and there is some evidence that they are chemotactically attracted to the archegonia. Following fertilization the zygote divides by a wall transverse to the axis of the archegonium. The outer cell, termed the suspensor, divides no further, but the inner continues to divide and gives rise to two regions of cells. The central region becomes the foot, while the inner differentiates into the root, first leaf, and stem apex of the embryo proper. This development results in the embryo being directed inwards, and the embryogeny is consequently said to be ***endoscopic***.

At first the apex of the embryo points vertically downwards, but expansion of the foot region pushes it to one side and the young sporophyte finally breaks out of the surface of the gametophyte. Before it becomes fully established, the young plant depends upon food materials absorbed from the gametophyte, probably through the foot. In some species of *Lycopodium* (e.g. in *L. cernuum*) the differentiation of the embryo is delayed and it emerges as a parenchymatous protuberance, termed a ***protocorm***. This eventually gives rise to one or more growing points, each of which yields a normal plant.

Phylloglossum

Phylloglossum is the only other living genus of the Lycopodiales. Its single species, *P. drummondii*, consists of an upright sporophyte, reaching 5 cm or less, with a basal whorl of leaves and a pedunculate strobilus (Fig. 5.13). During growth a lateral axis arises near the base of the plant and extends down into the soil, its tip eventually becoming transformed into a tuber. This tuber forms an organ of perennation, persisting through the dry period (when the remainder of the plant perishes) and giving rise to the following year's growth. The gametophyte of *Phylloglossum* is similar to that of *Lycopodium cernuum*, and there is an interesting resemblance between the protocorm stage in the development of the sporophyte and the perennating tuber. Although strikingly different from *Lycopodium* in habit, *Phylloglossum* is clearly not distant in the basic features of anatomy and reproduction. It is no doubt a form that has become specialized in relation to a particular kind of habitat.

Selaginellales In this Order, which includes a single living genus *Selaginella*, the habit resembles that of the Lycopodiales, but the microphylls are ligulate, the sporophylls are always aggregated into distinct strobili, and the spores are of two kinds, differing in size and in the sexes of the gametes they eventually produce.

There are more than 700 species of *Selaginella*, most of which are tropical, ranging from small epiphytes to large climbing plants. The stems of the sporophytes are much more branched than in *Lycopodium*, the branches and sub-branches often growing in the same plane and forming fern-like fronds. In many species, leafless rhizophores originate from the stem at points of branching and grow down towards the soil, dichotomizing

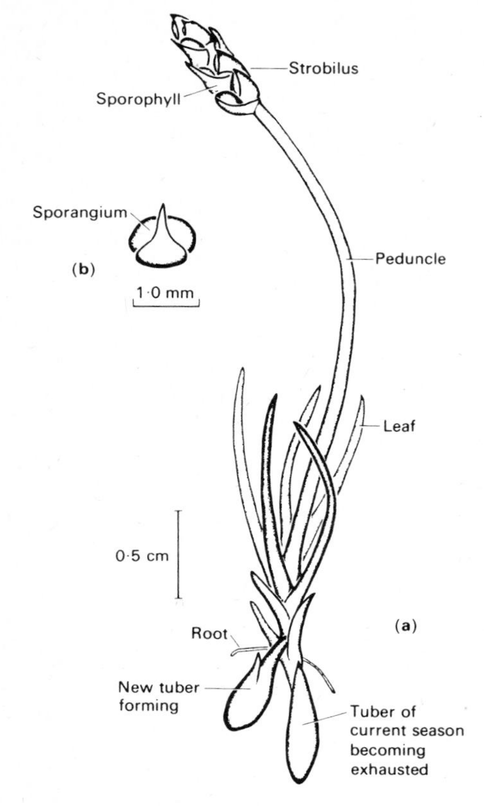

Fig. 5.13 *Phylloglossum drummondii.* (**a**) Habit. (**b**) Sporophyll and sporangium.

as they approach its surface. On contact, which apparently acts a stimulus, roots arise endogenously and penetrate the substratum. The leaves are usually spirally inserted, and in some species they radiate around the stem, as in *S. selaginoides*. On the lateral branches of other species, however, the leaves become coplanar with the branch system, so increasing the frond-like effect of the whole. This often involves heterophylly, the two lower rows of leaves becoming expanded in the plane of the branch system, and the upper leaves remaining small and adpressed. This is well seen in *S. kraussiana* (Fig. 5.14), now becoming common in some parts of southern England. The leaves of *Selaginella* resemble those of *Lycopodium*, but a minute, tongue-like ***ligule*** is inserted into the upper side of the leaf close to the axis. Also in some species of *Selaginella* the upper cells of the leaf

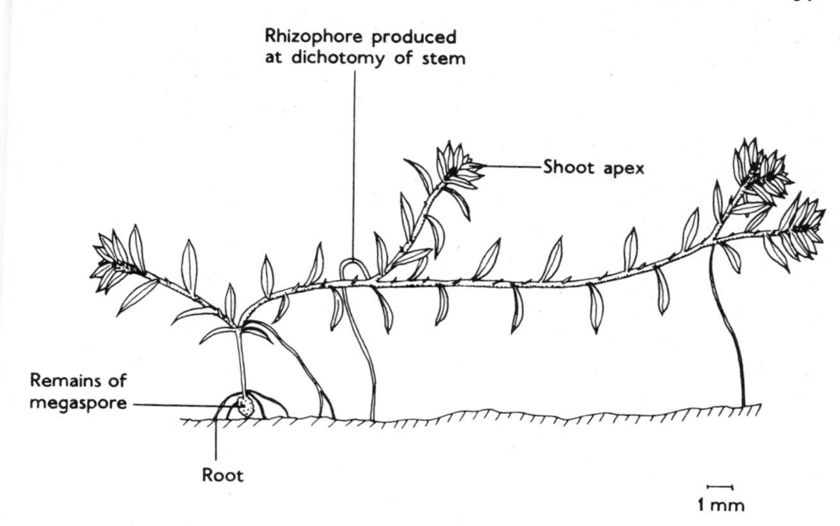

Fig. 5.14 *Selaginella kraussiana*. Young sporophyte.

contain a single large chloroplast, similar to that of *Anthoceros* (p. 120), but lacking a pyrenoid.

GENERAL FEATURES OF THE ANATOMY The anatomy of *Selaginella* differs little from that of *Lycopodium*, although occasionally the end walls of the tracheids fail to differentiate, thus giving rise in places to continuous channels simulating vessels. The endodermis is also highly peculiar, consisting of elongated, hypha-like cells which suspend the stele in the central cavity (Fig. 5.15). A ***trabeculate*** endodermis of this kind is unknown elsewhere in the Plant Kingdom, and its physiological implications are obscure. The stele of *Selaginella* is basically a protostele lacking internal parenchyma, but it is often ribbon-like instead of cylindrical, and, especially in aerial axes, several steles may ascend the stem together. In the rhizomes of a few species the xylem is in the form of a hollow cylinder lined on both surfaces by phloem. An endodermis also occurs both externally and internally. Such a stele, which is said to be amphiphloic, and is referred to as a ***solenostele***, clearly exhibits a more complicated pattern of differentiation than a protostele, but the factors controlling differentiation of this kind are still little known.

Rhizophores develop from meristematic areas often below the points of branching. These primordia seem to be to a certain extent indeterminate, and if a branch is removed the adjacent primordium will grow out to form a shoot instead of a rhizophore. However, if the branch is replaced by a

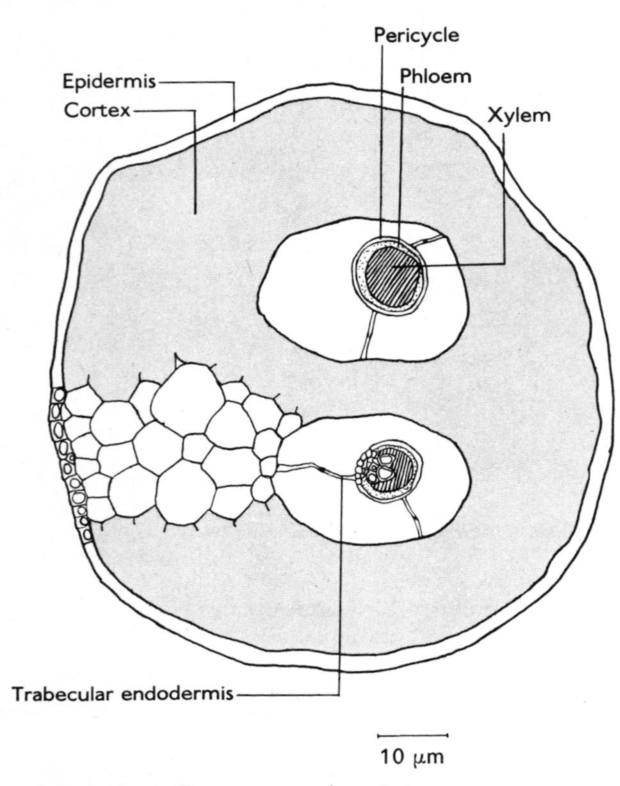

Fig. 5.15 *Selaginella* sp. Transverse section of stem.

source of the growth-regulating substance, indole-3-acetic acid, the primordium behaves normally. This was one of the first and most striking demonstrations of how growth-regulating substances, diffusing from one area to another in a plant, maintain the familiar pattern of morphogenesis.

REPRODUCTION In the strobili, which terminate the main or lateral axes, the sporophylls are usually in four ranks. Each sporangium lies in the axil of a sporophyll, between the ligule and the axis (Fig. 5.16). There are two kinds of sporangia, producing ***mega-*** and ***microspores*** respectively, located in different regions of the cone. The initial development of each kind of sporangium is the same, and closely resembles that of the sporangia of *Lycopodium*. Development diverges with the formation of the spore mother cells. In the microsporangia all the mother cells undergo meiosis and form microspores, whereas in the megasporangia all the sporogenous tissue, except (in most species) one mother cell, breaks down. This remaining mother cell yields a tetrad, but not all of these cells necessarily become spores. In some species, e.g. *S. sulcata*, only one haploid cell develops, so that the mature sporangium contains but one large spore. The number of megaspores in a megasporangium thus varies from about twelve to one, depending upon the species, but in all the megaspores are conspicuous for their size, their store of food materials built up at the expense of the degenerated sporogenous tissue, and the thickening and ornamentation of

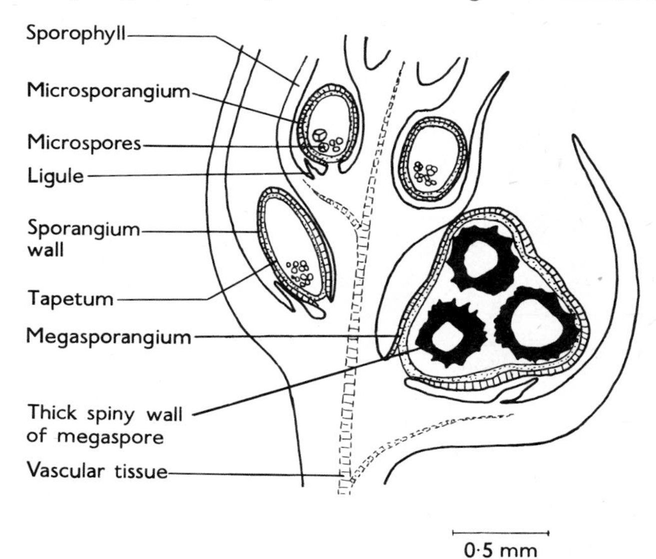

Fig. 5.16 *Selaginella kraussiana.* Longitudinal section of strobilus.

their walls. They are in consequence some of the most remarkable spores present in the Plant Kingdom; those of *S. exaltata*, for example, may exceed 1 mm in diameter.

Some nuclear divisions occur in the spores while they are still in the sporangia, but growth of the gametophytes, leading to rupture of the spore wall, does not resume until after the spores are shed. In the microspore an unequal mitosis produces a large antheridial cell and a small cell, termed a prothallial cell, which represents the sole development of the somatic tissue of the gametophyte. The antheridial cell continues to divide and develops into a normal antheridium, from which 128 or 256 biflagellate antherozoids are eventually liberated. In the megaspore the food material comes to occupy a central position, and free nuclear division occurs at its

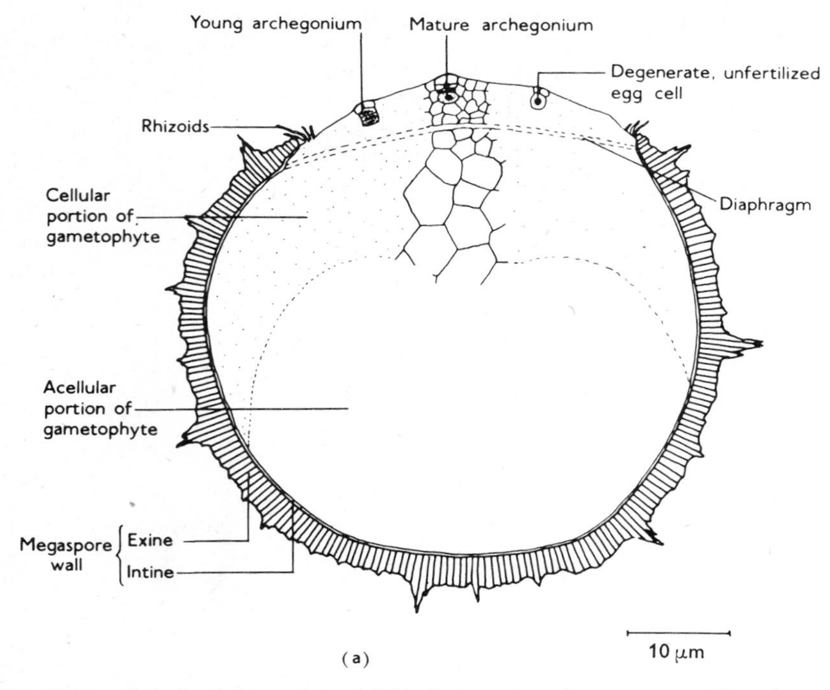

Fig. 5.17 *Selaginella kraussiana*. (**a**) Vertical section of megaspore with endosporic gametophyte.

periphery. Subsequently cell formation begins beneath the triradiate scar, which is eventually forced open by the general swelling, exposing a cap of gametophytic tissue (Fig. 5.17a). The somatic tissue of the female gametophyte is thus more extensive than that of the male, although chlorophyll is quite absent from both. There is some specific variation in the extent to which the cellular portion of the female gametophyte is delimited from the food supply below. In some species the boundary is imprecise, and cell formation gradually extends down into the lower region, but in other species a distinct diaphragm separates the upper cellular region from a largely acellular food reserve.

The female gametophyte, once exposed, protrudes as an irregular cushion bearing rhizoids at its margins and in the central region archegonia. The necks at the archegonia are very short, consisting of no more than two tiers of cells (Fig. 5.17b). Fertilization necessarily depends upon a microspore falling close to a megaspore, and a film of water being present when the gametangia are mature. The rhizoids around the female gametophyte may

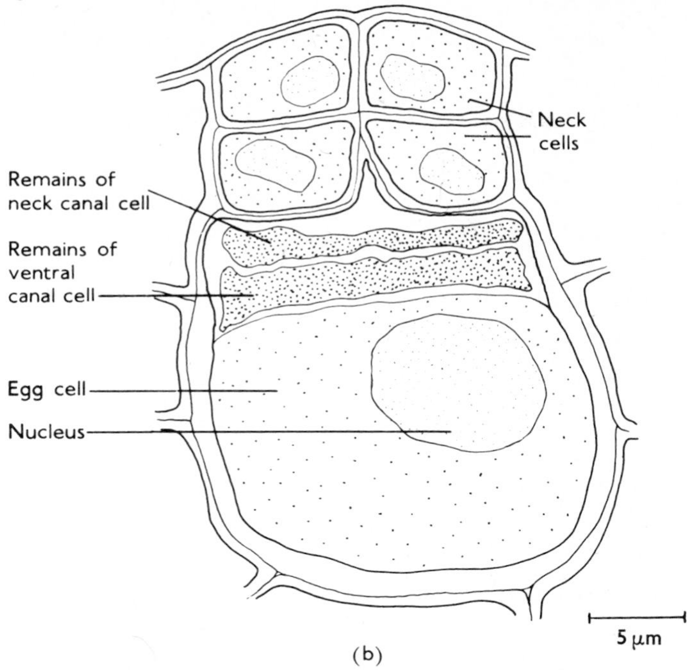

Fig. 5.17 Continued (**b**) Mature archegonium (from an electron micrograph).

in these conditions serve to retain a 'fertilization drop' above the archegonia in which the antherozoids congregate.

EMBRYOLOGY The embryogeny of *Selaginella* is like that of *Lycopodium* in being endoscopic, but differs in the greater development of the suspensor, and the wide variation in detail between species. In *S. selaginoides*, for example, elongation of the suspensor pushes the developing embryo down into the food reserve. In *S. kraussiana*, however, in which the food reserve is cut off by a diaphragm, the development of the suspensor is markedly less, but a curious downward extension of the archegonial canal carries the embryo through the diaphragm into the centre of the gametophyte. Apart from these features, there is also variation in the development of the foot region of the embryo, and in the relative positions in which the various parts of the embryo arise. In all species the embryo eventually emerges from the upper surface of the gametophyte (Fig. 5.18).

The substantial thickening of the megaspore wall in *Selaginella* probably protects the spore for considerable periods from desiccation and decay. Once conditions are favourable for germination, development is rapid and, because of the considerable food reserve of the megaspore, independent of an external supply of nutrients. Compared with the life cycle of the homosporous *Lycopodium*, there is considerably less time spent in the gametophytic phase. Since this, in view of the delicacy of the gametophytic tissues and their dependence on the maintenance of humid conditions, is the most vulnerable phase of the life cycle, the modifications that lead to its curtailment no doubt confer a considerable selective advantage on *Selaginella*. This is perhaps reflected in its numerous species.

In some species of *Selaginella* (e.g. *S. rupestris*) young plants emerge from the female regions of strobili. This has been regarded as following

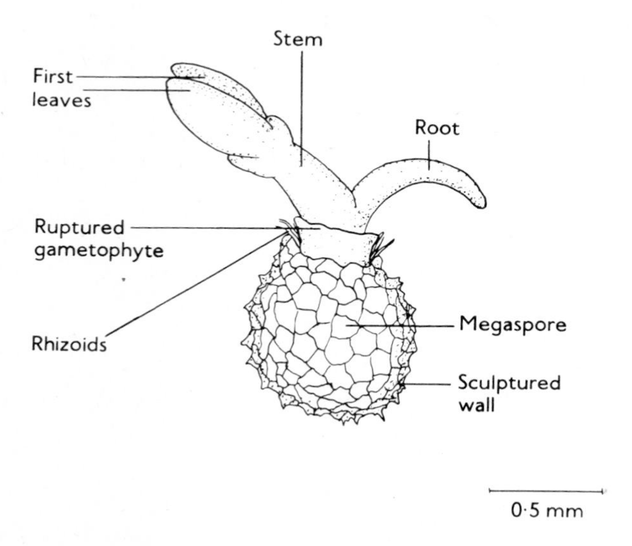

Fig. 5.18 *Selaginella kraussiana*. Sporophyte emerging from gametophyte.

from the lodging of microspores between the megasporophylls and the fertilization of an egg while the megaspore was still *in situ*, thus simulating an early step in the evolution of a seed. It is doubtful, however, whether this can be substantiated. In *S. rupestris*, at least, reproduction is apogamous, the megaspore producing an embryo directly without fertilization.[44]

Isoetales The members of this Order have a remarkable, rush-like habit, quite unlike that of any other lycopsid. They have short, fleshy, upright rootstocks, occasionally showing one or two dichotomies, and bearing a tuft of quill-like microphylls (Fig. 5.19). The microphylls are ligulate, and reproduction is heterosporous. All living representatives, con-

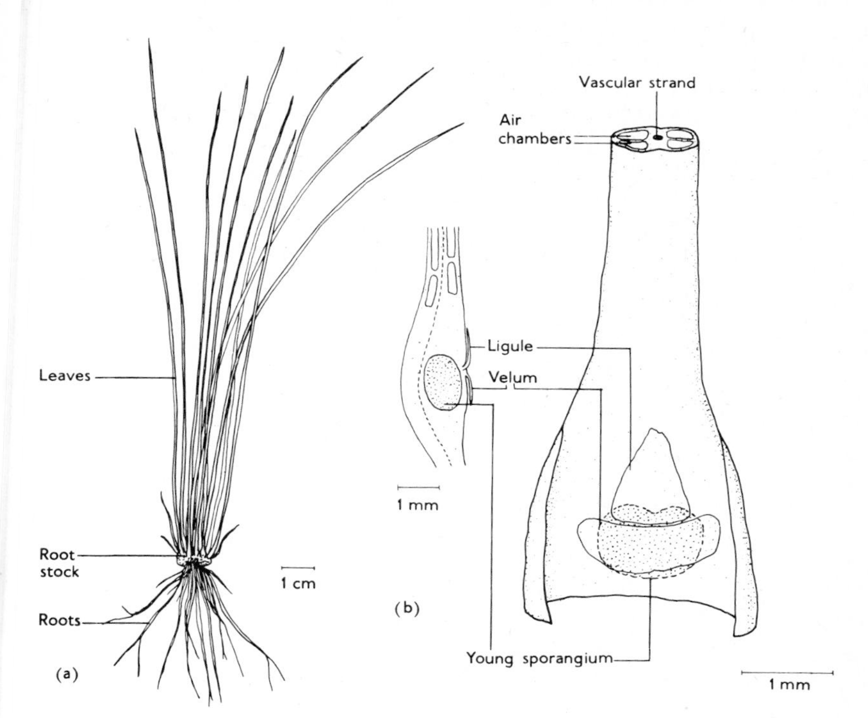

Fig. 5.19 *Isoetes echinospora*. (**a**) Habit. (**b**) Leaf base with young sporangium in face view and longitudinal section.

tained in the two genera *Isoetes* and *Stylites*, are aquatic, or plants of situations subject to periodic or seasonal inundation.

GROWTH HABIT AND ANATOMY *Isoetes* is widely distributed, three species occurring in Britain. The rootstock in all species is mostly below the level of the substratum, and is rarely branched. The leaves, which in some aquatic species may reach a length of 70 cm, are arranged in a dense spiral about a depressed apical meristem at the upper end of the rootstock. The roots arise from the lower end of the stock where the meristem is again depressed, here lying extended along a transverse cleft. The new roots are initiated at the base of the cleft (Fig. 5.20). Accompanying this singular morphology is an equally remarkable manner of growth. A cambial zone arises around the small amount of primary vascular tissue in the stock, but it contributes more to the cortex than to the stele. The activity of this cambium, in temperate species at least, is seasonal. In step with the addition of new material within, a girdle of outer tissue, complete with its decaying leaves above and roots beneath, sloughs away. Consequently, having reached its mature diameter (which may exceed its length), the stock remains more or less the same size, the new leaves and roots being carried up on to the shoulders of their respective meristems by the expansion of the products of the anomalous cambium.

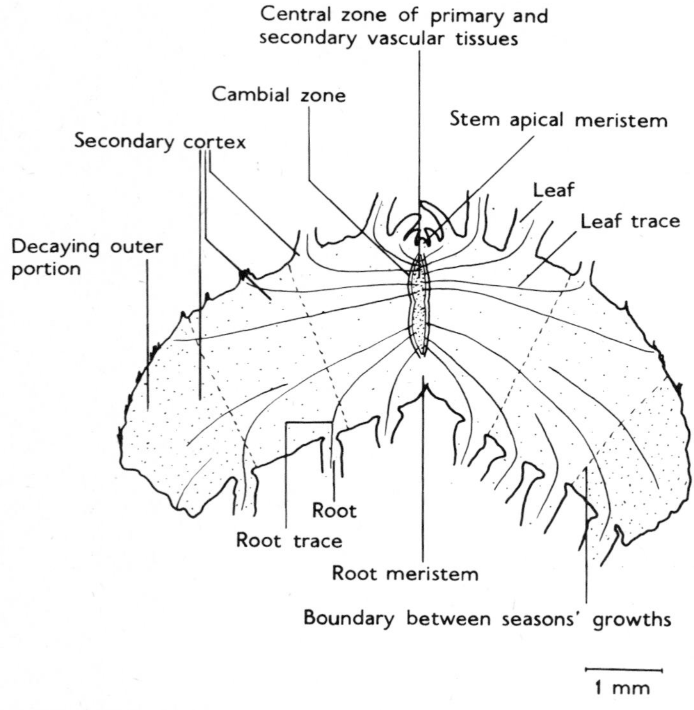

Fig. 5.20 *Isoetes echinospora*. Median longitudinal section of rootstock in plane perpendicular to that of the basal cleft.

The leaves contain a single vascular strand, often very tenuous, surrounded by four air canals, interrupted at intervals by transverse septa. These canals are especially striking in the aquatic species. The broadened leaf bases lack chlorophyll and overlap widely, forming a tight comal tuft. The anatomy of the stock presents a number of peculiar features. The primary xylem consists of more or less isodiametric tracheids, and at the base of the stele they are arranged in an anchor-like bifurcation lying in the same plane as the basal cleft of the stock. The tissue produced on the inside of the anomalous cambium differentiates as a mixture of tracheids, sieve cells and parenchyma.

The remainder of the tissue in the stock is parenchymatous, and no recognizable endodermis delimits the vascular tissue. The roots possess a single vascular strand surrounded by a cortex of two distinct zones: an outer fairly resistant to decay, and an inner of more delicate tissue with numerous air spaces.

REPRODUCTION Mature plants of *Isoetes* are usually abundantly fertile. The leaves first formed in a season's growth bear megasporangia, those following microsporangia, and the last formed are sterile. The sporangium is initiated much as in *Selaginella* between the ligule of the sporophyll and the axis, but distinctive features emerge as development proceeds. Part of the central tissue, for example, remains sterile and differentiates as trabeculae which divide the mature sporangium into a number of compartments (Fig. 5.19b). Also the ripe sporangium becomes enclosed in a thin envelope, called the velum, which grows up from the sporophyll, leaving a large pore (the foramen) on the adaxial side.

Spore formation in *Isoetes* is also peculiar. The megaspores are produced in tetrads with tetrahedral symmetry, and each spore consequently bears a triradiate scar. The microspores, however, are produced in tetrads in which each spore resembles a segment of a sphere produced by the bisection of a hemisphere perpendicular to its base. Each spore thus has only one linear edge, persisting in the mature spore as an elongated scar. Such spores are termed ***monolete***, in contradistinction to those with a triradiate scar, which are termed ***trilete***. *Isoetes* and *Stylites* are unique in the living Lycopsida in producing monolete spores, and they are some of the few plants in which both monolete and trilete spores are produced by the same individual.

The spores are liberated by the decay of the sporophylls. Their subsequent germination and development are similar to those of the micro- and megaspores of *Selaginella*. In *Isoetes*, however, the male gametophyte is wholly endosporic. The single antheridium contains only four antherozoids, differing from those of *Lycopodium* and *Selaginella* in being multiflagellate. They are released by rupture of the microspore wall. The female gametophyte is initially endosporic, but ruptures the megaspore at the site of the triradiate scar (Fig. 5.21), as in *Selaginella*. Chlorophyll again remains absent.

The first division of the zygote is slightly oblique. No suspensor is formed, but the embryogeny can still be termed endoscopic since the outer cell gives rise to the foot and the remainder of the embryo comes from the products of the inner cell. Differential growth causes the embryo to turn

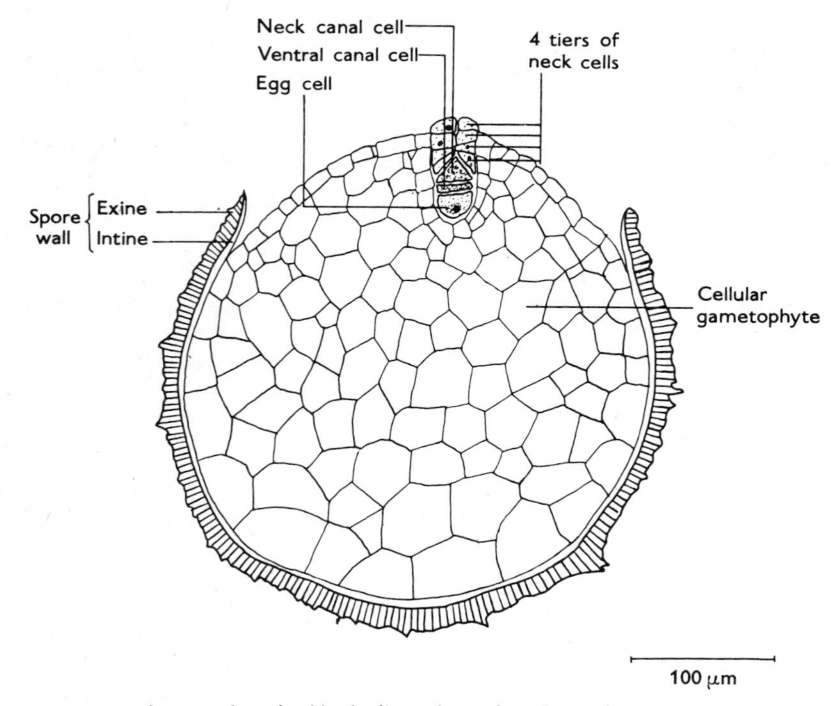

Fig. 5.21 *Isoetes hystrix*. Vertical section of endosporic gametophyte with archegonium.

round so that it is directed towards the upper surface of the gametophyte. It eventually breaks through, but the young plant remains for some time partially enclosed by a sheath of gametophytic tissue. Apogamy does not appear to have been detected in *Isoetes*, but the production of a bud in place of a sporangium is not uncommon.

Stylites[67]

Stylites closely resembles *Isoetes*, but the rootstock is dichotomously branched and may reach a height of 15 cm. The roots are confined to a single furrow which runs along the side of the stock. The plant has been found only in the High Andes of Peru, where it forms dense cushions by the sides of glacial lakes.

Although *Isoetes* and *Stylites* are like the remainder of the Lycopsida in essentials, they share a number of remarkable features unrepresented elsewhere in archegoniate plants. They appear to be the products of a line of lycopsid evolution that has been independent for a considerable period.

The early Lycopsida

The earliest recorded Lycopsida, the exact classification of which remains difficult, lived in the warm, humid climate of the Lower Devonian. An example is provided by the remarkable *Baragwanathia* (Fig. 5.22) from Australian rocks. This consisted of dichotomizing, aerial axes, some of

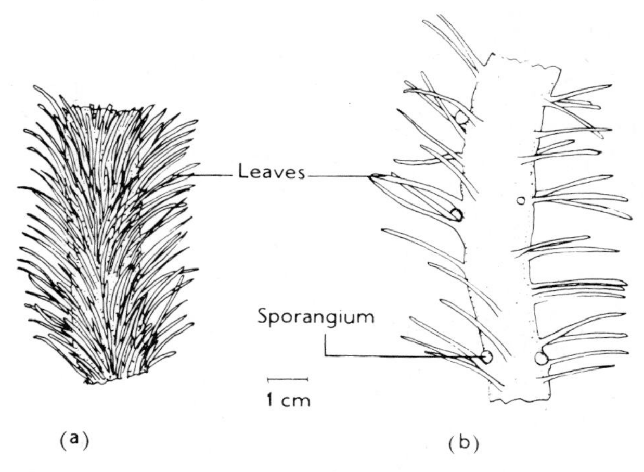

Fig. 5.22 *Baragwanathia*. (**a**) Vegetative shoot. (**b**) Shoot bearing sporangia. (After Lang and Cookson, *Phil. Trans. R. Soc.* **224 B**,421 (1935))

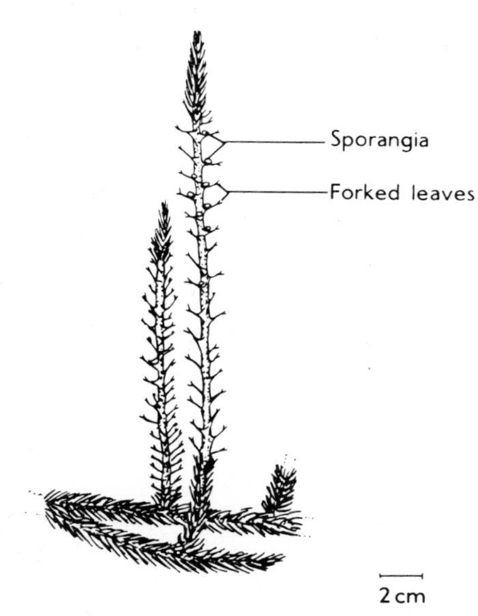

Fig. 5.23 *Protolepidodendron*. Habit. (After Kräusel and Weyland, from Delevoryas, *Morphology and Evolution of Fossil Plants*. Holt, Rinehart and Winston, New York, 1962)

which reached a diameter of 6.5 cm, bearing numerous overlapping, needle-like leaves about 4 cm long. The stem contained a core of tracheids, from which slender strands ascended into the leaves. Some of the shoots had fertile regions in which reniform sporangia, containing cutinized spores, lay amongst the leaves. Although the precise attachment of the sporangia is still unknown, the general resemblance of *Baragwanathia* to a lycopod is so striking that an affinity is clear. A similar plant (*Asteroxylon*) occurs fossilized in the Rhynie Chert.[53] *Protolepidodendron* (Fig. 5.23), from the Lower Devonian, apparently grew in a manner similar to *Lycopodium clavatum.* A prostrate, dichotomously branching stem gave rise to upright branching shoots, the whole system being clad in microphylls forked at the tip. The leaves of some of the erect branches bore unprotected sporangia on their adaxial surfaces.

Carboniferous rocks yield an abundance of lycopsid fossils, both of vegetative and reproductive structures and of isolated spores. Of the many forms they present, some were evidently herbaceous. Those having a general resemblance to living *Lycopodium* are placed in *Lycopodites*, but

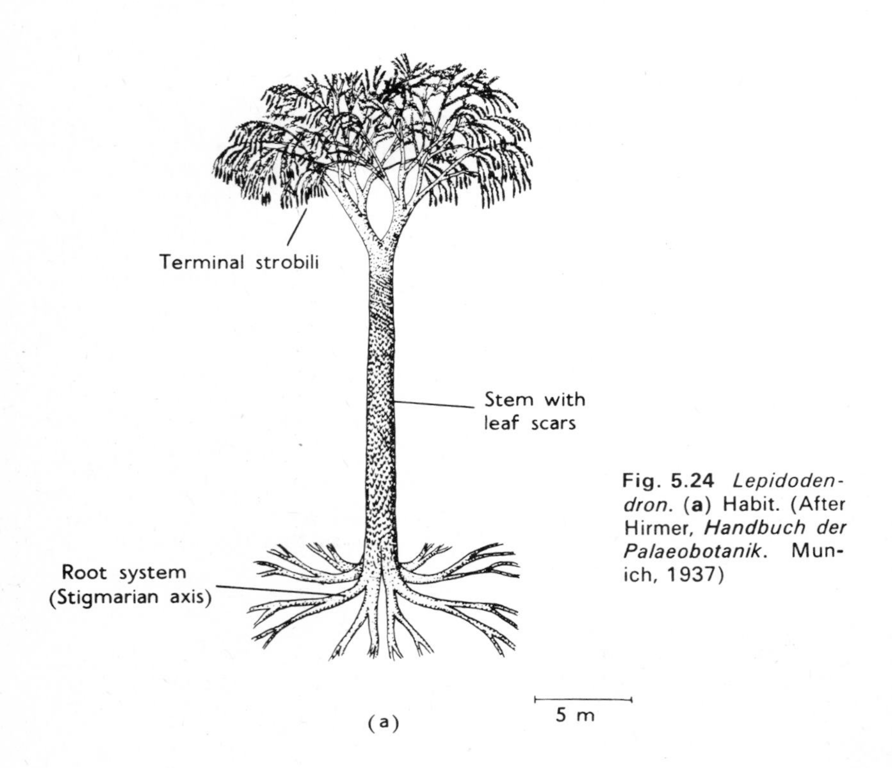

Fig. 5.24 *Lepidodendron.* (**a**) Habit. (After Hirmer, *Handbuch der Palaeobotanik*. Munich, 1937)

others, in which heterospory has been demonstrated, are believed to have been more like *Selaginella* and are placed in *Selaginellites*. The most impressive Lycopsida of the Carboniferous, however, were undoubtedly the arborescent Lepidodendrales, some of which achieved a height of 30 m. *Lepidodendron* (Fig. 5.24), for example, consisted of a trunk, 1 m or more in diameter at its base, which rose as a single column until it broke up by numerous dichotomies into the dense crown of branchlets. The upper parts of the tree bore simple ligulate microphylls, up to 20 cm long, triangular in cross-section, and arranged in regular spirals.

The trunk and lower branches of *Lepidodendron* and its relatives retained a characteristic pattern of diamond-shaped leaf scars. In each of these occurred a pit, which indicated the site of the ligule, and two lateral softer areas, one on each side of the vascular bundle. These are the remains of strands of aerenchyma (parichnos) which ran from the cortex of the stem to the leaf. The trunk was anchored at ground level by four radiating arms which, since they were first found detached and not immediately recognized, were named *Stigmaria*. The Stigmarian axes dichotomized freely and the smaller bore rootlets, anatomically similar to those of *Isoetes*, in spiral sequence.

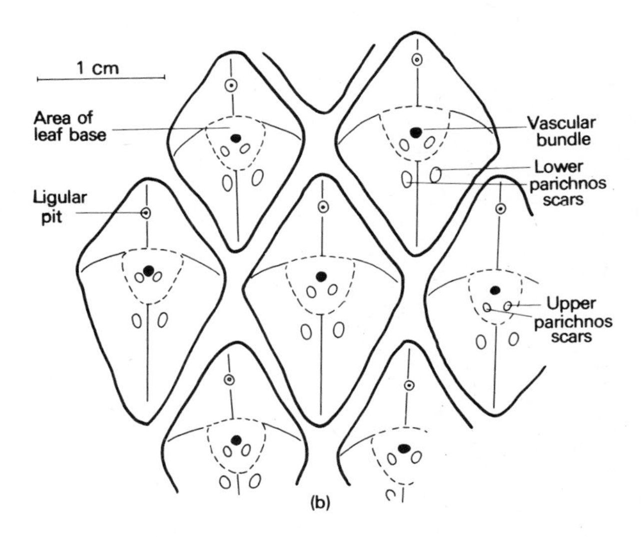

Fig. 5.24 Continued (**b**) Surface of cast of stem showing leaf bases. The thin-walled tissue of the parichnos strands has collapsed in the leaf cushions, producing the depressions referred to as secondary parichnos strands below the leaf scars.

Despite the girth of *Lepidodendron*, the anatomy of the trunk was comparatively simple, and its manner of growth consequently puzzling. Although a vascular cambium was present, it added only a narrow zone of secondary tissue to the primary stele. Additional secondary activity occurred in the outer cortex, resulting in a hard, sclerotic periderm which undoubtedly provided the principal mechanical support to the trunk. The inner cortex contained elongated cells, a so-called 'secretory tissue', which may have been a primitive form of phloem. The central zone of cortex, usually fragmentary or missing in the fossilized material, probably consisted of thin-walled aerenchyma, continuous with that in the Stigmarian axes and leaves. The curious anatomy, particularly the small amount of secondary xylem, has led to the view that the growth of these trees did not continue indefinitely, but was determinate.[4,29] It is envisaged that the plant first generated a massive apical meristem, the activity of which then produced an axis of considerable height before dichotomy began. At each dichotomy the apices became smaller, until eventually they ceased to be active.

In most Lepidodendrales the strobili, varying from 1 to 3.5 cm in width

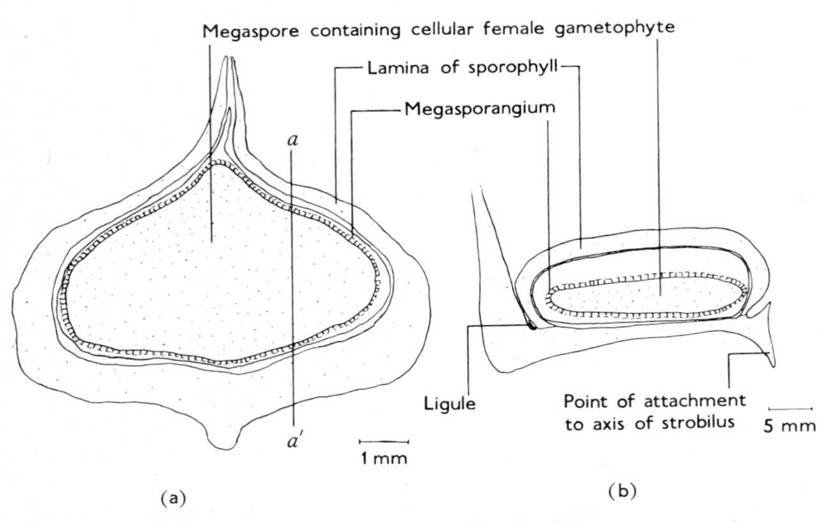

Fig. 5.25 *Lepidocarpon*. (**a**) Transverse section showing megasporangium enclosed in folded sporophyll. (After Arnold, *An Introduction to Paleobotany*. McGraw-Hill, New York, 1947) (**b**) Vertical section of region indicated by *a . . . a′* in (**a**). (After Hoskins and Cross, from Arnold, *An Introduction to Paleobotany*. McGraw-Hill, New York, 1947)

and 5 to 40 cm in length, were terminal on the branchlets. A few homosporous cones are known, but most were heterosporous, resembling the cones of *Selaginella* in general features and the placement of the sporangia in relation to the ligule. In a form known as *Lepidocarpon* (Fig. 5.25) the megasporangium was enclosed in an involution of the sporophyll, with a micropyle-like opening at the distal end. The female gametophyte was evidently retained in this structure, providing evidence that in the Carboniferous period some lycopsids produced a female organ approaching a rudimentary ovule. Germinating megaspores and microspores of the Lepidodendrales have occasionally been found in petrified material, and it is clear that reproduction was basically similar to that of *Selaginella*.

Pleuromeia (Fig. 5.26), the sole representative of the Pleuromeiales, is a little known plant found in Triassic sandstone. Its interest lies in its being morphologically intermediate between the Lepidodendrales and the Isoetales. A single trunk, little more than 1 m high, with spirally arranged leaf scars, rose from a four-lobed base which bore roots in the manner of Stigmarian axes. Above, the stem bore leaves and also terminal cones made up of curious obtuse sporangiophores. *Pleuromeia* appears to have been both heterosporous and dioecious.

The Jurassic and Cretaceous periods yield evidence of herbaceous forms very similar to *Lycopodium* and *Selaginella*. *Selaginellites hallei*, for example, was a small, heterophyllous plant, with leaves faintly denticulate at the margin, as in many species of *Selaginella*. There were four megaspores in each megasporangium, and each megaspore reached a diameter of about 500 μm, about 10 times the size of the microspores. *Nathorstiana*, from Cretaceous rocks of Germany, recalls at once the Isoetales. An upright stock, made irregular by leaf scars, reached a height of about 12 cm and bore a crown of needle-like leaves. The base was divided into a number of narrow vertical lobes from which the roots emerged. Unfortunately, nothing is known of the sporophylls.

The phylogeny of the Lycopsida

It is clear from the fossil record that the living Lycopsida are the relicts of a component of the Plant Kingdom which reached the apogee of its morphological complexity and floristic success in the Palaeozoic, about 200 million years ago. Besides the antiquity of the lycopsid kind of construction, the fossil record also indicates that the Lycopsida always had a distinctly heteromorphic alternation of generations. As in the living representatives, there is a conspicuous absence of morphological and anatomical similarities between the sporophyte and gametophyte generations, a situation which contrasts strongly with that in the Psilopsida. In the Lycopsida the gametophytes appear never to have been more than wholly parenchymatous plants of lowly status.

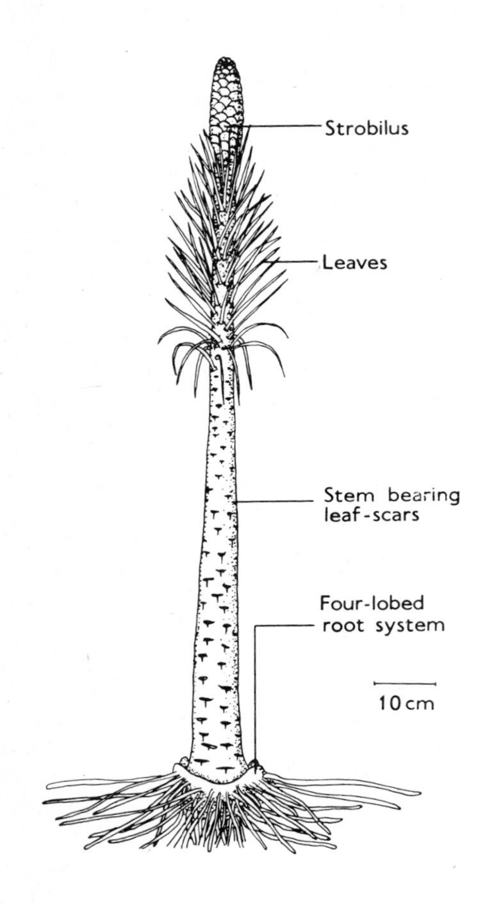

Fig. 5.26 *Pleuromeia sternbergi*. Reconstruction. (After Hirmer, from Andrews, *Studies in Paleobotany*. Wiley, New York, 1961)

The reduction of the gametophytes in the heterosporous Lycopsida to short-lived, almost wholly endosporic thalli can be regarded as a major step in the direction of becoming independent of humid conditions for the survival of this vulnerable generation. Had the Lycopsida acquired this valuable adaptation to terrestrial life earlier, and proceeded to the evolution of seeds, their spectacular decline at the end of the Palaeozoic might never have occurred.

SPHENOPSIDA

Sporophyte consisting of a monopodial branch system, some axes rhizomatous and bearing roots. Leaves microphyllous, borne in whorls. Vascular tissue of tracheids and phloem. Sporangia borne on sporangiophores, aggregated in terminal strobili. Gametophyte (in living forms) terrestrial. Antherozoids multiflagellate. Embryogeny exoscopic.

The only living sphenopsid is *Equisetum*, but the Sub-Division has a rich fossil record. Four Orders are recognized: the Equisetales (containing *Equisetum* and a number of fossil genera), and the extinct Calamitales, Sphenophyllales and Pseudoborniales.

Equisetales The striking feature of this Order is the jointed structure of the stem, and, in regions of uniform diameter, the regular alternations of the microphylls in the successive whorls. The stems are often conspicuously ridged, each ridge lying beneath a leaf. Consequently, provided the number of leaves in a series of whorls remains the same, the ridges also show regular alternation from one internode to the next.

Equisetum (Fig. 5.27), the horsetail, is a familiar sight in parts of the North Temperate zone, but is rarer in the Tropics and southern hemisphere, being absent altogether from Australia and New Zealand. About 15 species are now living, but others are known as Mesozoic fossils. The genus, or a form very closely similar, was widespread in Cretaceous times, and some species appear to have formed dense stands at the fringes of Cretaceous lakes. Moist habitats, such as river banks and marshy ground, are also favoured by most living species, but some are able to thrive in much drier places. *E. arvense*, for example, often flourishes on well-drained railway embankments. All species have a similar growth form. A perennial underground rhizome gives rise to green aerial shoots, and occasionally also to perennating tubers packed with starch. In temperate and arctic regions the aerial shoots die back at the end of the growing season and new shoots emerge in the following spring.

MORPHOLOGY AND ANATOMY OF *EQUISETUM* Although the height of the aerial system varies from a few cm in arctic and alpine species to as much as 10 m in the tropical *E. giganteum*, its morphology is strikingly uniform. Branches appear only at the nodes, and where several are present they too are whorled. The branch primordia are not, however, axillary, but they arise between the microphylls and eventually break through the sheath

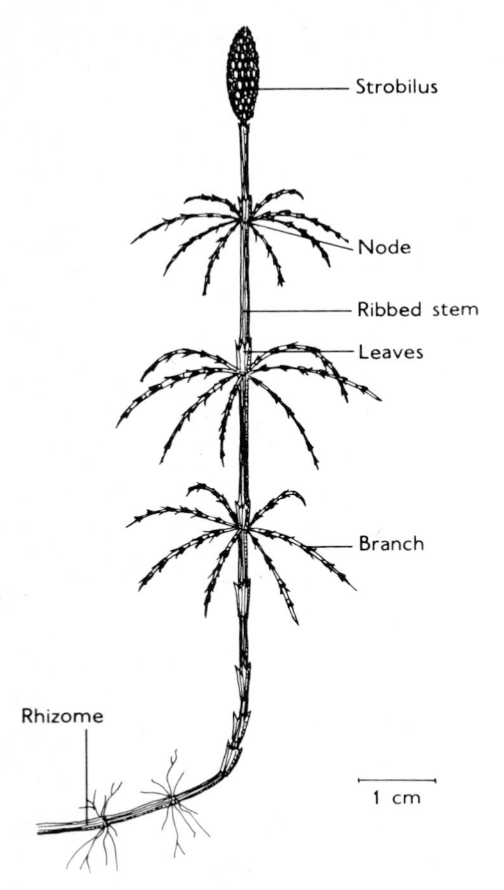

Fig. 5.27 *Equisetum sylvaticum*. Habit. Early summer condition, after production of vegetative branches, but before shedding of cone.

formed by their congenitally fused bases. In subterranean axes, roots emerge from directly below the sites of branch primordia. Branch and root primordia are in fact present at every node, but they develop only in appropriate environmental conditions, sometimes reproducible in the laboratory. In *E. arvense*, for example, green branches can be made to grow from the nodes of an etiolated, unbranched fertile shoot if they are enclosed in a moist chamber.[42]

The structure of the axes of *Equisetum* also shows little variation. The mature stems have a large central cavity, surrounded by a ring of vascular bundles (Fig. 5.28). These bundles are of the same number as the ribs on the outside of the internode (and hence as the leaves of the node above), and are also co-radial with them. An endodermis can usually be distinguished, either encircling the stele on the outside alone (as in *E. arvense*), or forming two continuous cylinders, one inside and one outside the ring of bundles (as in *E. hyemale*), or surrounding each bundle individually (as in *E. fluviatile*). In some species the position of the endodermis in the

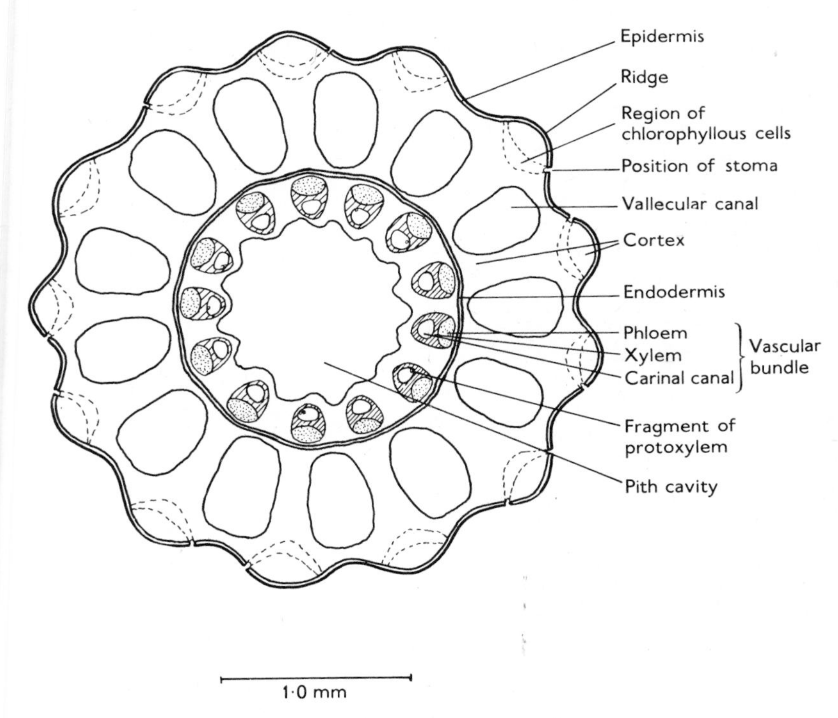

Fig. 5.28 *Equisetum arvense*. Transverse section of young stem.

rhizome is different from that in the aerial stem. Alternating with the vascular bundles, and lying between the endodermis and the periphery, are large longitudinal air chambers known as ***vallecular canals***.

The vascular bundles themselves contain very little lignified tissue. In a differentiated internode the protoxylem is represented solely by fragments of tracheids adhering to the sides of a cavity, termed the ***carinal canal***, present on the adaxial side of each bundle. The metaxylem differentiates as two groups of tracheids, one placed tangentially on each side of the phloem. The vascular anatomy of the nodes is highly peculiar. Here the tracheids, resembling those of the metaxylem, run horizontally and form a ring linking the bundles of the adjacent internodes. The root, branch and leaf traces also originate at this level, the leaf trace departing immediately above the entry of the bundle from the internode below.

The strength of the *Equisetum* stem depends principally upon the cortical ridges. These consist of sclerenchymatous cells reinforced by deposits of silica. The support that this rather tenuous outer framework can provide is, of course, limited. The taller species of *Equisetum* grow in

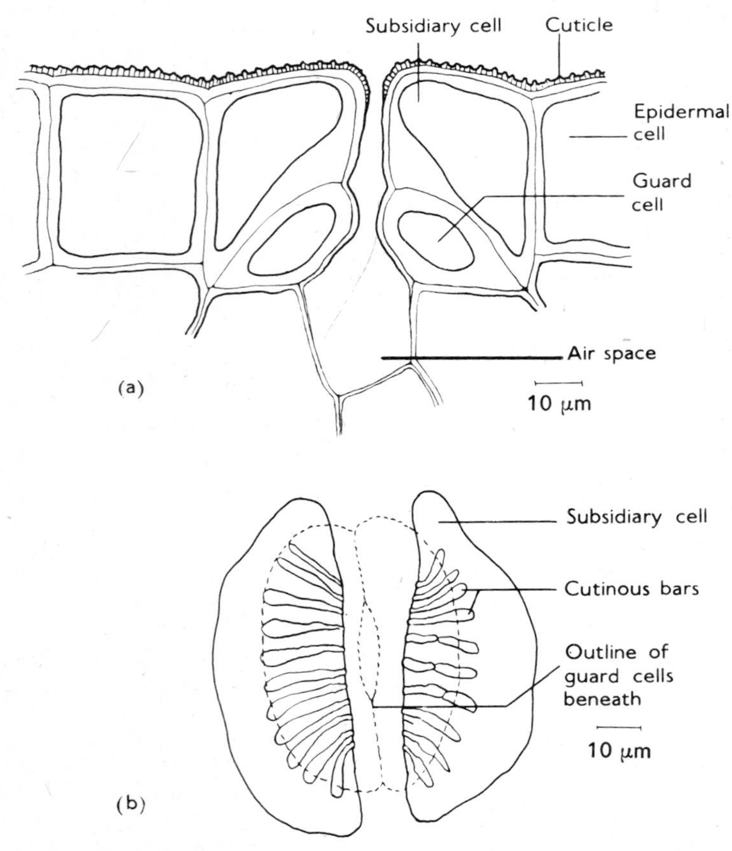

Fig. 5.29 *Equisetum* sp. Structure of stomata. (**a**) In vertical section. (**b**) In surface view partially macerated, showing the bars of cutin on the subsidiary cells. (After Hauke, *Bull. Torr. Bot. Club*, **84**, 178 (1957))

fact in groves and, since the rough, siliceous stems and branches do not readily slide over each other, the plants hold each other up.

The microphylls of *Equisetum* soon become scarious, and photosynthesis takes place predominantly in the surface layers of the stem. Apart from a few curious 'water stomata' or ***hydathodes*** in the adaxial epidermis of the tip of the microphyll, the stomata are confined to the valleys of the internodes, and thus lie above the vallecular canals. The stomata are often deeply immersed, merely a pore being visible externally. Each guard cell is flanked by a subsidiary cell, and each subsidiary cell bears transverse bars of cutin on its exposed surfaces (Fig. 5.29).

THE GROWTH OF THE AXES The growth of the axis of *Equisetum* provides a striking example of co-ordinated differentiation.[43] The axis is surmounted by a single tetrahedral apical cell. It is doubtful whether this cell itself undergoes division,[23a] but adjacent to its three posterior faces daughter cells are cut off in a regular, clockwise sequence. While the stem is growing there is no pause in the meristematic activity of the apical initials and their products are recognizable as three tiers of cells in the extreme apex. The primordium of a whorl of leaves first becomes visible as a ring at the base of the apical cone. While this ring is being superseded by another, leaf teeth initials become visible around the upper margin of the first. These initials are of limited growth and give rise to the free part of the microphylls, the basal ring meanwhile forming the sheath of fused bases.

The upper three or four nodal primordia remain close together, but the cells between them become organized as an intercalary meristem which surrounds, but does not cut across, the procambial strands. This meristem becomes noticeably active between the fourth and fifth nodes and remains active, with an increasing and then decreasing rate. Meanwhile, about five more nodes are initiated. This activity separates the whorls of leaves and generates the internodes, which reach their full development between about the ninth and tenth nodes.

The procambial tissue, the disposition of which foreshadows that of the vascular bundles in the mature axis, advances continuously, in step with the advance of the apex, at the level of the uppermost leaf whorl primordia. Differentiation of the procambial strand is not, however, a simple acropetal process. Protoxylem begins to appear at about the fourth node, and differentiation extends acropetally into the leaf and basipetally into the node below. Since this differentiation occurs during the time of maximum extension, most of the tracheids first formed in the internode are ruptured, and the area of weakness provided by the differentiating cells is pulled apart by the radial and tangential expansion of the stem, so yielding the carinal canal. Metaxylem begins to appear at the fifth node, and also differentiates basipetally. The rate of differentiation is, however, such that the descending strands do not fuse with the metaxylem of the node beneath until about the tenth node, when extension of the internodes is ceasing. The internodal metaxylem thus escapes rupture, although occasional small lacunae are formed within it, indicating that it is subjected to some stresses during the concluding phases of its differentiation. The differentiation of the phloem follows more or less the same course as that of the metaxylem.

In temperate and arctic habitats the aerial branches perish in the winter, and in this instance the growth of the shoot is a little different from that just described. Almost the whole of the next year's shoot overwinters in primordial form as a subterranean bud. In spring elongation of the internodes and further differentiation proceeds acropetally, so generating an aerial system of limited growth.

REPRODUCTION In all species of *Equisetum* the sporangiophores are aggregated into a terminal strobilus (Fig. 5.30) which terminates either a

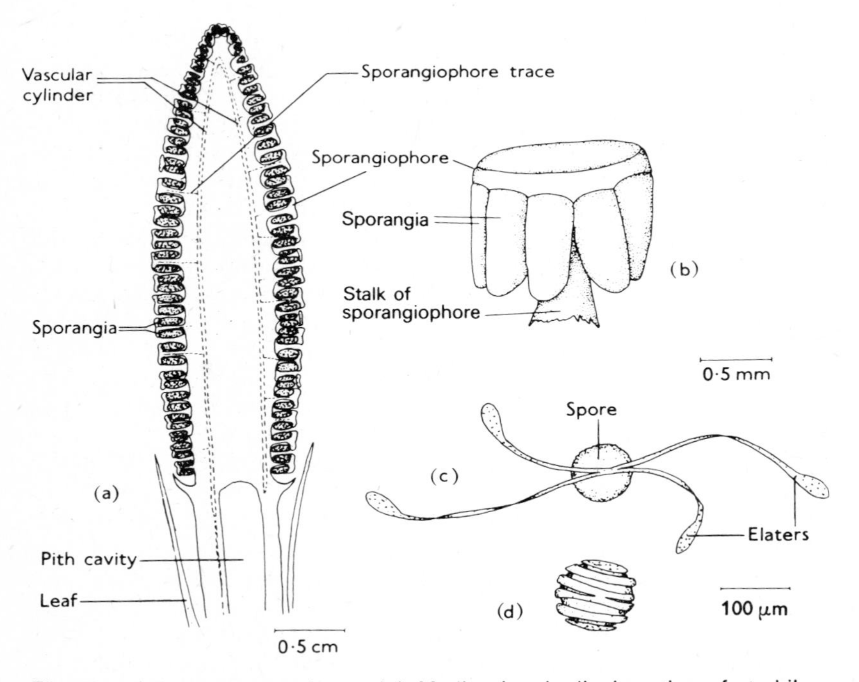

Fig. 5.30 *Equisetum maximum*. (**a**) Median longitudinal section of strobilus. (**b**) Peltate sporangiophore. (**c**) and (**d**) Spores with elaters in dry and moist condition respectively.

vegetative axis (as in *E. palustre*) or a specialized axis lacking pigmentation which appears early in the growth season (as in *E. arvense*). An intermediate state is shown by *E. sylvaticum*, in which the fertile shoots are at first colourless, but after release of the spores become green and branch. The vascular system inside the cone recalls that of a node, since it consists of a cylinder of metaxylem. Fine traces depart to the sporangiophores, and the cylinder is broken here and there by parenchymatous perforations which bear no evident relation to the departing traces. The sporangiophores, which are not necessarily arranged in whorls, are peltate, and are tightly packed so that heads acquire a polygonal outline. About 20 sporangia are pendent from the margin of the head, and so lie more or less radially in the intact cone.

A sporangium develops from a group of cells, in which a central archesporial tissue, surrounded by a wall several layers thick, can soon be distinguished. The inner layer of the wall functions as a tapetum, together with about a third of the archesporial tissue, the remainder being sporogenous. While the spores are maturing the spore coat becomes differentiated into several membranes, no doubt at the expense of the tapetal material. The innermost cellulose membranes are simple, but the outer are more complex and take the form of an X eventually becoming wholly free from the inner wall of the spore. The arms of the X are tightly wrapped around the spore as it lies in the sporangium, but on release, and after drying out, they respond to changes of humidity in the same way as the peristome teeth of the Bryales. They are thus referred to as elaters (or better as ***haptera***), and their quite violent movements assist distribution of the spores. The sporangium opens along a longitudinal stomium as the result of tensions arising during the drying out of the indurated and spirally thickened cells elsewhere in the wall.

Equisetum is usually considered to be homosporous. In *E. arvense* a large sample of spores falls into two classes whose mean diameters differ by about 10 μm, but this is an effect of drying.[26a] The spores, which contain chlorophyll, soon lose their viability if stored. In germination the cellulose wall bulges out to form a filament, and it has been shown that at the site of germination there is a large concentration of potassium ions.[54] This probably accounts for the increased plasticity of the wall at this place, since potassium ions are known to reduce the amount of cross-linking in cellulose microfibrils. The filament which emerges from the spore is transformed by division in a number of planes into a cushion of cells anchored by rhizoids. Subsequent growth is from a marginal meristem which forms a number of obliquely ascending lobes (Fig. 5.30), on the upper surface of which the sex organs appear. Growth in size is, however, limited, and the gametophytes rarely exceed 1 cm in diameter and 3 mm in height.

Culture experiments have shown that there are two different kinds of gametophytes. About half the gametophytes of a mass sowing of spores remain small and produce only antheridia, dying soon afterwards. The remainder are larger and longer lived. They first produce archegonia, and then, if none is fertilized, a crop of antheridia, followed by another of archegonia. Such gametophytes may last at least two years and in favourable circumstances produce several sporophytes. There is thus some evidence that *Equisetum* is heterothallic, but, since the proportions of the two kinds of gametophyte are related to the density of sowing there is the possibility that differences in competitiveness (deriving from chance variations in the cytoplasmic complements of the spores) may also influence the subsequent gametogenesis.[86]

The antheridia of *Equisetum*, which resemble those of the Lycopsida, yield numerous antherozoids, each furnished with a crown of flagella. The archegonia are formed mostly between the aerial lobes. The necks project, and at maturity the four distal cells become elongated and reflexed.

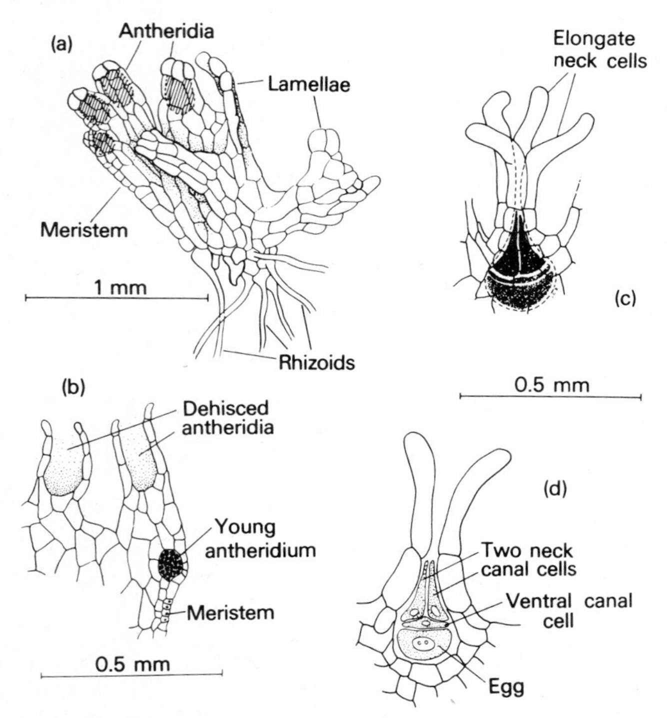

Fig. 5.31 (**a**, **b**) *Equisetum sylvaticum*. (**a**) Young male gametophyte. (**b**) Section through male branch. (**c**, **d**) *E. palustre*, archegonia. (**c**) General view. (**d**) Longitudinal section. (After Duckett, Ph.D. Diss. Cambridge University (1968))

Fertilization, dependent, as in the Lycopsida, upon the presence of a film of water, is followed by the division of the zygote in a plane perpendicular to the axis of the archegonium. The subsequent embryology is exoscopic, and the products of the outer cell give rise to the stem apex, the primordium of the first whorl of leaves, and, in a somewhat variable lateral position, a root apex.

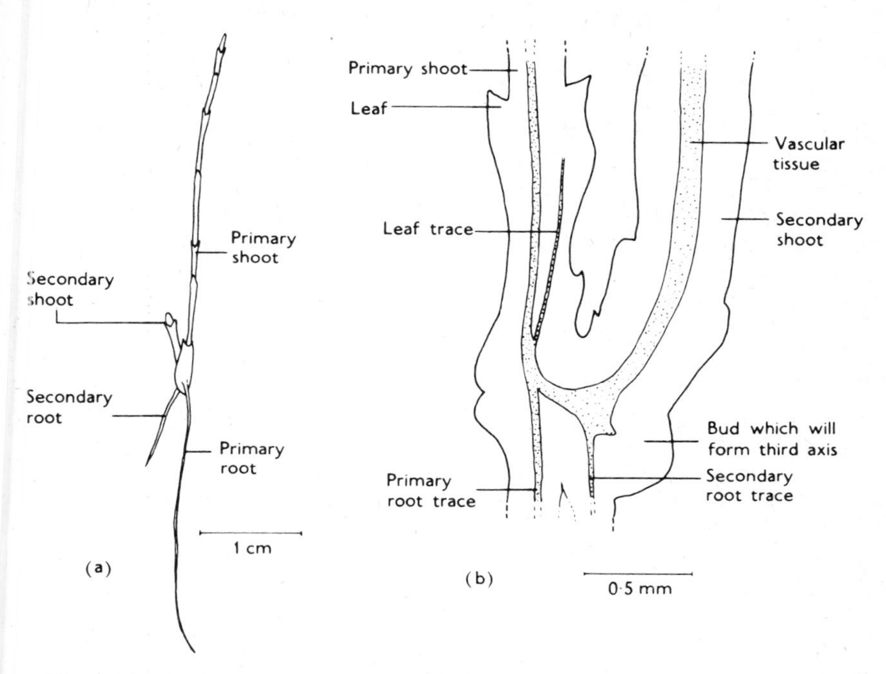

Fig. 5.32 *Equisetum arvense.* (**a**) Sporeling. (**b**) Development of mature form of plant. (After Barratt, *Ann. Bot.*, **34**, 201 (1920))

THE DEVELOPMENT OF THE SPOROPHYTE The development of the young sporophyte follows a curious course (Fig. 5.32), without parallel in other living archegoniate plants. The first axis, which contains a simple protostele, is of determinate growth and never increases in diameter. In *E. arvense*, for example, it produces about six whorls, each of about three leaves. As this shoot ceases to grow, a bud grows out from below the first whorl of leaves. This also produces an upright axis, of slightly greater diameter than the first and containing a protostele that shows a tendency towards medullation. This process is repeated two or three times until the axis reaches the diameter and structure characteristic of the mature plant. The rhizomatous growth habit is then initiated. The primary root of the embryo persists only a short time, and roots are produced freely from the nodes at and below the soil surface as the young plant becomes established.

The early Sphenopsida

The fossil record of the Sphenopsida parallels that of the Lycopsida, beginning early in the Palaeozoic and expanding in the Carboniferous, when the Sphenopsida must have formed a large part of the earth's vegetation. From the end of the Palaeozoic to the present time they have been of diminishing importance until today only a single genus remains.

The fossil sphenopsids which show the closest resemblance to *Equisetum* are the Calamitales. These occurred predominantly in the Carboniferous, reached the proportions of trees and were probably one of the main competitors of the arborescent lycopsids. The leaves were whorled, and, although in some of the early forms forked (e.g. *Asterocalamites*), in the later they were simple. The stems were conspicuously ridged, the ridges in some forms alternating from node to node, and in others lying superposed. The stomata of *Calamites* resembled those of *Equisetum*, transverse bars of cutin being present on the subsidiary cells (see (Fig. 5.29). Petrifactions also reveal that the vascular system of the Calamitales was basically similar to that of *Equisetum*. In the internodal region a pith cavity was present, and the protoxylem was associated with a canal. A cambium, however, arose between the primary xylem and phloem and contributed a considerable amount of secondary vascular tissue to the stem. Probably as a consequence of this radial expansion, no air canals were present in the cortex.

The strobili of the Calamitales, which terminated lateral branches, consisted of alternate whorls of sporangiophores and bracts, although the cone of one early form appears to have contained peltate sporangiophores alone. Since cones are often found detached, they are placed in ***form genera***, defined by the relative arrangements of the sporangiophores and bracts. The two form genera most widely represented are *Calamostachys* (Fig. 5.33) and *Palaeostachys*. The cones assigned to these form genera may, of course, have been produced by plants differing widely vegetatively. Both homosporous and heterosporous cones have been described.

Apart from the Cretaceous forms closely resembling living *Equisetum*, referred to earlier, there is evidence of herbaceous forms having also existed in the Palaeozoic. *Equisetites hemingwayi*, for example, from the Upper Carboniferous, is the remains of a fertile shoot very like *Equisetum* in the structure of the cone and order of size.

The Sphenophyllales, another Order of the sphenopsids,appeared in the late Devonian and were prominent in Carboniferous floras. They share little with the Equisetales and Calamitales except the whorled arrangement of the leaves. They were probably scrambling plants, supporting themselves on other vegetation in the manner of the familiar *Galium aparine* (cleavers or goosegrass). The leaves were wedge-shaped, with dichotomously branching venation, and were usually borne in multiples of three (Fig. 5.34a). The

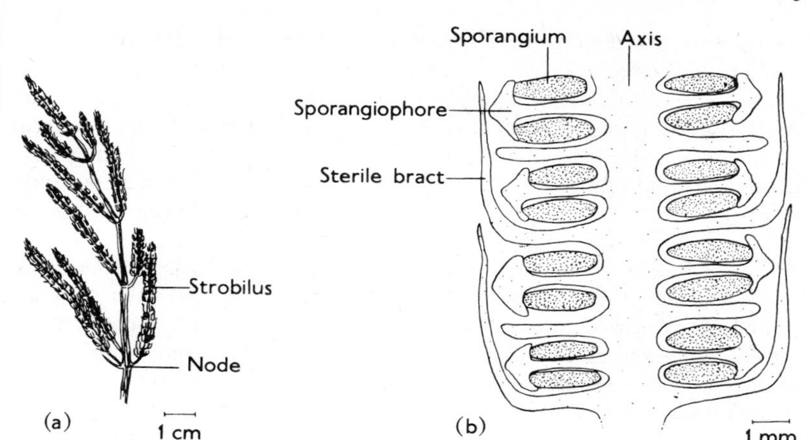

Fig. 5.33 *Calamostachys*. (**a**) *C. ludwigi*. Fertile shoot (from a compression). (After Weiss, from Andrews, Jr., *Studies in Paleobotany*. Wiley, New York, 1961) (**b**) *C. binneyana*. Longitudinal section of cone. (After Andrews, Jr., *Studies in Paleobotany*. Wiley, New York, 1961)

stems contained a solid core of primary xylem, triangular in transverse section, with protoxylem at the vertices. This primary xylem was surrounded by secondary in regular radial files, resulting in a vascular system strikingly reminiscent of that of a root. The strobili were terminal and frequently con-

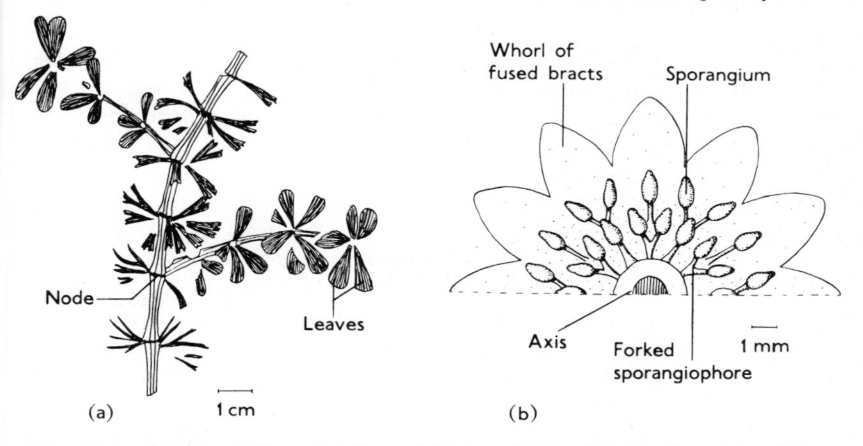

Fig. 5.34 *Sphenophyllum*. (**a**) *S. verticillatum*. Habit of vegetative shoot (from a compression). (After Potonié, *Lehrbuch der Pflanzenpalaeontologie*. Dümmler, Berlin, 1899) (**b**) *S.* (*Bowmanites*) *dawsoni*. Fertile whorl viewed from above. (After Hirmer, *Handbuch der Palaeobotanik*. Oldenbourg, Munich, 1927)

sisted of whorls of sterile bracts, each forming a cup-like sheath around the axis, on the adaxial side of which were attached the sporangiophores (Fig. 5.34b). The distal portion of the sporangiophore, which was free, branched and recurved, bore a number of sporangia. In other forms of cone the sterile part was less well developed, and the sporangiophore branching more complex. Most of the cones were homosporous, but distinct heterospory has been detected in one form. The Sphenophyllales disappeared at the beginning of the Triassic period.

The Pseudoborniales are based upon remains from the Upper Devonian of Bear Island. They appear to have borne their leaves, each of which forked once or twice, in whorls of four, and to have produced loose, terminal strobili, 30 cm or more in length. The sporangiophores were whorled, and each included a prominent sterile portion. Some of the stems reached a diameter of 10 cm, so the Pseudoborniales were probably bushes or small trees.

The origin of the sphenopsids

The bearing of the leaves or branches in whorls, and the articulate structure of the stem, although characteristic features of the Sphenopsida, are not of course confined to this Sub-Division of the Plant Kingdom. They are present in some of the algae (e.g. *Draparnaldia* (p. 41), *Batrachospermum* (p. 95) and in the flowering plants (e.g. *Galium*, *Casuarina*), and have clearly arisen a number of times in plant evolution. Nevertheless, the Sphenopsida appear to have specialized in this organization of the plant body from their beginnings and they may even have had a common origin in the remote past.

Unfortunately no fossils are known which show convincingly how the sphenopsids might have arisen. *Calamophyton* and *Hyenia*, both Middle Devonian in age, and consisting of axes bearing forking appendages, were once thought to be articulate plants close to the inception of sphenopsid evolution. This interpretation is now largely discarded,[6a] the transverse markings on the axes, formerly taken as an indication of nodes, now being regarded as nothing more than regular cross-fractures appearing during fossilization. In the absence of any firm evidence we may, however, reasonably speculate that the origin of the sphenopsids probably lay in a plant with a simple axial form like that of the psilopsids, but bearing lateral appendages. These appendages may have forked once or twice, those towards the tips of some of the shoots terminating in sporangia. We must suppose a tendency for these lateral appendages to be produced in whorls, followed by the consolidation of this pattern of development, with corresponding changes in vascular anatomy, in the genetic constitution, leading eventually to plants with a well-defined articulate morphology.

Although in some of the Carboniferous sphenopsids sporangia are produced in association with bracts, these complexes seem to have no affinity with the lycopsid sporophyll, a reproductive structure, as we have seen (p. 171), well established in early Devonian times. It appears unlikely, therefore, that the origin of the sphenopsids is to be sought amongst these plants.

Although the Sphenopsida themselves may have had a common origin, the relationships between the Calamitales and Equisetales are clearly closer than between the other Orders. There seems to have been a general tendency towards the sporangiophore becoming peltate, but this perhaps hindered the evolution of seed-like structures, of which there are only rare indications in the sphenopsids. The almost total elimination of the sphenopsids points to their limitations, which possibly lay principally in the reproductive mechanism. It is difficult to account for the survival of the Equisetales in preference to the other Orders. It may have been a consequence of only this Order having contained herbaceous forms sufficiently adaptable to meet the fluctuating conditions of the Mesozoic.

6

The Tracheophyta, II
(Pteropsida: Filicinae)

In the classification we have adopted, the fourth Sub-Division of the Tracheophyta, referred to as the Pteropsida, encompasses the diverse range of vascular plants known as the ferns, gymnosperms and angiosperms. This grouping, which gives the Pteropsida the same taxonomic status as, for example, the Sphenopsida, is justified by a number of basic features common to the higher archegoniate and flowering plants, but absent from the Sphenopsida, Lycopsida and Psilopsida. The general features of the Pteropsida may be summarized as follows:

PTEROPSIDA

Sporophyte usually branching freely, the branches bearing a distinct, often axillary, relationship to the leaves. The leaves often with an elaborately branched vascular system, the leaf trace commonly compound and leaving a parenchymatous gap at its departure from the stele. Vascular system consisting of tracheids or tracheids and vessels, and phloem. Position of sporangia various, sometimes enclosed in sporophytic tissue. Heterospory frequent, the female gametophyte in many forms contained within the sporophyte. Male gametes flagellate, or without any specialized locomotory apparatus. Embryogeny largely endoscopic, the embryo often retained and distributed within a seed.

The Pteropsida fall naturally into three Classes, namely the Filicinae, Gymnospermae and Angiospermae. The first two Classes have a rich fossil record, but that of the angiosperms is meagre and mostly recent. The Filicinae are thus some of the most ancient of the Pteropsida, and, despite their antiquity, they still remain plants of considerable morphological and ecological interest.

Filicinae[14,73]

Sporophyte herbaceous, leaves often large and resembling branch systems. Vascular system consisting of tracheids and phloem, usually lacking clearly defined secondary tissue. Sporangia borne on the leaves, but never on the adaxial surface of a microphyll. Heterospory rare. Gametophyte (of living species) usually autotrophic, antherozoids multiflagellate. Embryogeny mostly endoscopic.

Homosporous plants whose axes lacked secondary thickening, and which appear referable to the ferns, are found in rocks as old as the Devonian. The ferns were undoubtedly an important component of the succeeding Carboniferous vegetation, and, although many of these early forms died out with the Palaeozoic, others were able to adapt themselves to changing conditions. Consequently, the ferns, with about 10,000 living species, have remained an important and, particularly in humid regions, a conspicuous element of the world's vegetation. Apart from this relative floristic success, the ferns show the greatest range of growth forms amongst the archegoniate plants, and the greatest diversity in the morphologies of individual organs. The leaves, however, remain in most forms relatively large and branched structures, branching also being characteristic of their vascular systems. This kind of leaf (often called a frond), strikingly different from the microphylls of the Lycopsida, is termed a ***megaphyll***.

The ferns are conveniently considered as five Orders, namely the Cladoxylales, Coenopteridales, Ophioglossales, Marattiales and Filicales. The first two Orders are wholly extinct.

Cladoxylales The Cladoxylales are based upon a small number of peculiar, probably homosporous, plants from the Devonian and Lower Carboniferous. The axes (Fig. 6.1), the bigger reaching diameters of the order of 1.5 cm, branched irregularly, and bore small dichotomously branching leaves, rarely reaching 2 cm in length. Some of the terminal shoots were fertile, and here small fan-shaped sporangiophores took the place of leaves, each segment of the sporangiophore terminating in a sporangium.

The vascular system of the stem consisted of ascending, anastomosing plates of xylem, each in life probably surrounded by phloem and endodermis. Since the system has the appearance of falling into a number of independent sub-systems, it is said to be polystelic. The vascular supply of the branches was compound in origin, departing from several of the adjacent ascending plates. A curious anatomical feature was the presence of small parenchymatous areas within the xylem, especially towards the

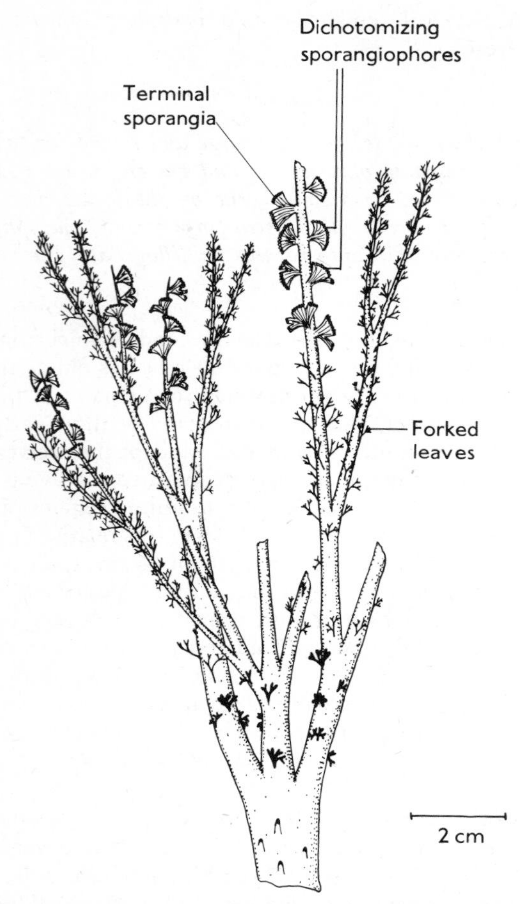

Fig. 6.1 *Cladoxylon scoparium*. Reconstruction. (After Kräusel and Weyland, from Delevoryas, *Morphology and Evolution of Fossil Plants*. Holt, Rinehart and Winston, New York, 1962)

periphery of the stem. Some species of *Cladoxylon* appear to have had rudimentary secondary thickening.

Coenopteridales The Coenopteridales, which were undoubted ferns, flourished in the late Palaeozoic, but apparently did not persist into the Mesozoic. They showed a wide range of habit, from tree ferns reaching a height of three or more metres to comparatively small epiphytes. The leaves were large and spreading, often resembling sprays of branchlets (Fig. 6.2). The branching was not always in one plane, and in one group,

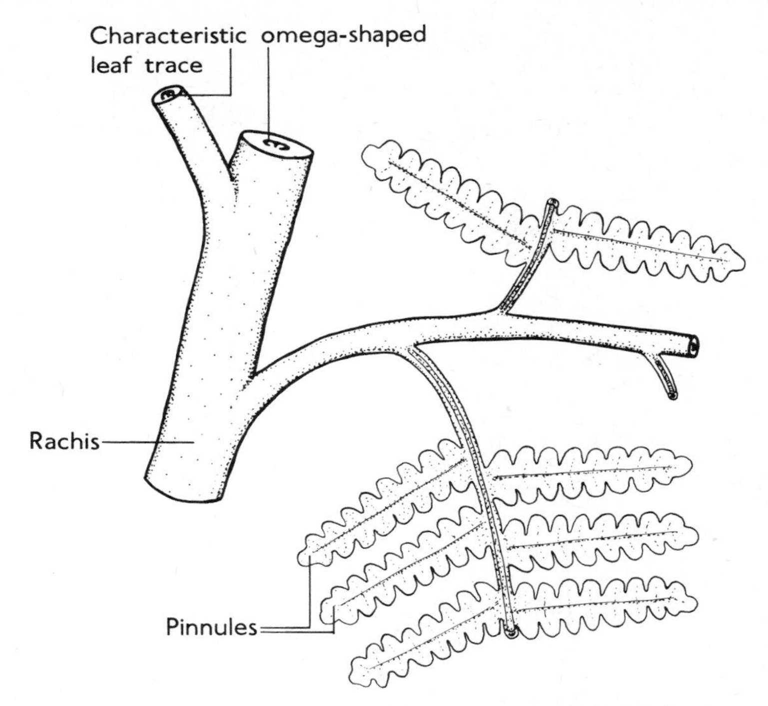

Fig. 6.2 *Botryopteris*. Reconstruction of portion of frond. (After Delevoryas and Morgan, from Delevoryas, *Morphology and Evolution of Fossil Plants*. Holt, Rinehart and Winston, New York, 1962)

the zygopterids, the branchlets were in a distinct quadriseriate arrangement. The morphological similarity of leaf and stem was sometimes reinforced by large decumbent fronds giving rise here and there to roots and shoot buds. An Upper Carboniferous species of *Botryopteris*, which must have formed low sprawling clumps, showed this kind of growth particularly well. In another form, *Metaclepsydropsis* (Fig. 6.3), from the Lower Carboniferous, the stem was rhizomatous and bushy fronds, elaborately branched in a quadriseriate manner, arose from it at intervals. That the complexity of the fronds appeared early in the phylogeny of this group of ferns is shown by equally remarkable fossils from the Devonian period.

THE VASCULAR ANATOMY The vascular system of the stem of the coenopterids was commonly a protostele, often with a parenchymatous medulla, and lacking any secondary thickening. The leaves, which were usually in a recognizable phyllotactic spiral, received a single vascular trace

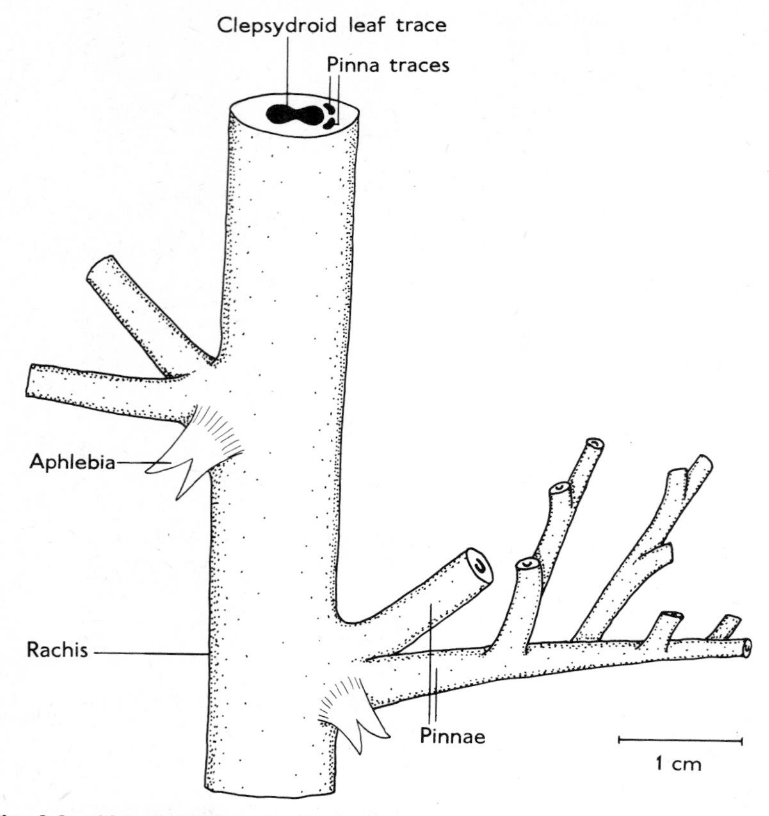

Fig. 6.3 *Metaclepsydropsis*. Reconstruction of portion of frond showing quadriseriate arrangement of pinnae.

which assumed in the petiole a definite and characteristic symmetry. Consequently the genera and species of these ferns are largely based upon the profile of the leaf trace in transverse section. In *Metaclepsydropsis*, for example, the section was hour-glass shaped (clepsydroid). The protoxylem lay towards each pole, each group associated with an island of parenchyma. Pinna traces arose from each pole in alternate pairs, in register with the quadriseriate branching. In *Stauropteris*, where the frond was similarly constructed, the petiole contained four groups of tracheids ascending in parallel. The petiolar trace of *Tubicaulis*, which produced complanate fronds, formed an arc in transverse section, with the convex surface adaxial. In early species of *Botryopteris* the petiolar trace was ovate in section, but in later species had the form of a ω, the curvature being abaxial and the protoxylem lying at the tips of the adaxial extensions. In many of the coenopterid fronds there were small branched emergences, each with an exiguous vascular supply, at the base of the rachis and at the sites of branching of the frond. These are referred to as ***aphlebiae***, and occur in some living ferns (e.g. *Hemitelia* and other tree ferns).

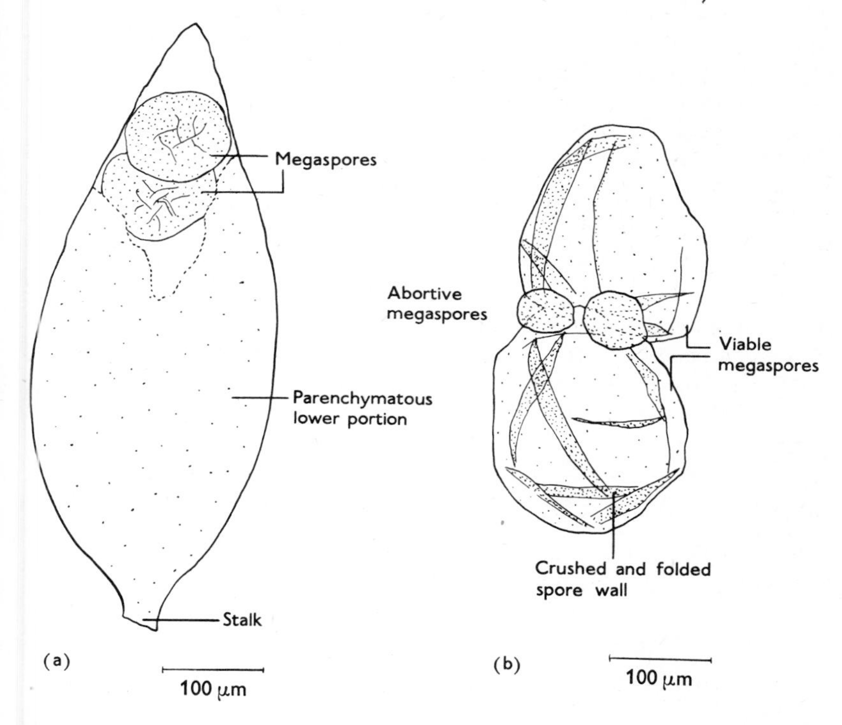

Fig. 6.4 *Stauropteris burntislandica*. (**a**) Megasporangium (from a specimen in the Oliver collection). (**b**) An isolated tetrad (*Didymosporites*). (After Chaloner, *Ann. Bot. N.S.* **22**, 197 (1958))

THE REPRODUCTIVE ORGANS In general in the coenopterids the sporangia terminated branchlets, either a basal or a terminal region of the frond being fertile. The sporangia of the Devonian *Rhacophyton* lacked any special opening mechanism, but in later forms the sporangia commonly possessed an annulus of thick-walled cells interrupted by a thin-walled stomium. In some species of *Botryopteris* the sporangia, which were about 0.75 mm in diameter, had a bounding wall only one cell thick, and in this they resembled the sporangia of many living ferns. Homospory was the predominant condition in the coenopterids, but a well-established example of heterospory is provided by the Carboniferous *Stauropteris burntislandica*. The megasporangia (Fig. 6.4) of this species, which were parenchymatous at the base, produced only one tetrad, consisting of two large and presumably functional spores about 200 μm in diameter, and two abortive spores.[19] In a related species, believed to be homosporous, spores germinating in a manner typical of ferns have been found petrified within sporangia. Apart from a few isolated instances of this kind, nothing further is known of the gametophyte generation of the coenopterids.

Ophioglossales The Ophioglossales form a small and morphologically peculiar Order of living ferns which, since they have no well-established fossil record, are of obscure origin. In all members the fertile region of the frond takes the form of a spike or pinnately branched structure, clearly set off from the vegetative portion. A feature that separates the Ophioglossales from other living ferns is that the fronds, instead of expanding from a closely coiled immature state (a process known as 'circinate vernation'), grow marginally from a more or less flat primordium. That of *Botrychium lunaria* when young shows a distinct kind of folding, the upper margins of the pinnules being covered by the lower margins of the pinnules above.

Of the three genera of the Class, *Botrychium* (moonwort) and *Ophioglossum* (adder's tongue) are fairly widespread, the former mainly in the north temperate zone and the latter in the Tropics. Both genera include species native to the British Isles. The third genus, *Helminthostachys*, is restricted to the Polynesian Islands in the South Pacific and a few regions in the Asian Tropics.

In *Botrychium* (Fig. 6.5), where the frond is annual, the vegetative and fertile parts are pinnately branched. *Helminthostachys* is basically similar, but the branches of the fertile part of the frond are very contracted. In *Ophioglossum* (Fig. 6.6) the sterile part of the frond, which is reticulately veined, is elliptic and entire or, in a few epiphytic forms, dichotomously lobed. The fertile part is never anything other than a simple spike.

THE VASCULAR SYSTEM OF THE RHIZOME The Ophioglossales are rhizomatous, growth taking place from a single apical cell. In *Botrychium* the rhizome of the young plant contains a medullated protostele (Fig. 6.7), but in the stele of older plants a parenchymatous area perforates the xylem anteriorly to the departing leaf trace. The endodermis remains wholly exterior, so we arrive at a stele intermediate between a protostele and a solenostele, often referred to as a siphonostele (see Fig. 6.20). A rudimentary solenostele does in fact arise in some species of *Botrychium* as a result of an endodermis appearing on the inside of the xylem cylinder. The stele shows a number of points of anatomical interest. The metaxylem tracheids, for example, bear bordered circular pits, found outside the Ophioglossales only in the gymnosperms and angiosperms. There is also cambial activity

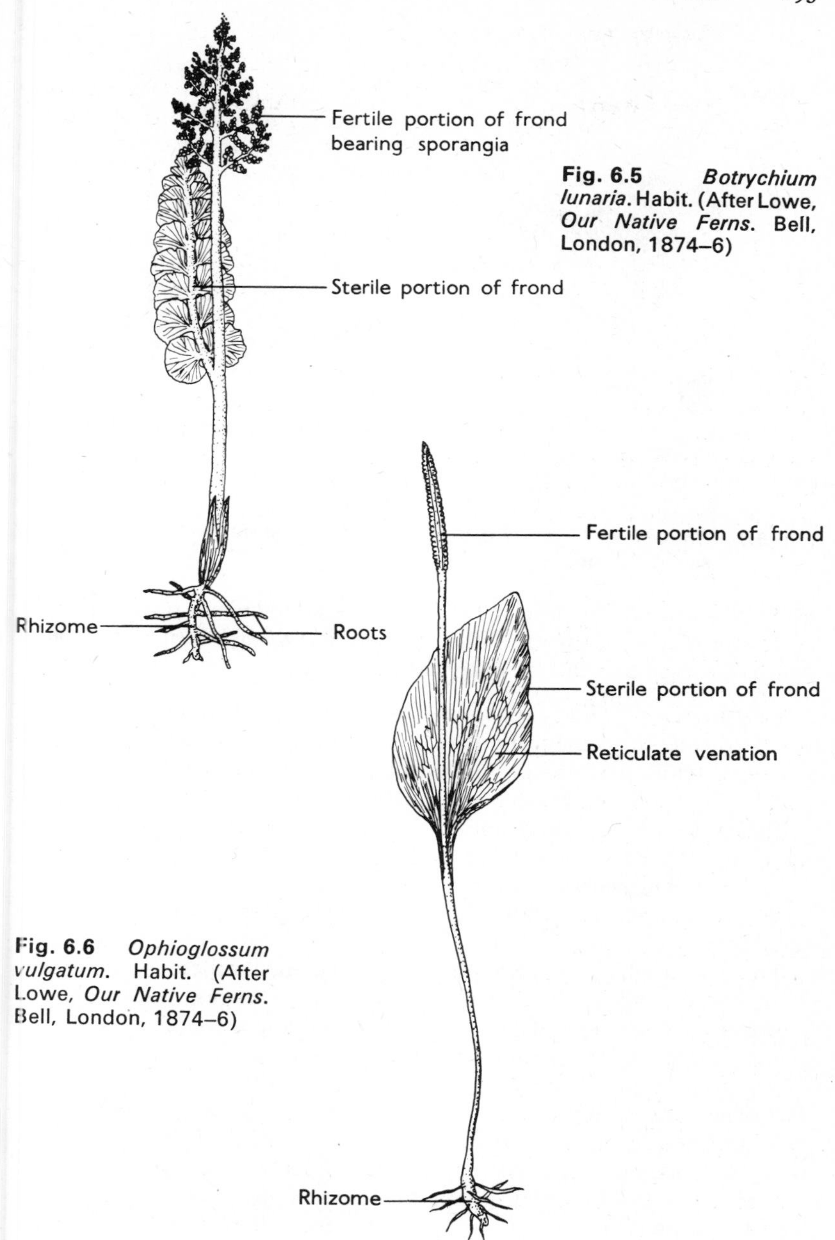

Fig. 6.5 *Botrychium lunaria*. Habit. (After Lowe, *Our Native Ferns*. Bell, London, 1874–6)

Fig. 6.6 *Ophioglossum vulgatum*. Habit. (After Lowe, *Our Native Ferns*. Bell, London, 1874–6)

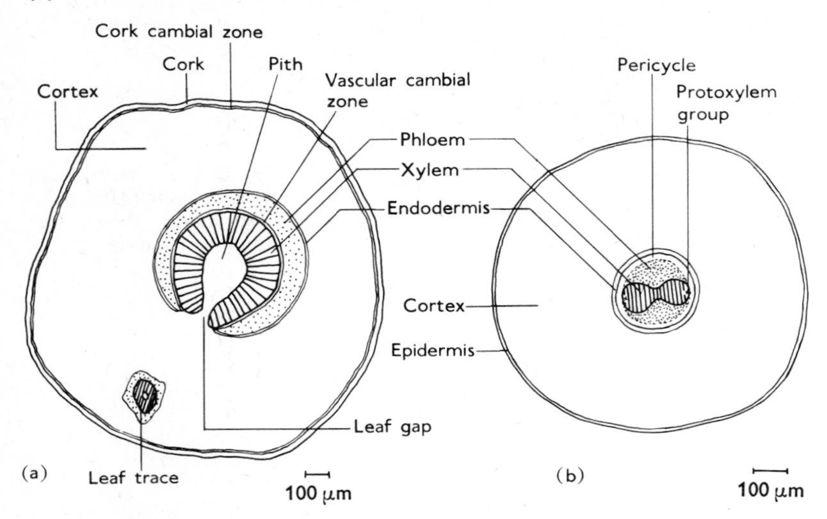

Fig. 6.7 *Botrychium lunaria*. **(a)** Transverse section of rhizome. **(b)** Transverse section of root.

leading to secondary vascular tissue, otherwise unknown in living ferns. Apart from this last feature, the anatomy of *Ophioglossum* and *Helminthostachys* resembles that of *Botrychium*.

REPRODUCTION The spherical sporangia of *Botrychium*, termed eusporangiate since they originate from a group of initial cells, arise in two ranks (Fig. 6.8) on the ultimate branches of the fertile part of the frond. The spore mother cells are enclosed in a tapetum, several cells in thickness, in whose disintegration products the spores mature. Even at maturity the wall of the sporangium is massive, and stomata interrupt its outer layer. The spores, a few thousand in number, are released by transverse dehiscence. *Botrychium*, like the Ophioglossales as a whole, is homosporous.

The gametophyte of *Botrychium* is a flattened tuberous prothallus, subterranean and invested with an endophytic fungus, presumably in mycorrhizal relationship (Fig. 6.9). The antheridia are sunken, and each yields over a thousand multiflagellate antherozoids. The archegonia, of quite normal construction, are partially immersed. The embryogeny of *Botrychium* is somewhat variable: in some species there is a suspensor and development is endoscopic, in others the suspensor is lacking and development is exoscopic. The first sporophytic organ to emerge is the root, infected from the first with the same endophytic fungus as the gametophyte. The young plant may remain subterranean in an immature condition for several years, and until the first leaf appears above ground the nutrition is presumably supplied wholly by the mycorrhiza.

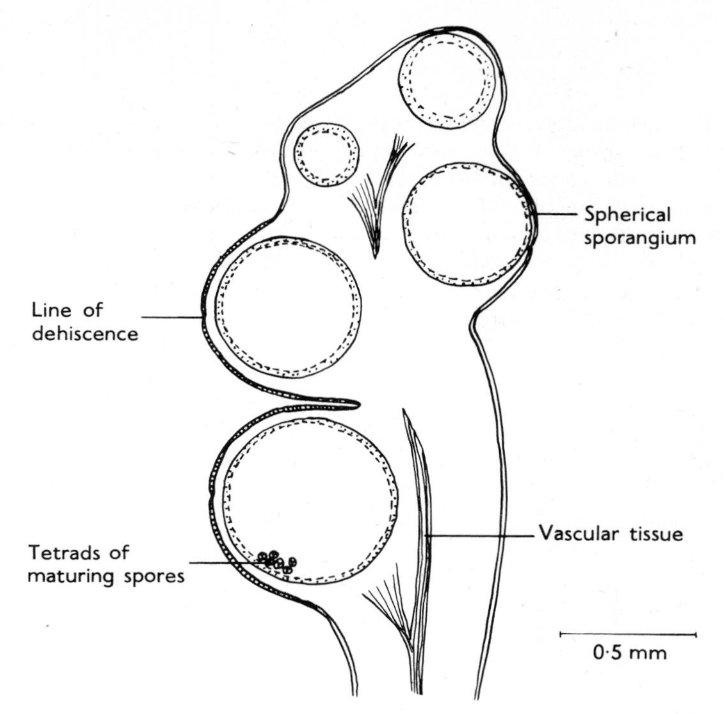

Fig. 6.8 *Botrychium lunaria*. Longitudinal section of fertile region of frond.

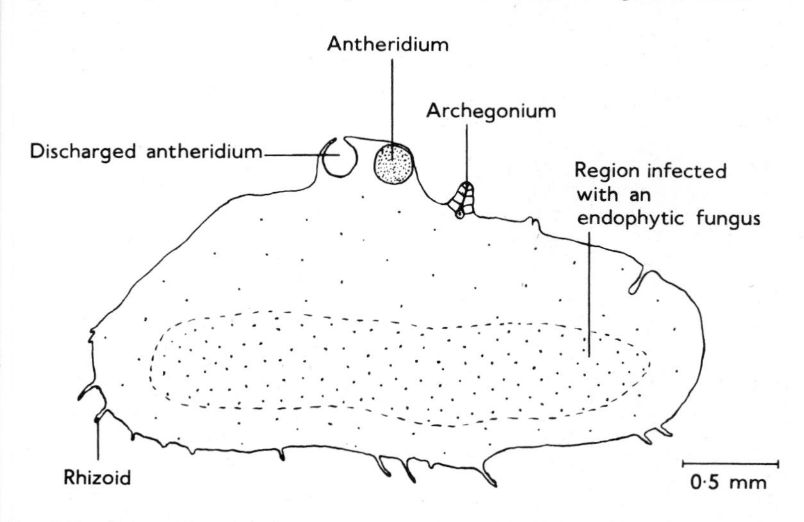

Fig. 6.9 *Botrychium virginianum*. Vertical section of gametophyte. (After Campbell, *The Structure and Development of Mosses and Ferns*. Macmillan, New York, 1905)

The reproduction of *Helminthostachys* is similar to that of *Botrychium*, but the dehiscence of the sporangia is longitudinal, and the embryogeny is regularly endoscopic. In *Ophioglossum* the sporangia, which occur as two rows partially embedded in the spike (Fig. 6.10), open by transverse clefts. Each contains numerous spores, in some species of the order of 15,000. The gametophyte of *Ophioglossum* is subterranean and cylindrical, sometimes approaching 5 cm in length. Both antheridia and archegonia are sunken. The embryogeny is exoscopic.

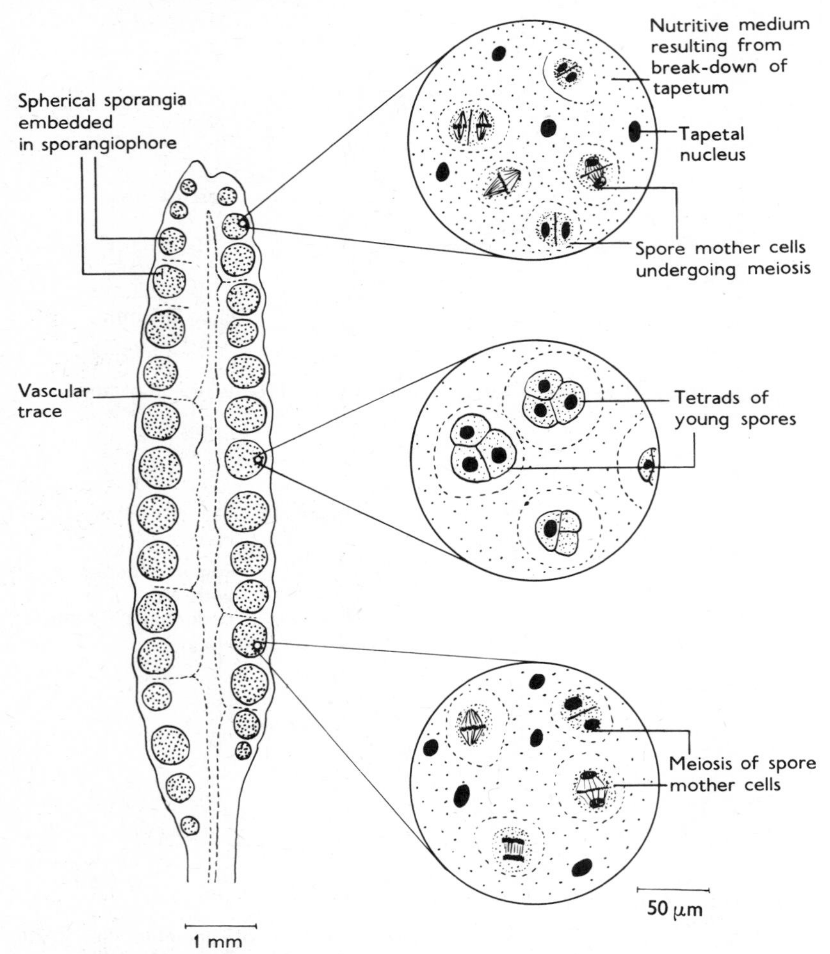

Fig. 6.10 *Ophioglossum vulgatum*. Longitudinal section of young fertile spike.

The phylogeny of the Ophioglossales

The Ophioglossales have no close relatives, and the evidence of distribution and comparative anatomy, particularly in relation to the massive sporangia and the stele, points to their being the relicts of an ancient lineage. The absence of a fossil history indicates that they were never very numerous.

Marattiales The Marattiales are another small Order of ferns, in this instance wholly tropical. Although not conspicuous in contemporary vegetation, they have a remarkably rich fossil record, and the Class can be recognized with certainty as far back as the Carboniferous.

MORPHOLOGY AND ANATOMY Of the living genera, the most common are *Angiopteris* and *Marattia*. Both have short upright trunks bearing large, pinnately branched and rather fleshy fronds (Fig. 6.11), sometimes

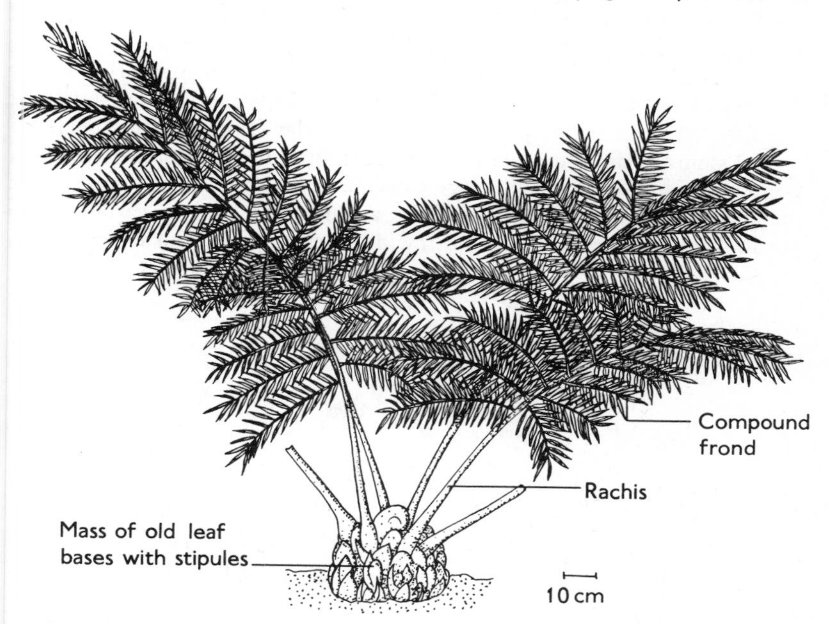

Fig. 6.11 *Angiopteris teysmanniana*. Habit of young plant. (After Bitter, in Engler and Prantl, *Die natürlichen Pflanzenfamilien*, I, 4. Engelmann, Leipzig, 1902)

reaching a length of 5 m, and showing circinate vernation. At the base of the petiole are two prominent stipules which persist after the leaf has fallen. *Christensenia*, a monotypic genus of the Indo-Malayan region, has a creeping rhizome with palmately divided fronds, which have the distinction of containing the largest stomata known in the Plant Kingdom. *Danaea*, a small genus confined to Tropical America, has one species with a simple, ovate frond, and another with a small pinnate frond in which the lamina is pellucid and filmy. These forms, although showing the diversity in the fronds of the Marattiales, are nevertheless unusual, and a massive angular construction is more characteristic of the fronds of the Marattiales as a whole. The laminae do, however, show differentiation into palisade and mesophyll, and the stomata are confined to the lower surface.

The stems of the upright forms grow not from a single apical cell, but from a more massive meristem, and in this they are unique amongst the living ferns. The leaves form an apical crown, and since each receives an extensive trace, consisting of several strands of vascular tissue, the form of the stele is highly complex. A transverse section of the stem shows a number of concentric cycles of partial steles (***meristeles***), and dissection reveals that the meristeles of each cycle anastomose freely, occasional anastomoses also occurring between adjacent cycles. Leaf traces originate from the outer cycle of meristeles. Root traces, which may arise at any depth, pass out obliquely into the cortex. An endodermis, although present in young plants, is usually absent from the stelar regions of the older.

The stems of the Marattiales also contain little if any sclerenchyma, but there is an abundance of mucilage canals and tannin sacs, as elsewhere in the plant (Fig. 6.12). These indicate a particular kind of carbohydrate metabolism that seems to have been widespread amongst the ancient ferns.

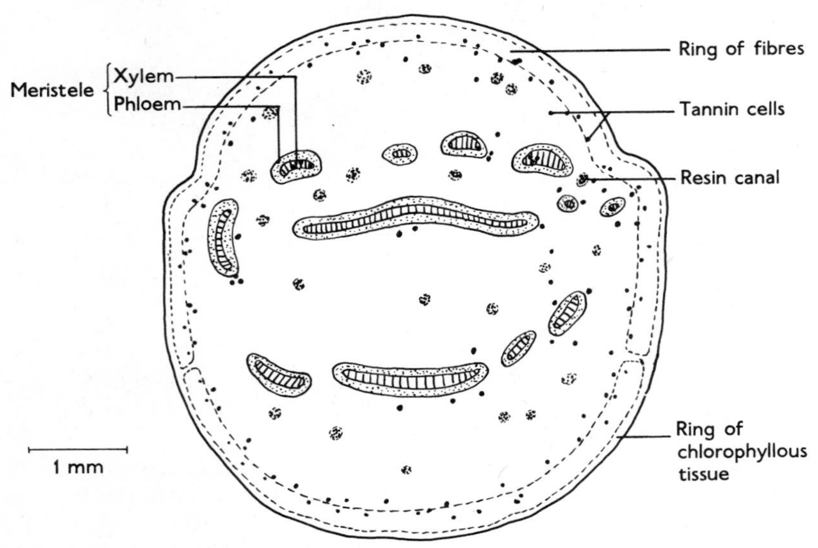

Fig. 6.12 *Angiopteris evecta.* Transverse section of secondary rachis.

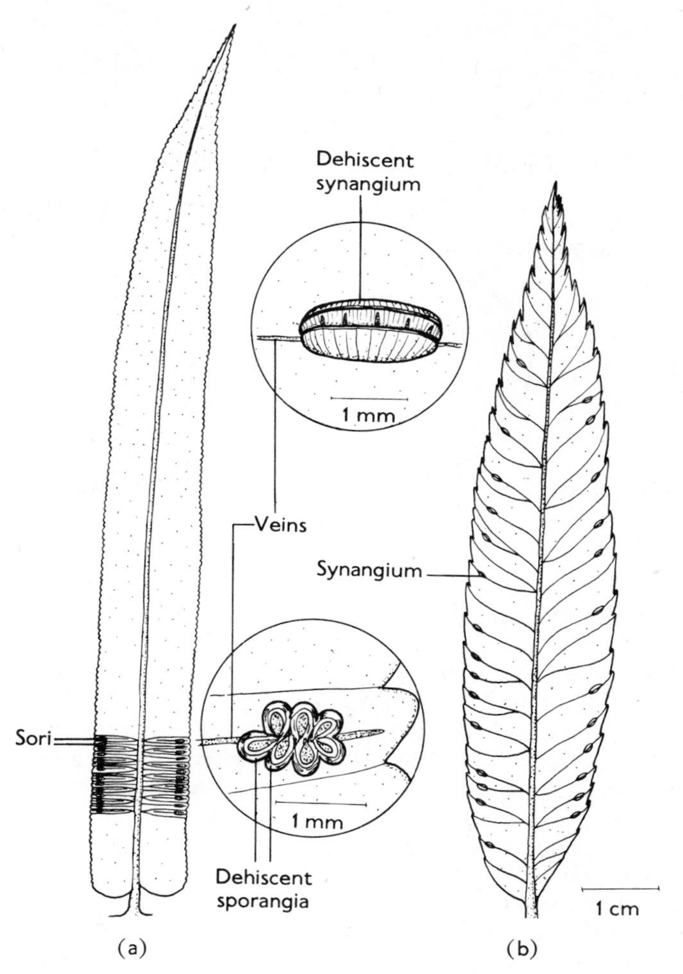

Fig. 6.13 Fertile pinnules of the Marattiales. **(a)** *Angiopteris.* **(b)** *Marattia.*

REPRODUCTION The fertile fronds resemble the sterile in most genera, and the sporangia, always eusporangiate in origin, are confined to the lower surface. In *Angiopteris* they arise in two ranks beneath veins towards the margins of the pinnules (Fig. 6.13a). The group is called a sorus and dehiscence of the sporangia, along a longitudinal stomium, is directed towards the mid-line of the sorus. In *Marattia* the fertile regions are similar, but the sporangia are congenitally fused into a synangium (Figs. 6.13b and 6.14). As the synangium matures and dries, it splits longitudinally into two valves (Fig. 6.15), and each compartment dehisces by a pore in the inner

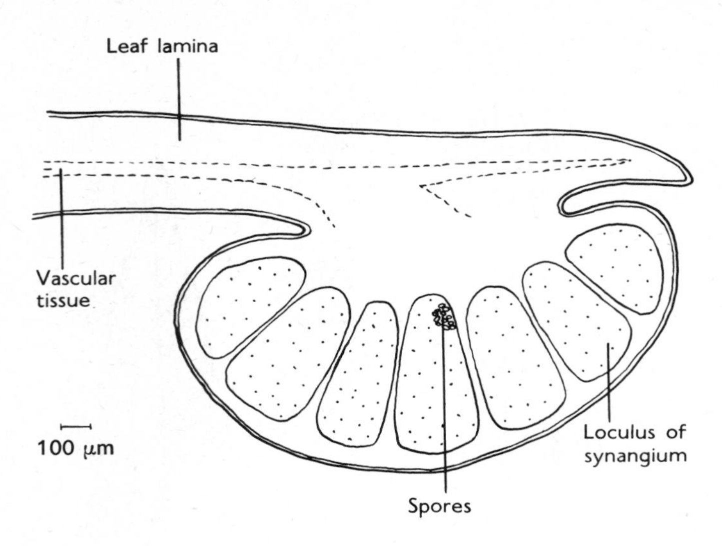

Fig. 6.14 *Marattia*. Vertical section of synangium.

face. The number of spores in each sporangium or synangial compartment in the Marattiales reaches a thousand or more.

Given warmth, moisture and light, the spores germinate rapidly, and, after passing through a brief filamentous phase, give rise to a green thallose gametophyte with apical growth, somewhat resembling a species of the liverwort *Pellia*. Although autotrophic, the lower cells contain an endophytic, and presumably mycorrhizal, fungus. The gametophyte may be long-lived, and old specimens reach a length of 3 cm or more. The antheridia are sunken, and occur on both surfaces, but the archegonia are confined to the median region of the ventral surface. The archegonia are also sunken, and the protruding neck cells form more of a cap over the egg than a neck (Fig. 6.16).

The first division of the zygote is by a wall transverse to the longitudinal axis of the archegonium. The subsequent embryogeny is endoscopic, and the embryo often emerges from the upper side of the gametophyte. A suspensor has been reported in *Danaea*, but is elsewhere lacking.

The fossil history of the Marattiales

The Marattiales are represented in the Carboniferous period by both vegetative and fertile material. *Psaronius*, for example, is the remains of a trunk surrounded by a mantle of descending roots (Fig. 6.17). The vascular tissue, which was wholly primary, formed a polycyclic array of anastomos-

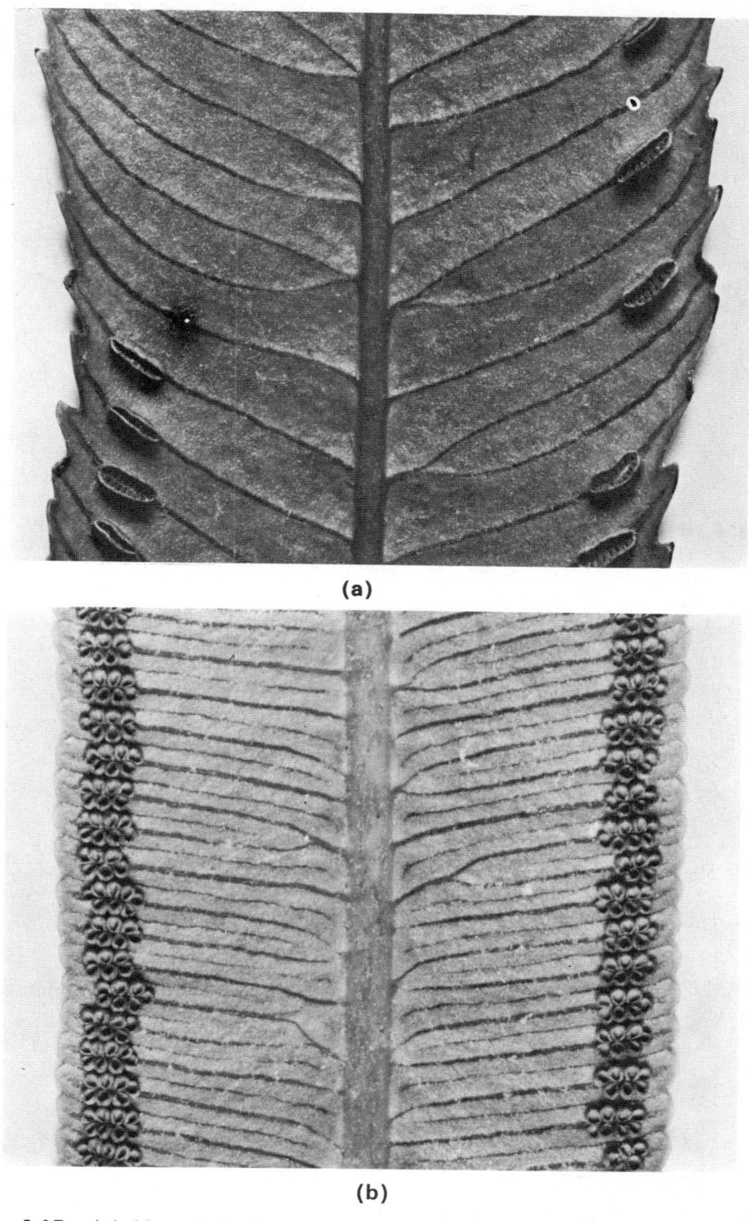

Fig. 6.15 (**a**) *Marattia fraxinea*. The abaxial surface of a fertile pinnule showing the arrangement of the synangia. The synangia have dehisced. ×5. (**b**) *Angiopteris evecta*. The abaxial surface of a fertile pinnule showing the grouping of the sporangia into sori, and the arrangement of the sori. The sporangia have dehisced. ×6.5.

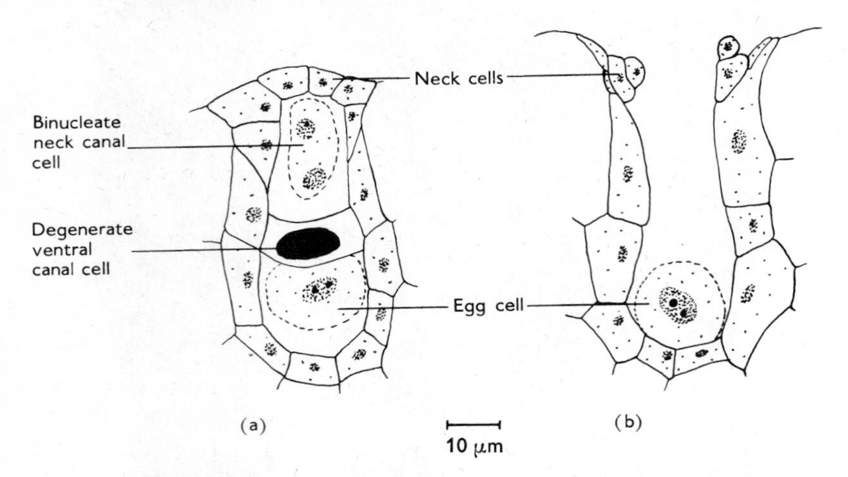

Fig. 6.16 *Angiopteris evecta*. Longitudinal section of archegonium. (**a**) Immature. (**b**) Prior to fertilization. (After Haupt, from Foster and Gifford, *Comparative Morphology of Vascular Plants*. Freeman, San Francisco, Copyright © 1959).

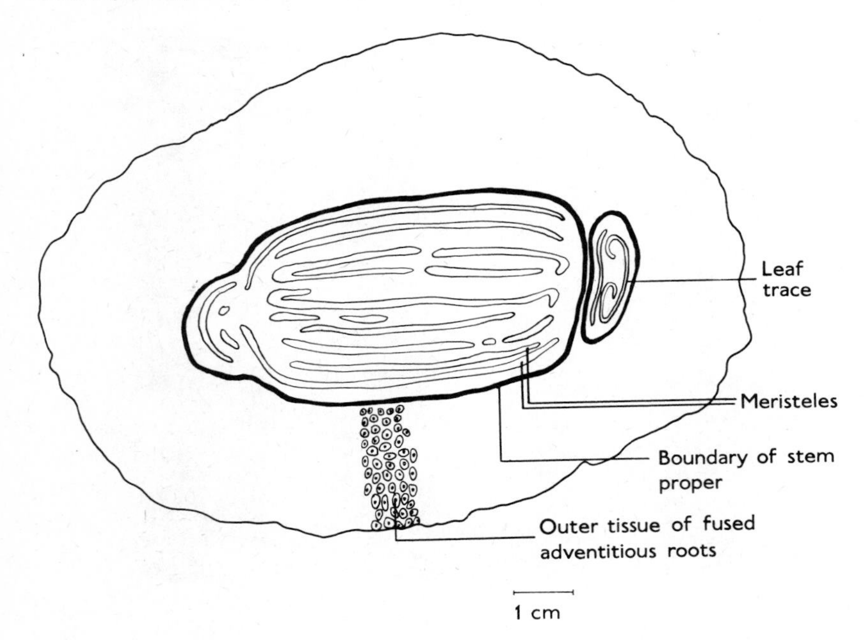

Fig. 6.17 *Psaronius conjugatus*. Transverse section of stem. From a specimen in the Oliver collection.

ing, band-like, meristeles. Morphologically and anatomically *Psaronius* is so suggestive of an arborescent *Angiopteris* that there seems little doubt of its affinity. Fertile material is represented by *Scolecopteris* (Fig. 6.18) and

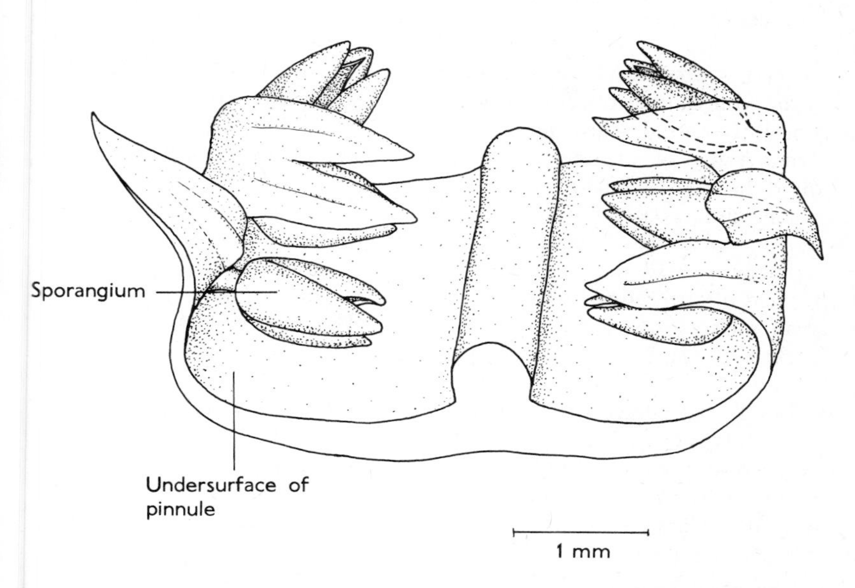

Fig. 6.18 *Scolecopteris incisifolia*. Reconstruction of portion of fertile pinnule. (After Mamay, from Andrews, *Studies in Paleobotany*. Wiley, New York, 1961)

Eoangiopteris, the sporangia of which were very similar to those of *Angiopteris*, although there were minor differences in the sorus. Fertile fronds of Marattiales, resembling those of various modern genera, are also found throughout the Mesozoic.

Filicales The Filicales number some 300 genera, and they include all the common ferns of temperate regions. Taken as a whole they show such a great diversity of growth form, morphology and anatomy that there are only two features, and these are at a cellular level, which distinguish them collectively from the remainder of the ferns. These are, first, the origin of the sporangium, and, second, the plane of the first dividing wall of the zygote. In the Filicales the sporangium develops from a single initial cell, a condition termed ***leptosporangiate***, in contradistinction to the eusporangiate condition of the Ophioglossales and Marattiales. Also in the Filicales, at least in those species which possess the typical heart-shaped gametophyte, the first dividing wall of the zygote is vertical or slightly oblique, parallel to the longitudinal axis of the archegonium, whereas in the Ophioglossales and Marattiales the first wall is perpendicular to the archegonial axis. The subsequent embryogeny of the Filicales is regularly endoscopic.

THE DISTRIBUTION AND ECONOMIC USES OF THE FILICALES Although world-wide in distribution, most Filicales are tropical, and they flourish in constantly humid, warm-temperate situations, such as are found within a certain range of altitudes on tropical mountains. In these situations are found, besides the herbaceous upright and rhizomatous forms of temperate regions, numerous epiphytes and arborescent and scandent forms. There is also an epiphytic species of south-east Asia in which the rhizome is curiously inflated and specialized to house colonies of ants, an adaptation (termed ***myrmecophily***) otherwise occurring only in flowering plants. A small number of wholly aquatic Filicales (Hydropterideae), all showing marked morphological specialization and heterospory, are widely distributed in fresh waters.

Few Filicales have any economic value. In some parts of the world the young leaves of certain species are eaten, and the sporocarps of *Marsilea* are used as a source of starch. Extracts and portions of some ferns have minor medicinal uses. *Adiantum* is a popular horticultural plant. Indians in the less stable parts of the Andes use trunks of tree ferns for building shelters in preference to timber, because of the tenacity of their sclerenchyma and their resistance to shattering by earthquakes.

THE GROWTH OF THE STEM The stems of the Filicales, with the occasional exception of some of the larger specimens of *Osmunda*, grow from a single, conspicuous, apical cell (Fig. 6.19), tetrahedral in shape. Daughter cells are cut off adjacent to its three posterior faces, although it is doubtful whether the apical cell itself divides. The meristematic activity in the apical cone, made evident by the anticlinal walls, diminishes towards the base. Below the apical cone, cell divisions are more generalized and variously directed. Leaf primordia arise in a definite phyllotactic sequence on the flanks of the apex, beneath the apical cone. A leaf primordium, first visible as a slight protuberance, soon develops its own apical cell. As the leaf primordia age and become separated by the expansion of the apex, bud primordia may be formed between them, but in some species buds do not appear at all so long as the apex is actively meristematic. Development of buds beyond the stage of primorida is in all cases rare in the region of developing leaves.

A curious situation is seen in *Pteridium* (bracken). The rhizomes of the mature plant are arranged in layers, the lowermost (up to 30 cm or more

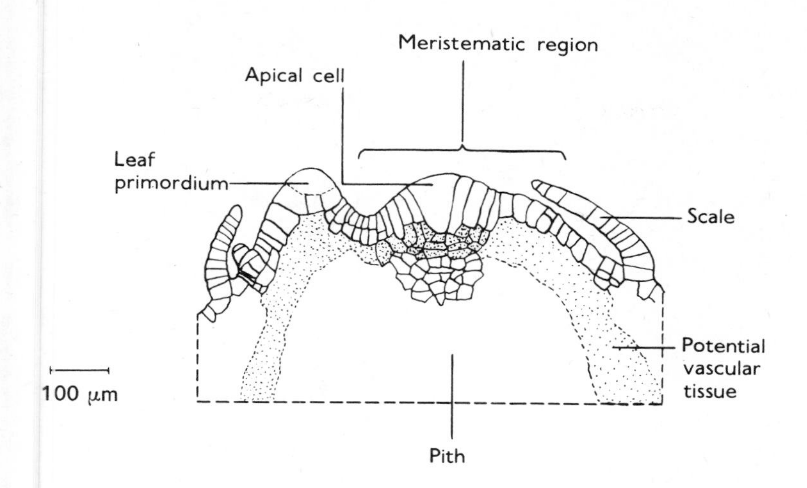

Fig. 6.19 *Dryopteris aristata*. Longitudinal section of stem apex. (After Wardlaw *Ann. Bot. N.S.* **8**, 173 (1944))

beneath the surface) consisting solely of 'long' shoots with extended internodes. Most of the fronds are borne on short stubby rhizomes near the soil surface. In the event of a 'front' of bracken invading a new area, the lowermost rhizomes head the advance.

Since the apices of many ferns are comparatively broad and accessible, they provide excellent material for experimental work on phyllotaxy.[80] The results indicate that the young leaf primordia, each a centre of meristematic activity, suppress growth in their immediate vicinity. Thus, if the position in which a leaf primordium is expected to arise is isolated by radial incisions from the neighbouring recently initiated primordia, then the new primordium develops with unusual vigour and outgrows the others. Similar experiments also confirm the fundamental similarity of stems and leaves in the Filicales. For example, tangential incisions on the anterior side of very young primordia that would normally yield leaves result in the production of stem buds instead. Incisions on the posterior side are without any effect. Consequently the determination of the sub-apical primordia appears to depend upon their being initially traversed by gradients of metabolites originating in the apical meristem. If a primordium is isolated from these gradients by an anterior incision it yields the radially symmetrical structure of a stem instead of the dorsiventral symmetry of a leaf.

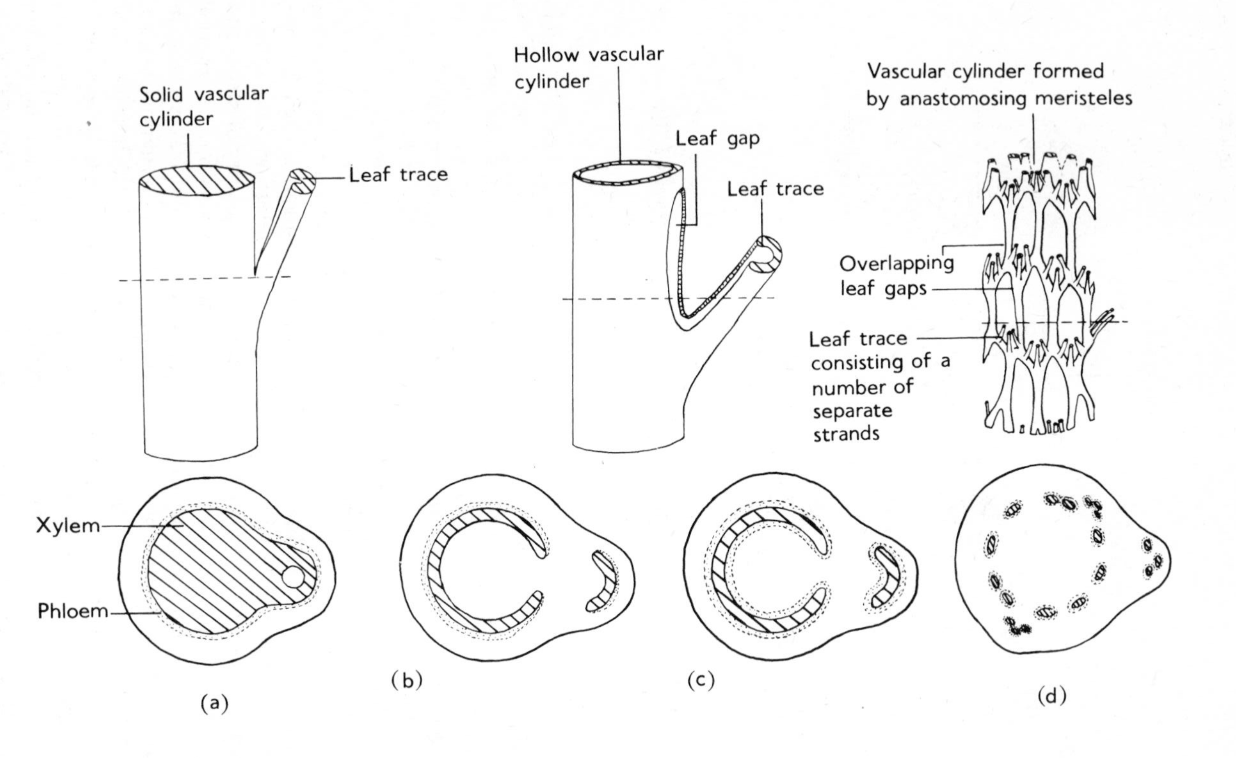

Fig. 6.20 Principal forms of fern steles. (**a**) Protostele. (**b**) Siphonostele. (**c**) Solenostele. (**d**) Dictyostele.

THE FORMATION AND MORPHOLOGY OF THE STELE The cells which yield the vascular tissue first become recognizable as a thin and distinctively staining layer immediately below the apical cone. This layer extends laterally, and beneath the leaf primordia becomes continuous with crescentic strands of similar cells which extend into each primordium. In descending the flanks of the apex these distinctive cells become transformed into procambial tissue, and this in turn is continuous with the vascular tissue of the mature shoot. The mature vascular tissue consists of tracheids and sieve cells. In a few instances (e.g. in the rhizome of *Pteridium*) the end walls of some of the metaxylem tracheids break down in development, yielding vessel-like channels. This recalls the situation in some species of *Selaginella* (see p. 157).

The form of the stele in the Filicales shows considerable variation (Fig. 6.20). In some species of *Gleichenia*, for example, the procambial tissue yields a solid core of tracheids from which the leaf traces depart without any break in the continuity of the xylem. Some other species of *Gleichenia* show a similar stele, but with medullation of the tracheidal core leading to the production of a siphonostele (Fig. 6.20b). In *Osmunda* (Fig. 6.21) the stele is basically a siphonostele, and the phloem and endodermis remain wholly external. The continuity of the xylem, however, is broken at the departure of the leaf traces, leaving a so-called 'leaf gap' which closes again anteriorly. Since, when dissected, the xylem (but not the stele as a whole) has the appearance of a cylinder of netting, *Osmunda* is said to have a dictyoxylic siphonostele.

There are a number of Filicales in which the stem possesses a solenostele (Fig. 6.20c). Leaf gaps are usually present, and sometimes other additional perforations unrelated to the departure of the leaf traces. The internal and external phloem and endodermis are in continuity around the margins of these gaps in the xylem. If the leaf gaps and other perforations are close together, as in *Dryopteris*, the stele becomes reduced to a ring of anastomosing vascular bundles (Fig. 6.20d), each with internal xylem and concentric phloem. This type of stele, which is of widespread occurrence, is termed a dictyostele. A complication, shown for example by the rhizome of *Pteridium* (Fig. 6.22) and the trunks of the tree fern *Cyathea*, is the presence of two or more concentric vascular systems, interconnected at intervals and usually all contributing to the leaf traces. These steles are said to be polycyclic. A point to be noted in passing is that steles are not always radially symmetrical. Those of ferns with creeping rhizomes, for example, are often markedly dorsiventral.

THE EXPERIMENTAL INVESTIGATION OF STELAR MORPHOLOGY The form of the stele in the Filicales has also been the subject of experimental

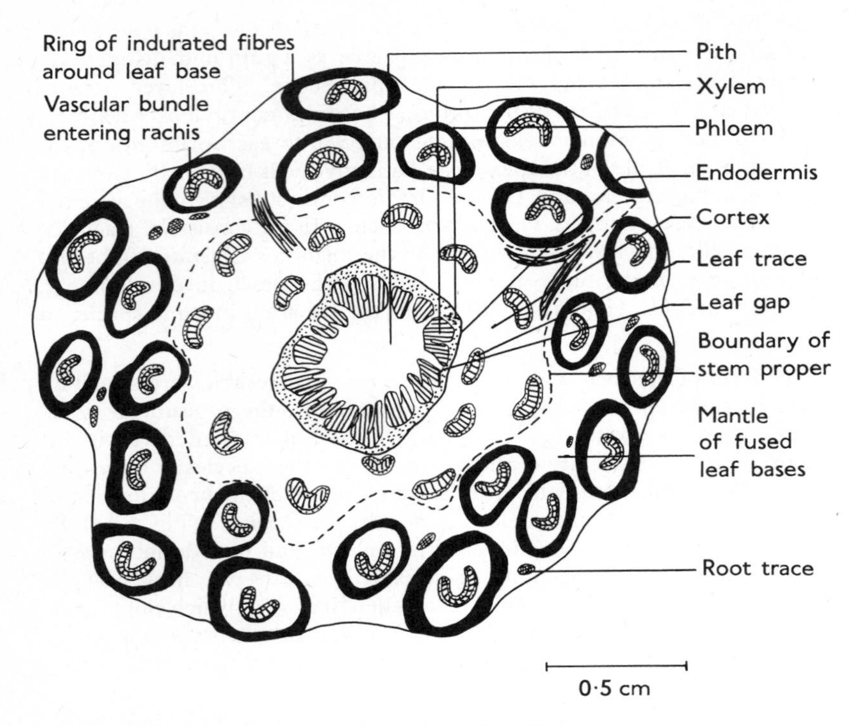

Fig. 6.21 *Osmunda regalis*. Transverse section of rootstock.

investigation. In *Dryopteris*, for example, if the apical region is isolated by vertical cuts, but left in contact below, it continues to grow and a solenostele differentiates behind it. As the apex expands and builds up a new crown, the stele gradually opens out to reform a dictyostele. In any one species, therefore, the size of the apex determines the form of the stele. This is also well shown in sporelings where a protostele is always present at the beginning. In protostelic species this merely increases in diameter as the plant develops, but in solenostelic and dictyostelic species the protostele of the sporeling becomes medullated, and phloem and endodermis appear within. The stele thus acquires its mature form in step with the increasing girth of the apex.[15] This relationship between size and form is clearly the consequence of physiological equilibria, but they are undoubtedly complex and have yet to be resolved.

OTHER ANATOMICAL FEATURES OF THE STEM In addition to the xylem, which is wholly primary, there are frequently bands or rods of scleren-

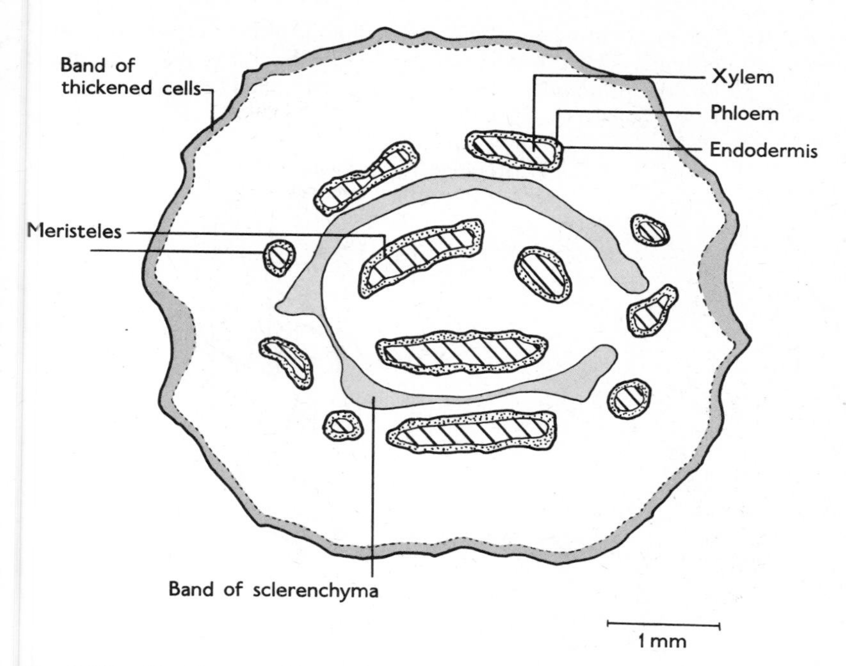

Fig. 6.22 *Pteridium aquilinum.* Transverse section of rhizome.

chyma in the stem contributing to its rigidity. In the tree ferns, for example, some of which may reach a height of 10 m, mechanical stability is dependent almost entirely upon the extremely tough girdle of sclerenchyma in the outer cortex and in association with the leaf bases. In those Filicales which are believed, on the basis of fossil evidence, to be relicts of very ancient groups (e.g. *Gleichenia*), the parenchymatous tissue of the stem often contains resin sacs and mucilage canals. Among the Filicales believed to be more recent in origin, however, these features are less evident, but in these ferns strikingly asymmetric deposits of cellulose, usually coloured brown by phlobaphene, frequently occur in the cells on the outside of the endodermis.

ROOTS The roots of the Filicales are in no way peculiar. They are all adventitious, in arborescent forms often being produced even in the aerial regions and providing a mantle of stubby outgrowths between the leaf bases. They show a distinct apical cell, but this is believed to be quiescent, divisions being confined to the cells at its flanks. They resemble the roots of other Pteropsida in possessing a root cap and, in many species, in producing root hairs. The xylem is commonly diarch, and in many epiphytic forms all but the protoxylem often remains unlignified.

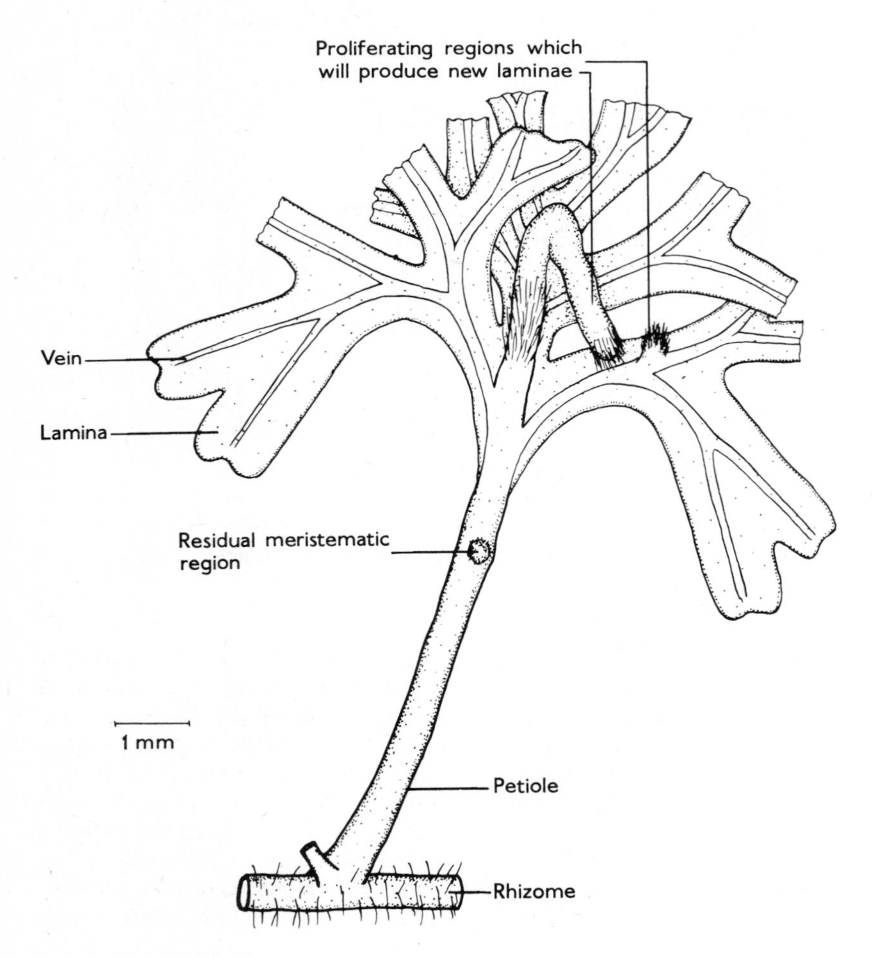

Fig. 6.23 *Trichomanes proliferum*. Frond, showing indefinite growth. (After Bell, *New Phytol.* **59**, 53 (1960))

THE MORPHOLOGY AND ANATOMY OF THE VEGETATIVE LEAVES The leaves of the Filicales retain an apical cell during their development from the primordium. Some form of pinnate branching is usually present in the mature leaves. True dichotomous branching occurs very rarely (the frond of *Rhipidopteris peltata* provides one of the few examples), but cymose branching, superficially resembling dichotomy, is shown by the leaves of several species of *Gleichenia*. In a few Filicales meristematic areas are retained in the differentiated leaf, and these subsequently grow out to form either additional leaves (as in *Trichomanes proliferum* (Fig. 6.23)) or new plantlets (as in *Asplenium mannii* and *Camptosorus radicans*). These forms illustrate how in the living ferns, as in the extinct (see p. 189), the leaves sometimes display features suggestive of stems. All parts of the young leaf show circinate vernation, and the extension of the rachis and the unrolling of the pinnae clearly involve considerable co-ordination of growth in space and time. There is some evidence that this is dependent upon the diffusion and varying relative concentrations of auxins in the expanding leaf. The expanding leaves of some ferns, for example of the tropical *Dryopteris decussata*, are enveloped in mucilage, possibly with some protective function.

The lamina of the leaf is commonly differentiated into palisade and mesophyll, but the texture is very variable and in some species a thick cuticle on the upper surface gives the leaf a surprising harshness. 'Filminess', the possession of laminae only one cell in thickness, is found in *Leptopteris* and throughout the family Hymenophyllaceae. Filmy ferns are necessarily confined to situations of continuously high humidity. They are often able to thrive in light intensities far below those tolerated by flowering plants, and more akin to those of bryophyte communities, with which they are frequently intermixed. In some tropical epiphytes the sterile leaves are of two forms, one remaining photosynthetic and the other soon dying, but persisting in a rigid, scarious condition as a collector of humus and moisture. *Platycerium* (Fig. 6.24) provides a striking example of this kind of habit. In other epiphytes of similar situations the leaves, borne on a short upright rootstock, are stiff and tightly overlapping, so forming a funnel which traps rain and organic matter. *Asplenium nidus* provides a typical example of these 'nest ferns'. The material at the base of the funnel is freely penetrated by absorptive rootlets. The leaves of the Hydropterideae are wholly peculiar. In *Salvinia*, for example, the surface of the leaf is made unwettable by a covering of waxy hairs, and in *Azolla* the minute leaves contain cavities inhabited by the blue-green alga *Anabaena azollae*.

THE FERTILE LEAVES AND THE NATURE OF THE SPORANGIA The fertile leaves of the Filicales are often quite similar to the sterile (as in *Dryopteris*), but dimorphy is not uncommon. In *Blechnum spicant*, for example, both the sterile and fertile fronds are simply pinnate, but in the fertile the sterile part of the lamina is strikingly reduced. The sporangia arise from single initial cells (except in *Osmunda* where a few additional cells are

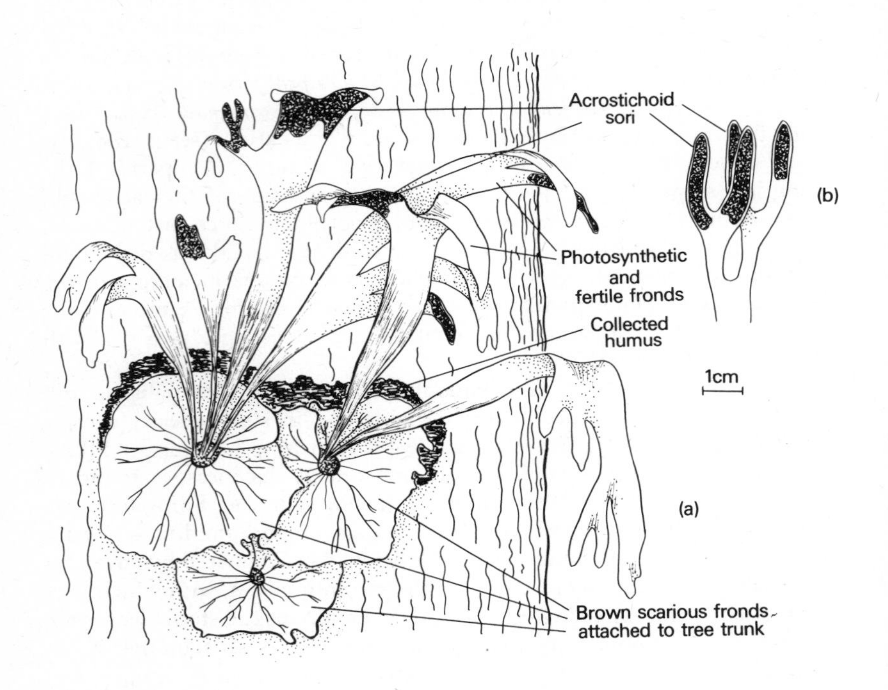

Fig. 6.24 *Platycerium*. (**a**) Habit, showing the two forms of leaves. (**b**) Lower surface of fertile portion of frond.

involved), either at the margin or on the lower surface of the leaf. The mature sporangium has a distinct stalk, the structure of which ranges from a broad multicellular stump to a delicate and relatively long column of cells. The wall of the capsule is only one cell thick (again with the exception of *Osmunda*, where a thin inner layer may also be present), and it always contains several indurated cells and a well-defined stomium. In *Osmunda* the indurated cells are grouped laterally (Fig. 6.25), and a linear stomium extends from them over the apex of the sporangium. In other Filicales the indurated cells are arranged in a single band (***annulus***) which encircles the sporangium, either transversely near the apex of the sporangium (as in *Aneimia* (Fig. 6.26) and other Schizaeaceae), or obliquely (as in *Gleichenia* (Fig. 6.27) and the Hymenophyllaceae), or vertically (as in *Dryopteris* and most common temperate ferns). The annulus is interrupted by the stomium. Where the annulus is vertical, it is also interrupted by the stalk of the sporangium, the stomium then lying just in front of the stalk. The

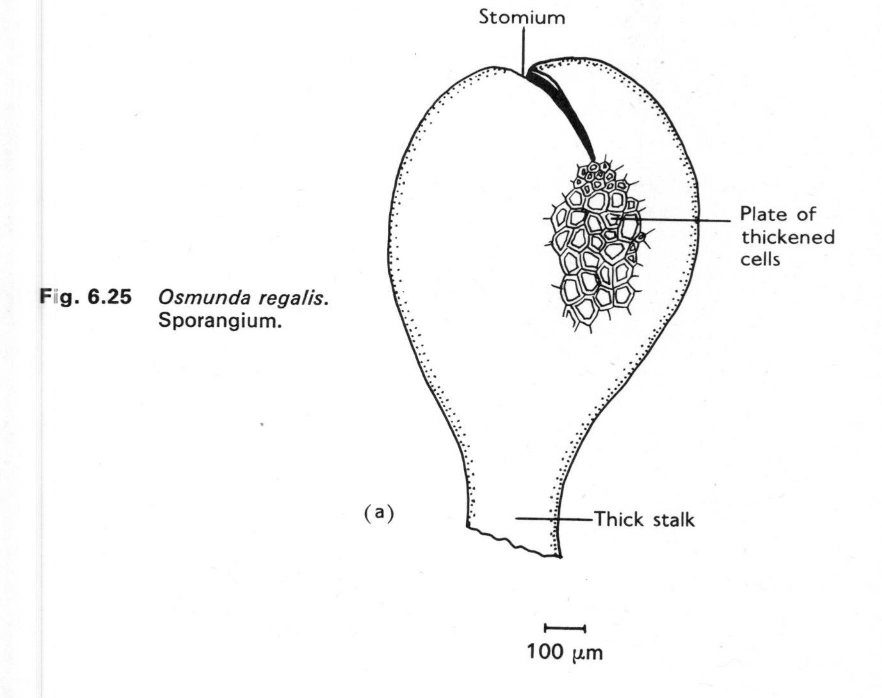

Fig. 6.25 *Osmunda regalis*. Sporangium.

sporangium dehisces at the stomium as a consequence of tensions set up in drying. Although this process is of a general nature in *Osmunda*, it is more precise in those Filicales where the sporangia have annuli, particularly where the annulus is vertical. Here, as the cell sap in the annular cells diminishes by evaporation, differential thickenings in the cells (Fig. 6.28a) cause an increasing tangential tension which tends to reverse the curvature of the annulus. The stomium eventually breaks, and the upper part of the sporangium gradually turns back as if on a hinge (Fig. 6.28b). Tension in

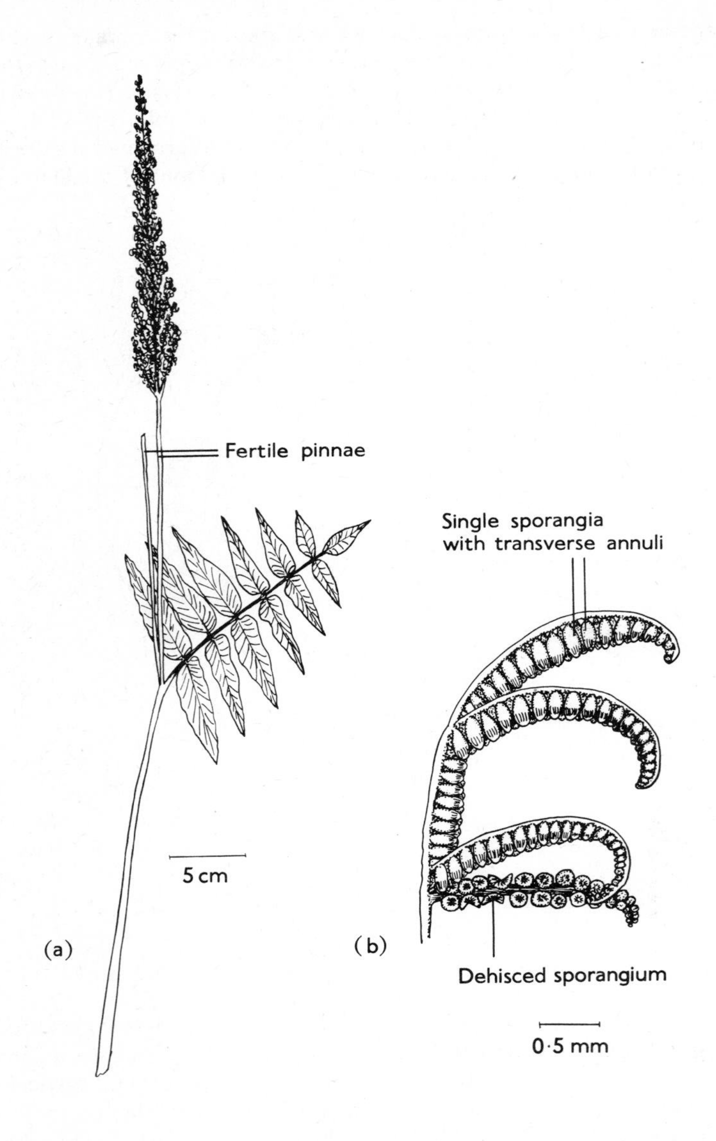

Fig. 6.26 *Aneimia phylliditis*. **(a)** Fertile frond. **(b)** Portion of fertile region.

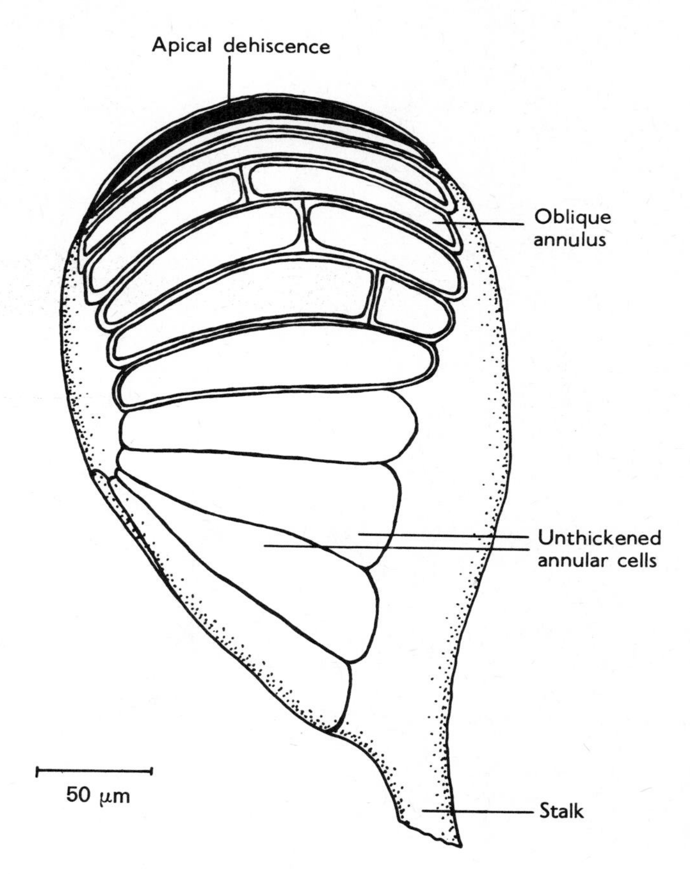

Fig. 6.27 *Gleichenia*. Sporangium.

the cells of the annulus ultimately reaches a critical level; at this point a gas phase suddenly develops, water remaining in the annular cells becoming vapour. The tension is immediately released, and the upper part of the sporangium flies back to more or less its original position (Fig. 6.28c). These two movements effectively disperse the spores.

THE ARRANGEMENT OF THE SPORANGIA In most Filicales the sporangia arise in distinct groups, called ***sori***, usually beneath or near the extremities

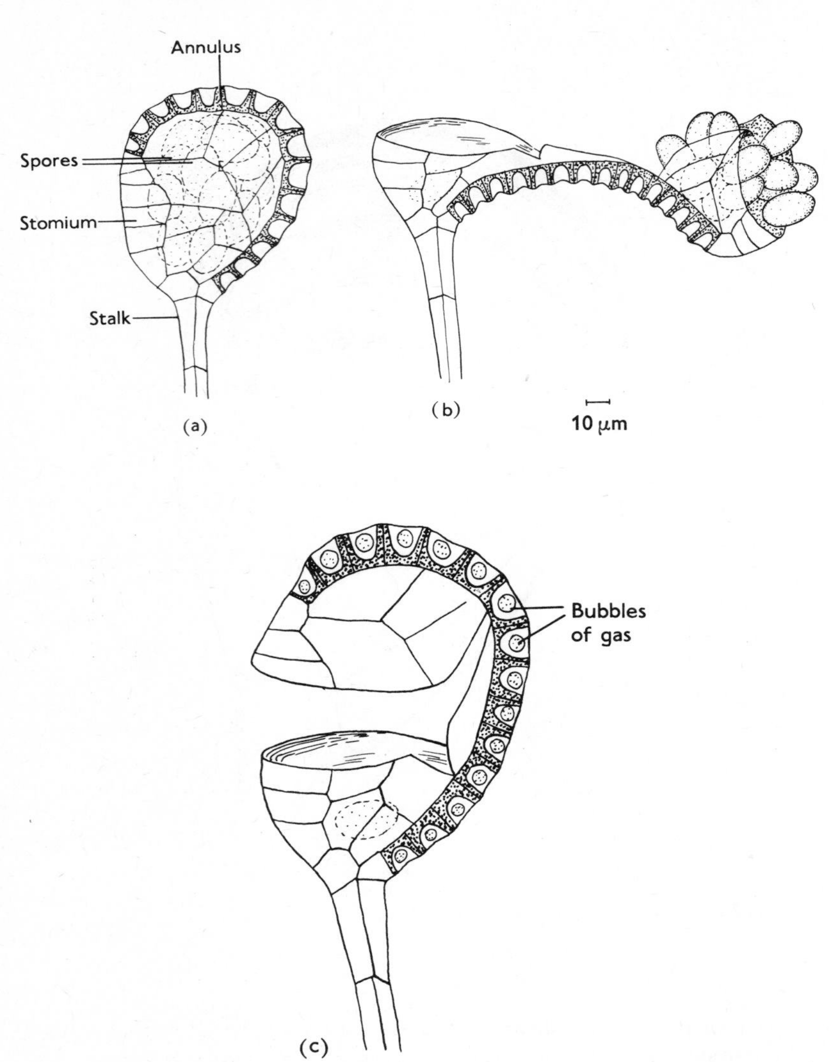

Fig. 6.28 *Dryopteris*. Stages in dehiscence of sporangium. (After Ingold, *Spore Discharge in Land Plants*. University Press, Oxford, 1939)

of veins, these two positions being called superficial and marginal respectively. Sometimes the sporangia are produced in a continuous line, referred to as a ***coenosorus***, well shown, for example, by *Pteridium*. The sorus is often partly or wholly covered by an outgrowth of the lamina, called an

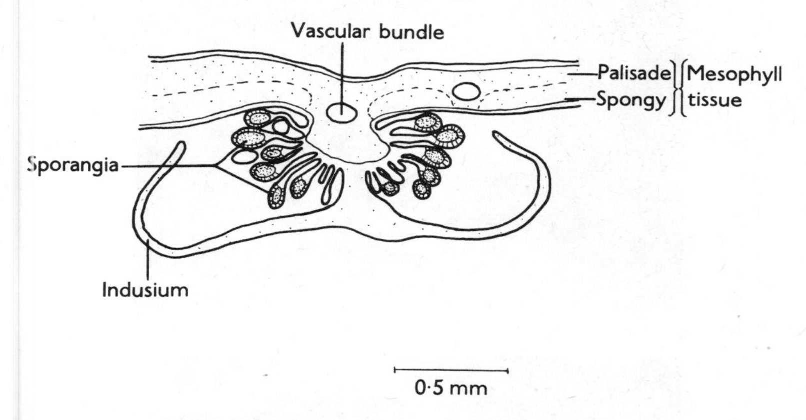

Fig. 6.29 *Dryopteris filix-mas*. Vertical section of sorus.

indusium (Fig. 6.29), which adopts a characteristic form. In *Dryopteris* for example, the indusium is reniform (Fig. 6.30), and in *Polystichum* peltate. In *Onoclea* the sori are protected by a rolling up of the fertile pinnules, a development carried to an extreme in the water ferns. In the family Marsileaceae, for example, the fertile segments of the frond are concrescent and form an indurated ***sporocarp*** at the base of the petiole. In some species of *Polypodium*, where an indusium is typically absent, the sori are immersed in the lower surface of the lamina. A few ferns show the so-called 'acrostichoid' condition in which the sporangia arise as a continuous felt on the lower surface of the fertile frond (e.g. *Platycerium*; see Fig. 6.24b). In some ferns (e.g. *Gleichenia*) the sporangia in a sorus are all of the same age (Fig. 6.31a), in others they are produced in spatial and temporal sequence on an elongating receptacle (as in the Hymenophyllaceae (Fig. 6.31b)), yielding a so-called 'gradate' sorus, and in yet others the sporangia are produced over a period but intermingled, leading to a so-called 'mixed' sorus (Fig. 6.31c) (as in *Dryopteris* and most common temperate Filicales).

THE DEVELOPMENT OF THE SPORANGIA AND SPORES In the development of the sporangium a cluster of spore mother cells becomes surrounded by a two-layered tapetum which developmentally is part of the wall tissue and not of the archesporium, as in eusporangiate sporangia. Some 2^6 to 2^8 spores are produced in each sporangium, the higher numbers being charac-

(a)

(b)

Fig. 6.30 The abaxial surfaces of fertile pinnules of *Dryopteris filix-mas*. (**a**) The sori in a young condition, before dehiscence of the sporangia. (**b**) Three weeks later, most of the sporangia having dehisced. $\times 6$.

teristic of the families believed to be more primitive. The spores of the homosporous Filicales (those of the heterosporous are considered later) are usually of the order of 40 μm in diameter. Some ferns (e.g. *Dryopteris borreri*) produce spores about 80 μm in diameter, but in these, as a consequence of a peculiar kind of sporogenesis, the nuclei contain an unreduced number of chromosomes. The outer wall of the spore (the exine), the material of which is largely derived from the tapetum, is sometimes deposited in a characteristic pattern of bars and ridges. This is especially true of the Schizaeaceae, and it provides a feature that has been very useful in identifying fossil forms. In some ferns the spores are monolete instead of trilete. Monolete spores often have an additional translucent investment, called a perispore, formed from the remains of the tapetum.

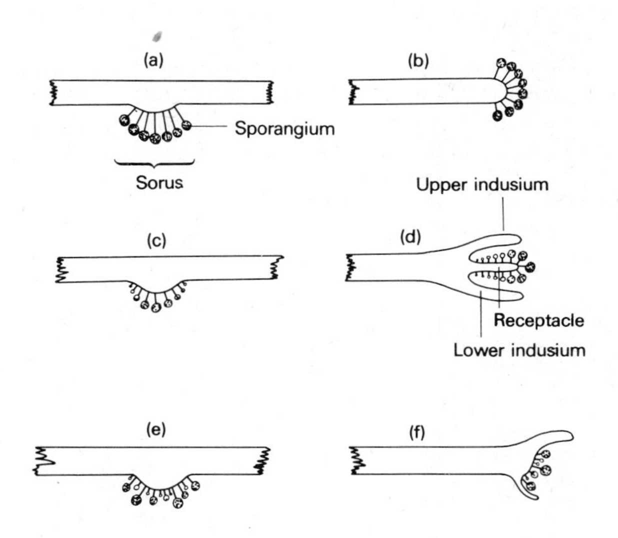

Fig. 6.31 Principal forms of sorus. (**a**, **b**) Simple, the sporangia all of the same age. (**a**) Superficial. (**b**) Marginal. (**c**, **d**) Gradate. (**c**) Superficial. (**d**) Marginal. (**e**, **f**) Mixed. (**e**) Superficial. (**f**) Marginal.

THE GAMETOPHYTE GENERATION Except for the Osmundaceae and sporadically in some other families, where the spores contain chlorophyll and are short-lived, the spores of most Filicales remain viable for two to three years. Germination, for which moisture and light are necessary, leads to the production of an alga-like filament, the growth of which is predominantly apical. In some Filicales (the Hymenophyllaceae, for example, and some Schizaeaceae), the gametophyte remains filamentous and the sex organs are borne on lateral cushions of cells. In most, however, the cells at the apex of the filament soon begin to divide in a number of directions and so form a cordate (heart-shaped) gametophyte (Fig. 6.32), of which *Dryopteris* provides a familiar example. The gametophytes of the Filicales are usually intolerant of all but the most humid and temperate conditions, and those of only a few species are able to survive desiccation or prolonged sub-zero temperatures. In a few tropical species the gametophyte is subterranean and closely similar to that of *Psilotum*[11a] (see pp. 145, 228).

Many cordate gametophytes produce only antheridia in the first stages of growth, but subsequently the production of antheridia declines or even ceases, and archegonia then appear in sequence on the lower surface of the gametophyte behind the apical meristem. The necks of the archegonia, consisting of about six tiers of cells, usually project conspicuously and are often slightly recurved (Fig. 6.32b). The ordered and symmetrical development of the cordate gametophyte of the homosporous Filicales has been shown to depend upon a stream of auxin diffusing posteriorly from the

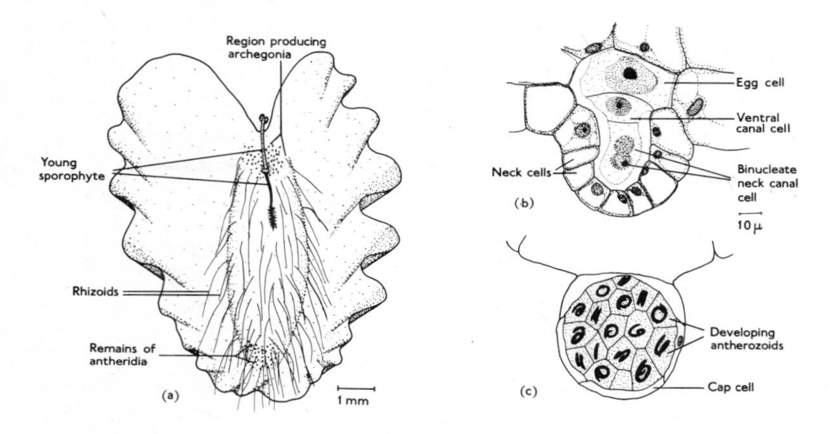

Fig. 6.32 (**a**) *Pteris ensiformis*. Lower surface of gametophyte, bearing young sporophyte. (**b**) *Pteridium aquilinum*. Median longitudinal section of young archegonium. (**c**) *Pteridium aquilinum*. Mature antheridium.

apical meristem. Beyond a certain size the system becomes unstable. The gametophyte then proliferates irregularly, and produces principally antheridia. In ferns such as *Dryopteris borreri*, where the spores contain nuclei with the same number of chromosomes as the parent, germination leads to a cordate gametophyte which is at first male, producing antherozoids capable of fertilizing the eggs of related sexually perfect species. Subsequently a young sporophyte is produced directly, without any fertilization or nuclear fusion, from the sub-apical region of the gametophyte where archegonia are normally found. This is an example of apogamous reproduction. The converse, the direct production of a gametophyte from a sporophyte without reduction of chromosome number, is also known to occur in certain conditions. This is termed apospory.

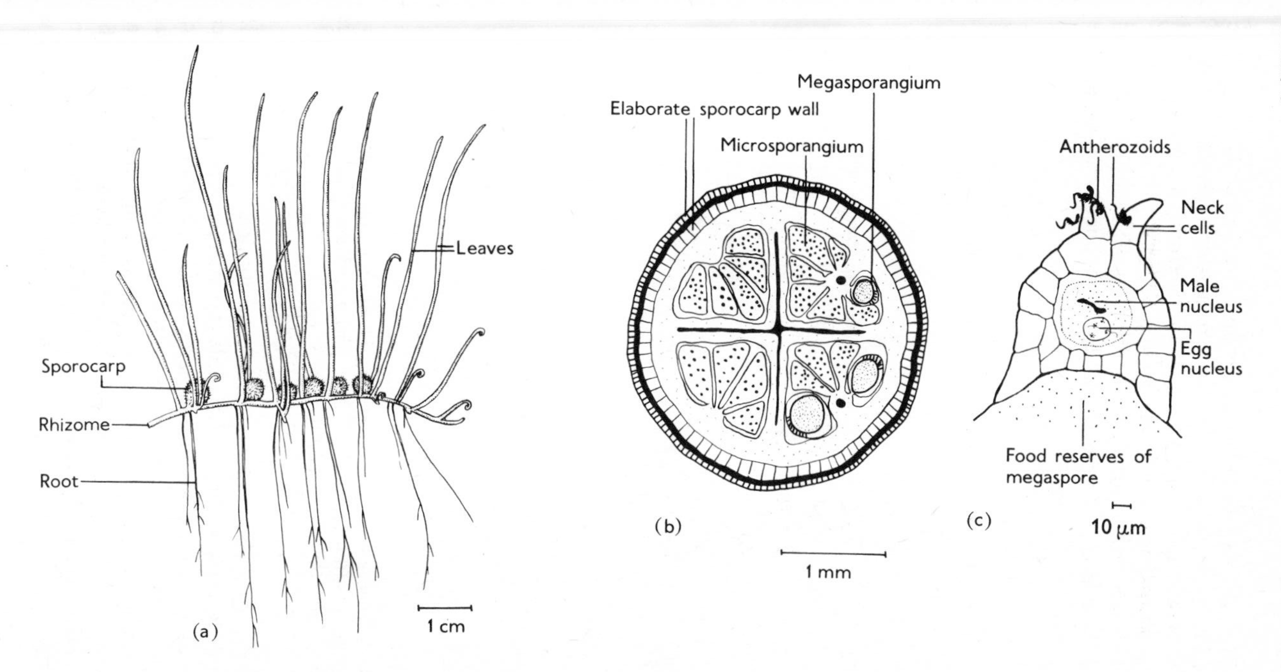

Fig. 6.33 *Pilularia globulifera*. (**a**) Habit. (After Hyde and Wade, *Welsh Ferns*. National Museum, Cardiff, 1940) (**b**) Transverse section of sporocarp showing the four sori. (**c**) Longitudinal section of inseminated archegonium.

THE HETEROSPOROUS REPRODUCTION OF THE HYDROPTERIDEAE In the heterosporous Hydropterideae, where reproduction resembles that of *Selaginella* and *Isoetes*, some sporangia produce single megaspores and others up to 64 microspores. In *Pilularia* (Fig. 6.33), which is representative of the Marsileaceae, the spores are liberated by the eventual decay and rupture of the sporocarp. Germination begins at once; the megaspore

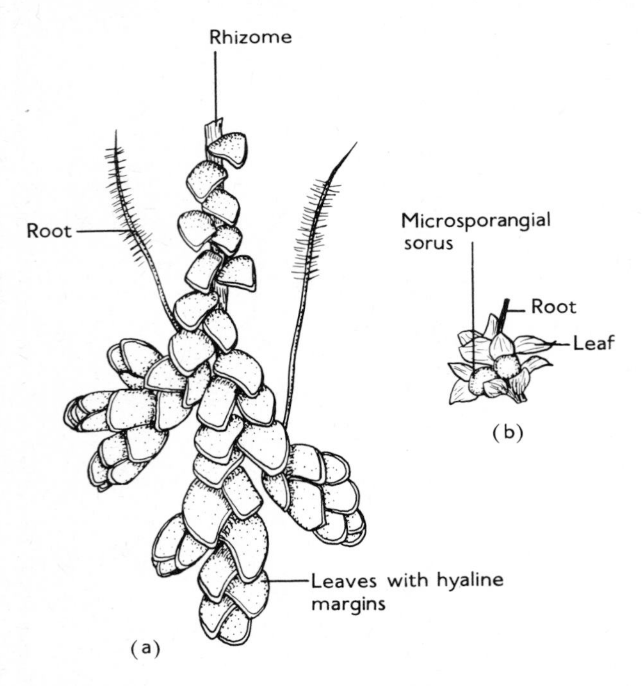

Fig. 6.34 *Azolla filiculoides.* (**a**) Habit. (**b**) Lower surface of shoot showing microsporangial sori. (After Campbell, *The Structure and Development of Mosses and Ferns.* Macmillan, New York, 1905)

rapidly gives rise to a single archegonium surrounded by a few somatic cells (which may develop chlorophyll), while the microspores each produce a single antheridium containing 16 antherocytes. In *Azolla* (Salviniaceae) (Fig. 6.34) the megaspore becomes surmounted by four frothy ***massulae*** (Fig. 6.35) formed from the tapetum, and these give the liberated megaspore some buoyancy, but it is doubtful whether they keep it afloat indefinitely. A variable number of massulae are formed in the microsporangium (Fig. 6.36). Each includes a number of microspores at its periphery, and is

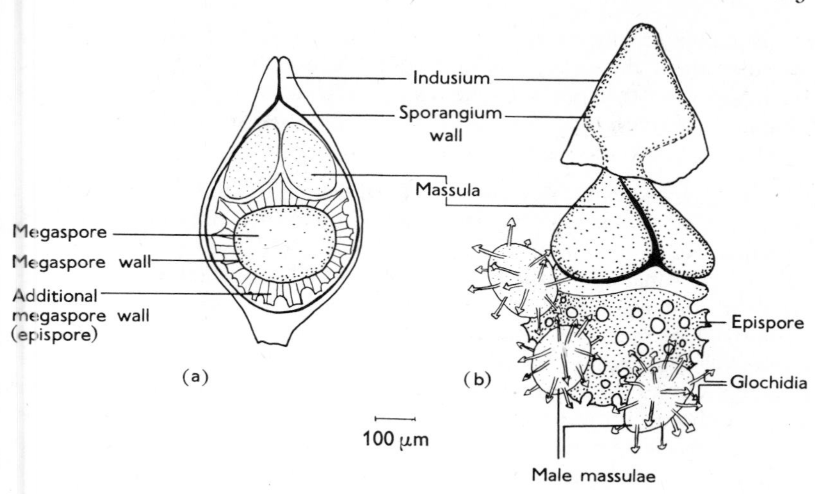

Fig. 6.35 *Azolla filiculoides.* (**a**) Longitudinal section of megasporangium. (**b**) Liberated megaspore with male massulae attached. (After Strasburger, *Ueber Azolla*. Abel, Leipzig, 1873)

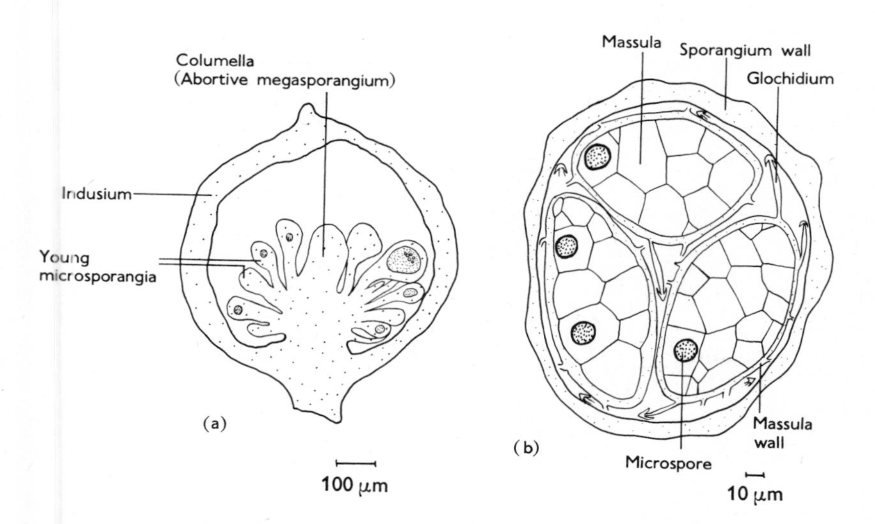

Fig. 6.36 *Azolla filiculoides.* (**a**) Longitudinal section of young microsporangial sorus. (After Campbell, *The Structure and Development of Mosses and Ferns*. Macmillan, New York, 1905) (**b**) Transverse section of mature microsporangium. (After Smith, *Cryptogamic Botany*, II. McGraw-Hill, New York, 1955)

furnished externally with peculiar anchor-like ***glochidia***. These male massulae hook themselves to the female, and the complex then sinks. Germination of the spores, in the main similar to that of the spores of the Marsileaceae, then follows. The cutinized glochidia of *Azolla* can still be recognized in lake deposits of considerable antiquity.

FERTILIZATION AND EMBRYOGENY Fertilization in the Filicales depends, as with most other archegoniate plants, upon the presence of water. Moisture causes the mature archegonia to open, and also converts the contents of the canal above the egg to mucilage. This mucilage, possibly because it contains traces of malic acid (known to possess chemotactic properties), attracts the antherozoids and effectively confines them to the region of the archegonia. Because of the rapid development of the gametophytes in the heterosporous Hydropterideae, fertilization may occur within 12 hours of the germination of the spores, but in most homosporous ferns there is a delay of some weeks before eggs are produced.

Division of the zygote follows about 48 hours after fertilization. The first vertical wall is succeeded by a horizontal so that in lateral aspect the zygote appears divided into quadrants. These quadrants indicate in a general way the course of the subsequent embryogeny. The upper anterior region, for example, goes to form the apex of the new sporophyte, the lower anterior the first leaf, the lower posterior the first root, and the upper

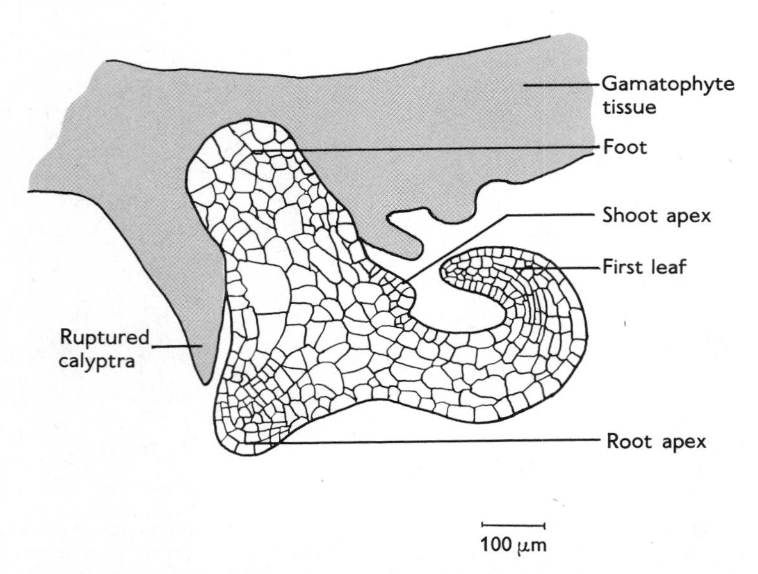

Fig. 6.37 *Pteridium aquilinum*. Vertical section of embryo.

posterior the foot (Fig. 6.37). There is no suspensor. Following fertilization, and probably a consequence of the movement of growth-regulating substances, the growth of the gametophyte diminishes and eventually stops. At the same time the cells of the archegonium immediately above the fertilized egg proliferate, forming a conspicuous cap (calyptra). Experiments have shown that this calyptra, probably by exerting mechanical pressure on the developing embryo, plays an important part in determining normal embryogenesis, recalling the situation in the mosses (p. 131). If it is removed from above a very young zygote, the zygote gives rise to a mass of parenchymatous tissue before producing differentiated growing points.[82] With cordate gametophytes, an intact apical meristem, probably in consequence of the auxin it produces, is also essential for normal embryogenesis. If this meristem is destroyed, differentiation of the embryo is markedly slower and the emergence of the first root very much delayed.[50]

The evolution of the Filicales

The evolutionary relationships of the living Filicales can be studied at two levels. First, by examining the interrelationships of living species and sub-species we can obtain a picture of the most recent evolution. Second, by comparing the morphology and anatomy of living species with the fossil, we can obtain a general picture of the development of the contemporary Filicales.

The study of the living species involves breeding experiments and the recognition of chromosome homologies,[57] particularly of the large number of polyploid ferns now known to exist. In this way it has been discovered, for example, that *Dryopteris filix-mas* ($4n$) is an allotetraploid probably derived from a hybrid between two diploid ancestors, one similar to *D. abbreviata* and the other so far unknown. This kind of evolution, involving considerable speciation, but little profound anatomical and morphological change, has clearly been widespread in recent geological time.

The study of the fossil Filicales has given valuable information about the evolutionary status of the present-day families. It is clear, for example, that the Osmundaceae have a very long history. A transverse section of the stem of the late Permian *Thamnopteris* is very similar, even to the extent of the arrangement of the sclerenchyma and the packing of the leaf bases, to that of *Osmunda*. The record of this family, which includes preserved fronds and sporangia, continues through the Mesozoic to the present. The Schizaeaceae also have a well-established fossil history extending back to the Palaeozoic, and that of the Gleicheniaceae is similar. Other families first appear in the Mesozoic. Nevertheless, despite these evidences of antiquity, most of the living Filicales either have no fossil record, or no record extending back further than the Tertiary. This is particularly true of the large family Polypodiaceae, and we must suppose that these ferns are comparatively recent, probably having evolved towards the end of the Cretaceous period and subsequently.

Comparison of the living Filicales with the fossil, quite apart from tracing particular lineages, also reveals those features which can be regarded in a general way as primitive. Protostelic and solenostelic vascular systems, the simultaneous production of the sporangia in the sori, short thick sporangial stalks, the indurated cells of the sporangial wall aggregated laterally or arranged in a transverse annulus, and a large number of spores in each sporangium are all features of the early Filicales. Conversely, dictyosteles, the production of mixed sori, long and delicate sporangial stalks, vertical annuli, and low spore numbers are all features of Filicales with little or no fossil record.

On the basis of these criteria it is possible to assign the families of living ferns to three grades according to their evolutionary advancement. It is also possible to arrange them in two series, according to whether the sporangia are marginal or superficial in origin, but recent research has thrown considerable doubt upon the significance of this feature. Nevertheless, it is tentatively retained here, and the position of some of the more important families and genera in this double classification is shown in Table 6.1. The Osmundaceae are considered as contributing to both

Table 6.1

Features of evolutionary significance	Position of origin of sporangia: Marginal	Position of origin of sporangia: Superficial
Steles commonly dictyosteles. Sori mixed. Sporangia with vertical annuli interrupted at stalk	*Pteridium* *Davallia*	*Dryopteris* *Polystichum* *Asplenium* *Athyrium* *Polypodium*
Steles dictyosteles or solenosteles. Sori gradate. Sporangia with oblique annuli	Hymenophyllaceae	Cyatheaceae
Steles protosteles or solenosteles. Sori simultaneous. Indurated cells of sporangial wall aggregated, or in transverse or oblique annuli	Schizaeaceae *Osmunda* (Osmundaceae)	Gleicheniaceae *Todea* (Osmundaceae)

(Side arrows, reading from bottom row to top row: 2^{7-8} --- Number of spores/sporangium ---> 2^{5-6}; ------ Reduction in number of cells forming sporangial stalk ------>)

series, since the sporangia are marginal in *Osmunda*, but superficial in *Todea*. This classification does not of course imply that living families and genera have evolved from each other; it merely illustrates relative primitiveness. The arrangement is substantiated by the fossil record which is of greater duration in respect of the Filicales at the bottom of the table than of those at the top.

THE ORIGIN OF THE FILICINAE AND THE MORPHOLOGICAL NATURE OF THE MEGAPHYLL

Since the Cladoxylales, Coenopteridales, Marattiales, Filicales, and possibly the Ophioglossales were all in existence together towards the close

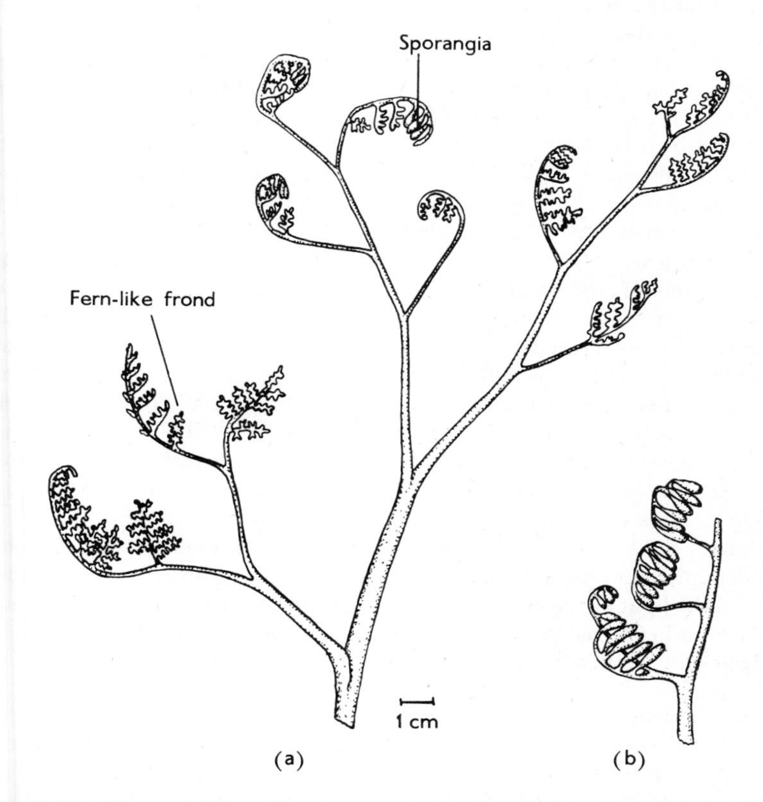

Fig. 6.38 *Protopteridium hostimense*. (After Kräusel and Weyland, from Delevoryas, *Morphology and Evolution of Fossil Plants*. Holt, Rinehart and Winston, New York, 1962)

of the Palaeozoic era, it seems clear that these Orders represent parallel lines of evolution within the Filicinae, only three of which have survived. There are sufficient similarities between the Orders, however, to suggest a common ancestor in some earlier period. A significant feature of many Filicinae is the close resemblance, both in appearance and behaviour, between leaves and branch systems. This has given rise to the view that the origin of the Filicinae should be sought in some early *Psilophyton*-like form.

Although *Psilophyton* itself was predominantly axial in construction, at least one species, *Psilophyton goldschmidti* (from the Lower Devonian of Norway), bore repeatedly forked lateral branches which, although they lacked any development of lamina, were comparable to the fronds of ferns. *Svalbardia* (Middle Devonian) appears to have been a frond-like branch system, possibly borne on an early member of the Filicinae. Another Middle Devonian genus of considerable importance is *Protopteridium* (Fig. 6.38), a branching plant which seems to have been truly intermediate between the psilophytes and the Filicinae.[47] Some of the ultimate branchlets, which may even have shown circinate vernation, were flattened and resembled pinnules, while others ended in sporangia. A number of protostelic axes, possibly belonging to these plants are also known, and they show some anatomical detail suggestive of the Coenopteridales.

Further grounds for accepting the axial nature of fern fronds has come from a study of *Stromatopteris*, a rare fern of New Caledonia. Fronds and branches arise from the creeping rhizome in an almost indistinguishable manner. Moreover the frond, which consists of a rachis with a row of pinnules on each side, is morphologically similar to the shoot of *Tmesipteris* (p. 144). This resemblance between a fern and the Psilotales extends even more strikingly to the gametophyte and embryo.[11b] The sporangia and sori of *Stromatopteris* are undeniably fern-like, but the features in common between this fern and the Psilotales are so numerous and intrinsic that an affinity seems undoubted. The difference appears one of degree. In the Psilotales the branches produced by the rhizome fit more readily into the concept of shoot, those of *Stromatopteris* (which is regarded as a primitive fern in the light of the criteria discussed earlier (p. 226)) into that of frond. In most other ferns, of course, the distinction between shoot and frond is much clearer, but, as we have seen (pp. 189, 210), indications of the shoot-like nature of the megaphyll are encountered throughout the Filicinae.

The evidence at present available is, therefore, fully in accord with the idea that the frond of the Filicinae originated in a lateral branch system, and that the Filicinae evolved from some early group of psilophytes in which the plant body became differentiated into main and lateral axes.

7

The Tracheophyta, III
(Pteropsida: Gymnospermae)

The gymnosperms,[74] although a diverse Class with possibly more than one origin from the earliest land plants, are with few exceptions archegoniate and almost entirely arborescent. Their fossil record rivals that of the ferns in richness and variety.

Gymnospermae

Sporophyte usually arborescent; branching and leaves various. Secondary vascular tissue always present, consisting of tracheids (in a few forms also of vessels) and sieve cells. Sporangia borne on specialized structures, probably of axial origin. Heterospory general. The megasporangium enclosed within a specialized tissue (nucellus) of the sporophyte, this in turn surrounded by a distinctive sheath (integument), perforated at the apex by a narrow channel (micropyle), the whole termed the ovule. Neither male nor female gametophytes autotrophic. Fertilization by multiflagellate antherozoids, or by male cells with no specialized means of locomotion, occurring within the ovule, either before or after its being shed. Embryogeny endoscopic, the embryo remaining contained within the seed developed from the ovule.

All the early seed plants, with the exception of the specialized *Lepidocarpon* and *Miadesmia* of the Lycopsida (see p. 171), are referable to the Gymnospermae. Gymnospermy, a term which implies the bearing of naked seeds, unenclosed in any carpellary structure, can thus be legitimately regarded as the most primitive form of the seed habit.

The first plants recognized as gymnosperms are found in the late Devonian period. These are probably early members of the two Orders

Pteridospermales and Cordaitales, both now extinct. Other Orders of the Gymnospermae we shall consider here are the Coniferales, Ginkgoales and Gnetales, all containing living representatives, and the extinct Bennettitales and Caytoniales.

Pteridospermales (Cyacadofilicales) Investigators of the fern-like fronds found in Carboniferous rocks soon became aware that not all these were in fact referable to ferns. Some were undoubtedly associated with seeds, and others with stems in which there were secondary thickening and other anatomical features not found in the Filicinae. Nevertheless, the habit of these plants was probably something like that of the Marattiales. The leaves were megaphyllous, compound and pinnately branched, and borne on stems of varying height.

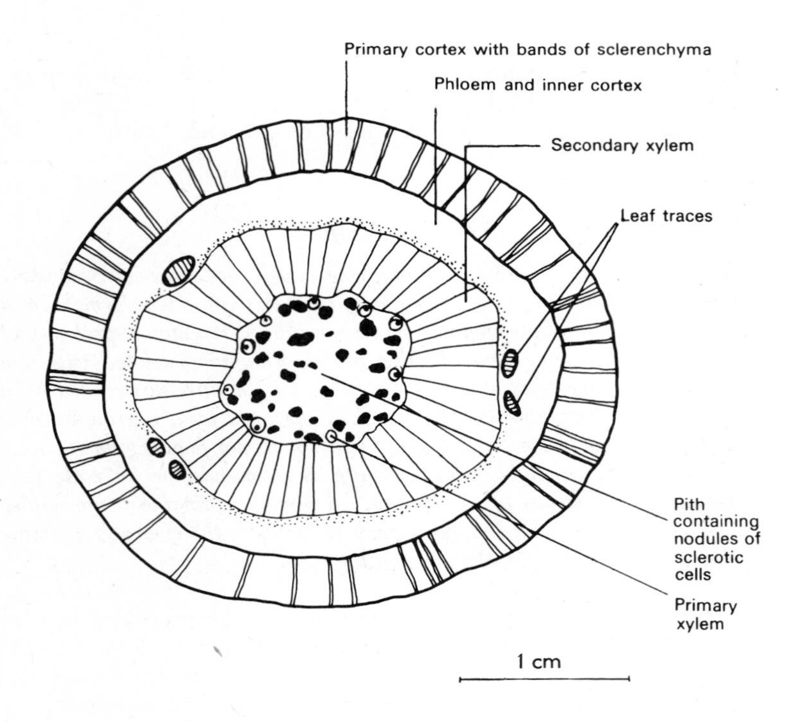

Fig. 7.1 *Lyginopteris oldhamia.* Transverse section of stem. The glandular epidermis has been lost. The leaf trace divides into two strands in the cortex, but these reunite to form a V-shaped trace at the base of the petiole.

VEGETATIVE FEATURES One of the best-known pteridosperm stems of Carboniferous age is *Lyginopteris*. This ranged in diameter from about 0.5 to 4.0 cm, and occasionally branched. There was a central pith, which contained nodules of thick-walled cells (similar to the groups of stone cells in the flesh of a pear), surrounded by a ring of primary xylem strands (Fig. 7.1). Exterior to these was a relatively large amount of secondary xylem. The tracheids of the secondary xylem, like those of the metaxylem, contained bordered pits, but in the somewhat smaller tracheids of the secondary xylem they were absent from the tangential walls. A girdle of phloem, rarely well preserved, lay outside the xylem. A characteristic feature of *Lyginopteris* was the anatomy of the outer cortex. This contained radially elongated bands of fibres which anastomosed freely and clearly gave considerable mechanical support to the stem. The outer surface of the stem was furnished with peculiar multicellular glands.

The leaves of *Lyginopteris* (originally described as *Sphenopteris*) were borne in a 2/5 phyllotaxy and when young showed circinate vernation. In mature leaves, which sometimes reached a length of 50 cm, the rachis dichotomized at about half its length, but the remainder of the branching was pinnate and the ultimate segments were narrow pinnules. All surfaces of the frond bore glands similar to those of the stem.

THE REPRODUCTIVE ORGANS The seeds (originally described as *Lagenostoma*) terminated axes which were probably branches of otherwise normal fronds. Each seed (or ovule) was partially enclosed in a cup formed by a number of glandular and basally fused bracts (Fig. 7.2a). This structure, called a cupule, in some forms contained more than one seed. The seed itself was an upright, radially symmetrical structure, about 0.5 cm long

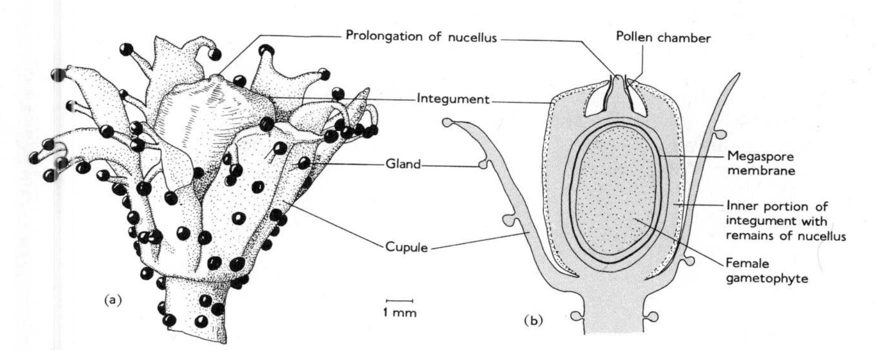

Fig. 7.2 *Lagenstoma lomaxi.* (a) Reconstruction of seed. (After a drawing by Oliver in the Department of Botany, University College, London, W.C.1) (b) Longitudinal section of seed. (After Walton, *An Introduction to the Study of Fossil Plants.* A. and C. Black, London, 1940)

and a little less broad. The central part (Fig. 7.2b), the nucellus, possibly a specialized archesporial tissue, was surrounded by an integument of two layers, the outer of which contained a sclerenchymatous sheath. A single vascular bundle entered the base of the seed and divided symmetrically into nine which ascended the inner fleshy part of the integument. The upper part of the integument around the micropyle was shallowly lobed, the lobing corresponding to the intervals between the ascending veins. A female gametophyte, surrounded by a distinct membrane and bearing archegonia on its upper surface, lay within the nucellus. This gametophyte probably developed from a megaspore, formed as one of a tetrad within the nucellus, the remaining megaspores degenerating.

A peculiar feature of pteridosperm seeds was the form of the upper part of the nucellus. The central part of the nucellus was in some prolonged as a column, and this was surrounded by a sheath of similar tissue, a cylindrical space being left around the column. Since pollen grains have been found in this chamber, it is thought that it was here that the microspores germinated.

The male reproductive organ of *Lyginopteris* is not yet known with certainty. It is, however, very probable that it consisted of a small ovate

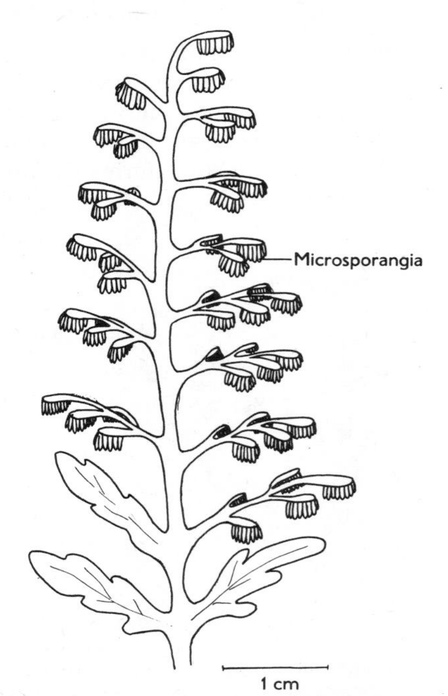

Fig. 7.3 *Crossotheca* sp. Reconstruction of fertile shoot. (After Andrews, *Studies in Paleobotany*. Wiley, New York, 1961)

plate, about 2 mm in length, terminating a branchlet in a somewhat peltate manner (Fig. 7.3). About six bilocular sporangia, each about 3 mm long and 1.5 mm wide, were attached to one face of the plate. These microsporangiophores were borne on a branch system (the whole being known as *Crossotheca*) which may have formed part of a *Lyginopteris* frond.

The pollen grains, which bore triradiate scars, were presumably distributed by wind. By analogy with the pollination of living gymnospermous ovules, it is thought the grains were trapped by a drop of sugary fluid which protruded from the micropyle, and that subsequent absorption of this drop drew the grains down into the nucellar chamber. No seeds have have been found which indicate the form of the male gametophyte, and whether fertilization was brought about by antherozoids or by unspecialized male gametes remains unknown. It is curious that no seeds have come to light containing embryos, and this has led to the view that fertilization probably occurred after the seed was shed.

The later pteridosperms

Lyginopteris is representative of a wide range of Carboniferous pteridosperms, but towards the end of the Palaeozoic other, more complex, forms became prominent. These were more like tree-ferns in habit, and the leaves had pinnules conspicuously larger than those of the earlier pteridosperms. The seeds, about six times the size of those of *Lyginopteris*, were more clearly in association with foliar organs, and the microspores, which were monolete, were produced in large cup- or spindle-shaped synangia. The stems had a complicated stelar structure. Vascular tissue differentiated from a number of successive and often discontinuous cambia, so that it became interspersed with much parenchyma.

The origin of the pteridosperms

The pteridosperms probably had their origin in the same group of axial plants as that which yielded the ferns, but were the consequence of a trend in which the evolution of megaphylls was accompanied by increasingly pronounced heterospory, and by the formation of secondary vascular tissue within the stem. *Archaeopteris*, a pinnate frond of late Devonian age (Fig. 7.4), is a well-known example of an early, fern-like, heterosporous plant. The recent discovery that at least one species of *Archaeopteris* was borne on a stem with pronounced secondary thickening, and that the rachis had a stem-like anatomy,[7a] is in keeping with the kind of origin envisaged for the pteridosperms (see also p. 283).

The radial symmetry of many pteridosperm seeds (on account of which the ovules are termed radiospermic) suggests that they may have been derived from a tassel of megasporangia (resembling the male organ *Crossotheca*) in which a central megasporangium was closely surrounded by a ring of similar megasporangia. Sterilization of the outer ring, but not of the centre, would then have resulted in an integumented megasporangium. Although no intermediate stages in such a transformation have yet come to light, it is significant that in one of the earliest pteridosperm seeds from the Lower Carboniferous (*Salpingostoma*) the integument consists of a ring of finger-like processes, each containing a vein, fused only in the basal region.

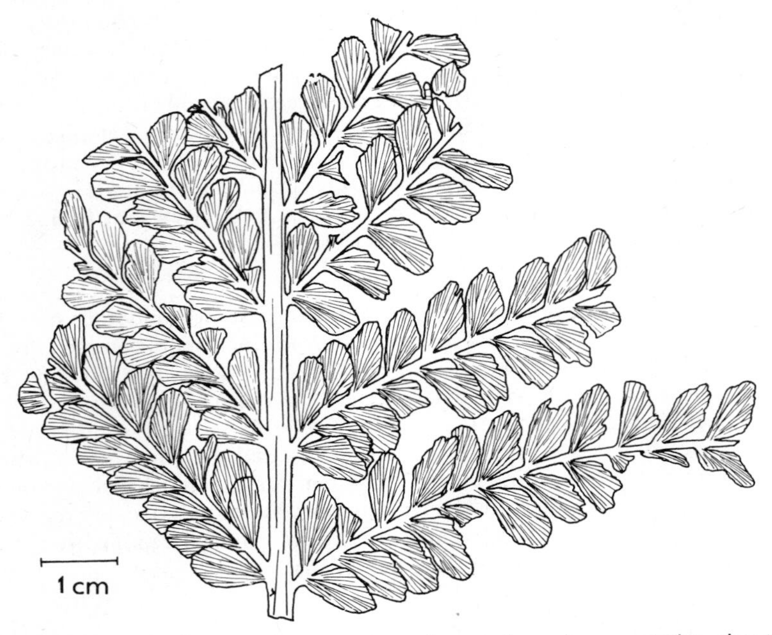

Fig. 7.4 *Archaeopteris* sp. Compression of a portion of a vegetative shoot. (After Andrews, *Ancient Plants and the World They Lived In*. Comstock, New York, 1947. Used by permission of Cornell University Press)

Progressive fusion of components of this kind might then have led to the entire integuments of the pteridosperm seeds of the Upper Carboniferous, the shallow lobing often seen in the micropylar region of these integuments being the only remaining indication of their compound origin.

Another seed which can be interpreted in this way is *Archaeosperma* (Fig. 7.5). This is of additional interest, not only for its great age (it comes from the Upper Devonian of Pennsylvania), but also because of the undoubted occurrence of three aborted spores (Fig. 7.5b) above the functional megaspore of the seed. The arrangement and markings of these spores clearly indicate that the symmetry of the tetrad was tetrahedral, and the situation in the nucellus of *Archaeosperma* thus bears a striking resemblance to that in the megasporangium of a heterosporous fern, such as *Pilularia* (p. 222).

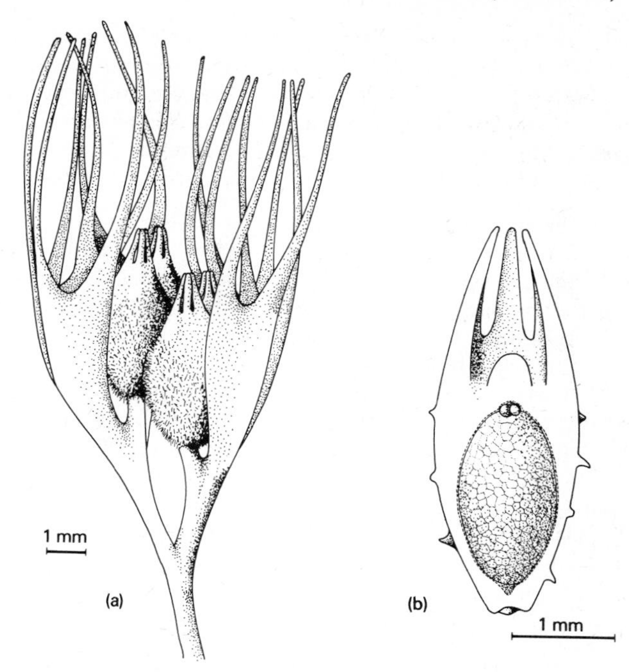

Fig. 7.5 *Archeosperma arnoldii.* (**a**) The two-seeded cupules were borne in pairs, the two pairs of seeds facing each other. (**b**) Reconstruction of an individual seed showing the three abortive megaspores at the top of the one which is functional. The integument is deeply lobed above. (After Pettitt and Beck, *Contr. Mus. Paleont. Univ. Mich.* **22**, 139 (1968))

The descendants of the pteridosperms

It seems beyond doubt that some pteridosperms persisted into the early Mesozoic, although the remains become much less frequent after the close of the Palaeozoic era and of an unfamiliar form. There is also evidence that the evolution of the pteridosperms in the northern and southern hemispheres diverged at this time, but the two floras remained in contact in certain regions of Africa. The early Mesozoic rocks of these regions have yielded a number of puzzling plants, almost certainly derived from the pteridosperms and probably of very great importance in the evolution of the later seed plants. Some of them had leaves with reticulate venation and an appearance strikingly like that of the leaves of some modern flowering plants.

Cycadales The cycads are seed plants with upright, naked ovules, resembling in general features and size those of the later pteridosperms. The fossil evidence indicates that the cycads came into prominence at the beginning of the Mesozoic era.

Although in Mesozoic times the cycads were distributed as far north as Siberia and Greenland, they are today confined to tropical or sub-tropical regions in both the Old and New Worlds. There are nine genera in all, the commonest being *Zamia*, *Macrozamia*, *Cycas*, *Encephalartos*, *Dioon* and *Ceratozamia*, but only *Cycas*, extending eastwards from the Malagasi Republic into Polynesia, has anything approaching a wide distribution.

HABIT The sporophyte is upright and has either a short, stocky stem, often with a large portion below ground (Fig. 7.6), or is much taller, reaching heights of up to 15 m. Below ground is a massive tap root which bears, together with normal roots, others which are negatively geotropic and which break up at the soil surface into coralloid masses. These contain endophytic fungi and blue-green algae.

Apart from stature, all cycads have a similar growth form. The thick stem, usually unbranched, bears an apical rosette of large, pinnate leaves. The rate of growth is very slow, and although there are no recognizable annual rings in the stem, the age of any specimen can be calculated approximately from the rate of leaf production and the number of leaf bases. A specimen of *Dioon* only 2 m high was estimated in this way to be about 1,000 years old.

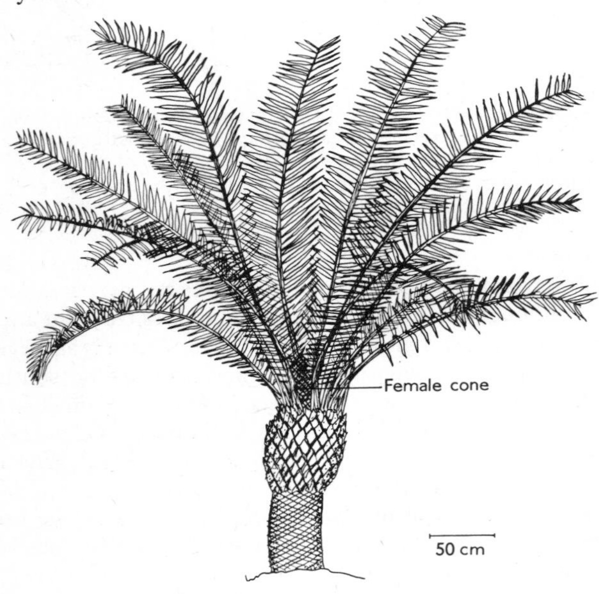

Fig. 7.6 *Encephalartos hildebrandtii*. Habit of plant bearing female cone. (After Eichler, from Eichler in Engler and Prantl, *Die natürlichen Pflanzenfamilien*, II, 1. Engelmann, Leipzig, 1889)

THE VEGETATIVE ANATOMY The stem grows from a massive apex in which there is generalized meristematic activity, and considerable centrifugal expansion, as well as growth in length. Behind the apical initials a core of central tissue, which soon becomes distinguishable from the peripheral, differentiates into the vascular tissue and pith. The peripheral zone becomes cortex. The primary vascular tissue consists of a ring of bundles with endarch protoxylem, and these surround an extensive pith. The secondary xylem is traversed by wide parenchymatous rays and the radial walls of its tracheids are furnished with several series of circular, bordered pits (except in *Zamia* and *Stangeria* where the pits are of the narrower kind characteristic of ferns). The first cambium is of limited activity, and it is followed by others which arise successively outside the vascular cylinder. These cambia are of diminishing activity, the last producing merely a few concentric bundles lying out in the cortex. The stele is thus highly parenchymatous, and the main mechanical support of the stem comes from its armour of sclerenchymatous leaf bases. Mucilage canals, tannin cells, and cells containing crystals of calcium oxalate occur in the pith and cortex of mature stems.

The leaves are of two kinds, foliage leaves (or fronds) and scale leaves, sequences of each following each other in a definite phyllotactic sequence. The scale leaves cover the apex and the upper part of the stem (Fig. 7.7), often disintegrating to form a fibrous sheath below. The foliage leaves in some genera show circinate vernation (as in *Cycas*), but in others a vernation similar to that seen in the fern *Botrychium*. In all cycads the leaves are pinnately branched, but are twice pinnate only in *Bowenia*. The leaf trace, which is horse-shoe shaped in section in the petiole, has a complex origin in the stem. Some strands arise opposite the insertion of the leaf and girdle the stem obliquely upwards into the leaf base. In an individual bundle of the trace much of the metaxylem is adaxial to the protoxylem, but characteristically a few tracheids, often separated by parenchyma, lie on the abaxial side adjacent to the phloem ('centrifugal xylem').

The venation of the pinnae is various, but any branching is dichotomous and open. In section small patches of transfusion tissue (anatomically intermediate between parenchyma and tracheids) are often present on each side of the xylem. In *Cycas*, where each pinna has only a midrib, a sheet of similar cells extends from the midrib to the margin ('accessory transfusion tissue').

The pinnae have a leathery texture. A conspicuous cuticle is usually present, and an epidermis (often accompanied by a hypodermis), palisade and mesophyll are well differentiated. The cell walls of the lower epidermis are straight or slightly sinuose, and the stomata, although usually sunken, are surrounded by a simple ring of subsidiary cells (***haplocheilic*** stomata). These epidermal features, which remain clearly evident in fossil material,

Fig. 7.7 *Zamia* sp. Young plant with female cone. Although produced terminally, the female cone becomes pushed to one side by the sympodial growth and appears to be lateral. Photograph by Frank White. Approx. × ½.

are of great value in distinguishing extinct Mesozoic forms from superficially similar contemporary plants (see Fig. 7.14).

THE REPRODUCTIVE STRUCTURES The mega- and microsporangiophores of the cycads are aggregated into separate strobili borne on different plants. The female cone either terminates the main axis (in which case subsequent growth is sympodial (Fig. 7.7)) or it is lateral, according to the genus. The situation in *Cycas* is exceptional for here the main axis, having given rise to a sequence of megasporangiophores, continues to be active and reverts to the production of normal vegetative leaves.

The female cones vary in compactness and in the number of ovules borne on each megasporangiophore. At one extreme stands *Cycas*, in which the female cone consists of a loose aggregate of megasporangiophores, the distal, sterile portions of which are more extensive than in any other cycad. Several pairs of ovules are attached in the proximal region (Fig. 7.8a), the micropyles of the ovules being directed obliquely outwards. At the other extreme are *Zamia* and *Encephalartos*, in which small, peltate sporangiophores (Fig. 7.8b) are tightly packed in a distinct ovoid cone. Each sporangiophore bears two ovules, the micropyles of which are directed towards the centre of the cone. Both the cones and ovules in the cycads generally are of extraordinary size. In *Encephalartos* female cones have been recorded weighing as much as 45 kg, and in *Macrozamia* the ovules reach a

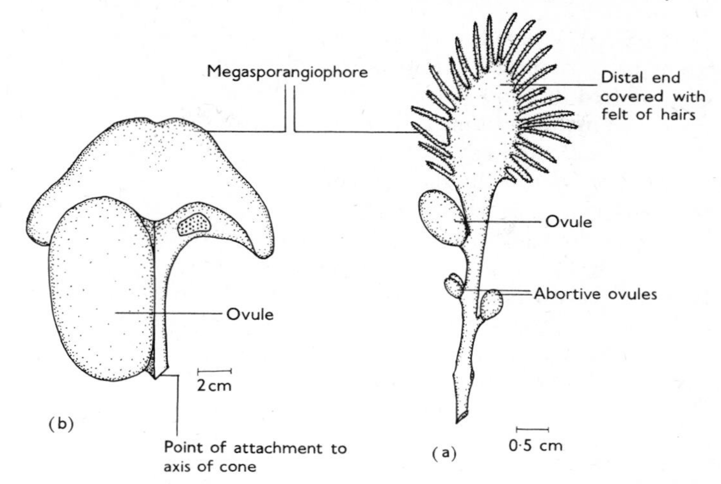

Fig. 7.8 (**a**) *Cycas revoluta*, megasporangiophore. (**b**) *Encephalartos hildebrandtii*, megasporangiophore.

length of 6 cm. The whole of the female reproductive system is thus on a much larger scale than in any other living plants.

The male cones of the cycads (Fig. 7.9) are also either terminal or lateral. Where terminal, subsequent growth is always (including that of

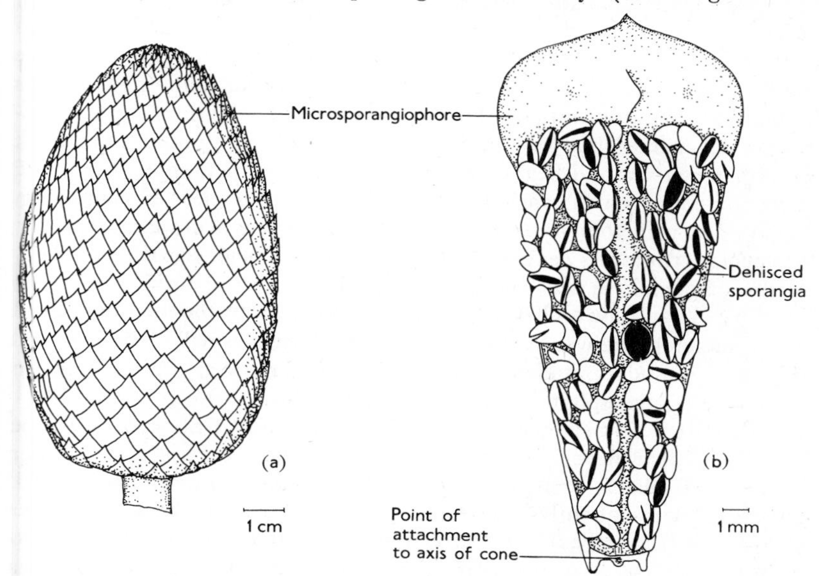

Fig. 7.9 *Cycas* sp. (**a**) Male cone. (**b**) Microsporangia, viewed from below.

Cycas) sympodial. Where lateral, growth is monopodial and the cones may be present in considerable numbers. In *Macrozamia*, e.g. 20–40 cones may be produced in rapid sequence around the lower part of the apex. There is much more uniformity in the structure of the male cones than in that of the female, although again considerable variation in size. In some species of *Encephalartos* the male cones reach a length of 50 cm, but in *Zamia* only 5 cm. The microsporangiophores are in the form of scales, closely appressed during growth and when mature covered on their lower surfaces with several hundred microsporangia (Fig. 7.9b), resembling in appearance and their lateral dehiscence the sporangia of the fern *Angiopteris*. The scales show little development of a distal sterile region. The sporangia, about 1 mm in length, are grouped in sori, each consisting of 3–4 sporangia. Their origin is eusporangiate and their formation almost simultaneous. In *Encephalartos* considerable temperatures, some 15°C above the ambient, have been recorded in male cones at the time of meiosis,[49] no doubt a consequence of the intense respiratory activity throughout the cone at this stage of development. The microspores of the cycads are uniformly trilete, and in most species each sporangium produces some hundreds of spores. Germination begins in the sporangium, and each spore when shed already contains three cells, namely a single prothallial cell, a generative cell and a tube nucleus (Fig. 7.11a).

THE DEVELOPMENT OF THE GAMETOPHYTES AND FERTILIZATION The ovules of the cycads are very similar to those of the later pteridosperms. They are distinctly radiospermic and the integument is differentiated into sclerenchymatous and fleshy layers. In the immature ovule the megaspore mother cell, which is deeply embedded in the nucellus, undergoes meiosis and gives rise to four megaspores in a linear tetrad. The three outer megaspores degenerate, but the inner or chalazal megaspore remains viable. It germinates *in situ* and begins to form the female gametophyte, the initial development of which consists of a sequence of free nuclear divisions (Fig. 7.10a). The gametophyte enlarges with the expanding ovule and eventually a vacuole forms at its centre. Wall formation then begins at the periphery of the gametophyte and continues towards the centre until it becomes wholly cellular. The archegonia (Fig. 7.10b) appear in the micropylar end of the gametophyte where the cells are comparatively small. A distinct and thickened boundary, known as a 'megaspore membrane', forms between the haploid gametophyte and the diploid nucellus.

Although at first sight unfamiliar, the archegonia of the cycads can be seen from their development to be quite similar to those of the lower archegoniate plants. A single initial cell divides into an outer primary neck cell (which subsequently gives rise to one tier of neck cells) and an inner central cell (Fig. 7.10c). The latter rapidly expands, and then divides to form the egg and a small superficial ventral canal cell, which degenerates as the egg becomes ready for fertilization. Maturation of the egg involves

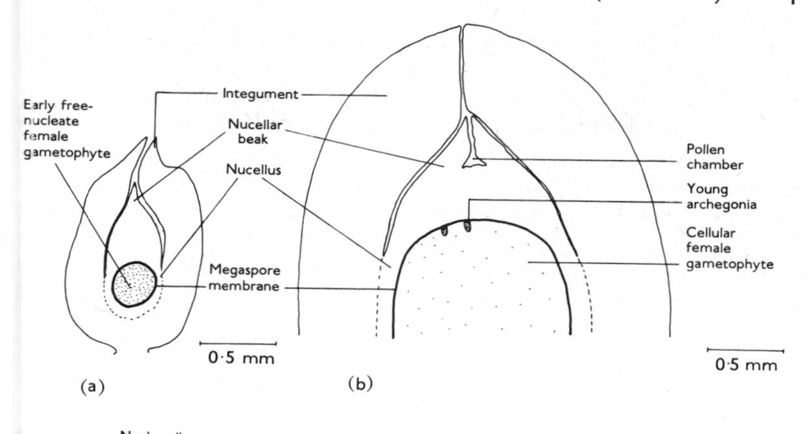

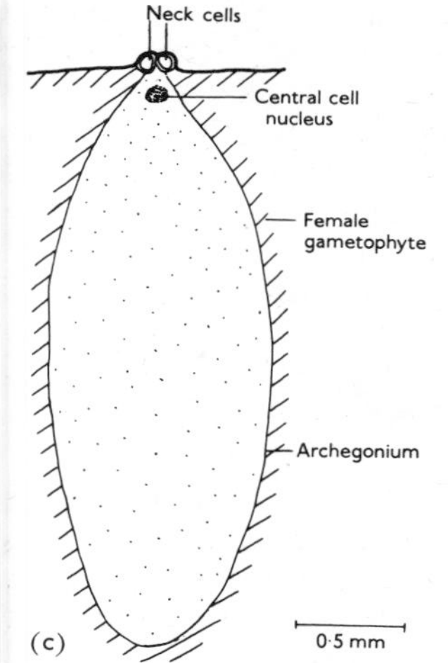

Fig. 7.10 *Cycas* sp. Longitudinal sections of ovule showing (**a**) free-nucleate female gametophyte. (**b**) Initiation of archegonia. (**c**) Almost mature archegonium. (All after Swamy, *Am. J. Bot.* **35,** 77 (1948))

considerable cytological activity. In the egg of *Zamia*, for example, small bodies, which appear in the light microscope as refractive droplets, seem to stream away from the surface of the large central nucleus into the cytoplasm.[16] The cycads possess the largest egg cells known amongst land plants. Their diameters are of the order of 3 mm or more, and even the nucleus may reach 0.5 mm.

Pollination, which occurs during the closing stages of the growth of the female gametophyte, is brought about by the microspores, which are distributed by wind (and in some species possibly by insects), being caught in a sugary 'pollination drop' secreted at the orifice of the micropyle. Probably as a result of resorption of this drop, the microspores reach the surface of the nucellus, the upper part of which has now broken down to form a shallow pollen chamber (Fig. 7.10b). Here the pollen undergoes renewed germination, each grain putting out a tube from the distal part of the grain (i.e. on the side away from the centre of the original tetrad) laterally into the nucellus. This tube, which is of limited growth, has a purely haustorial function, and only the tube nucleus enters it. In the presence of the pollen, and possibly accelerated by enzymes secreted by it, the upper part of the nucellus continues to break down, until all that remains above the mature archegonia is a small pool of fluid. The development of the male gametophyte meanwhile continues. The generative cell, the only cell to show further activity, divides, giving rise to a body cell and a stalk cell

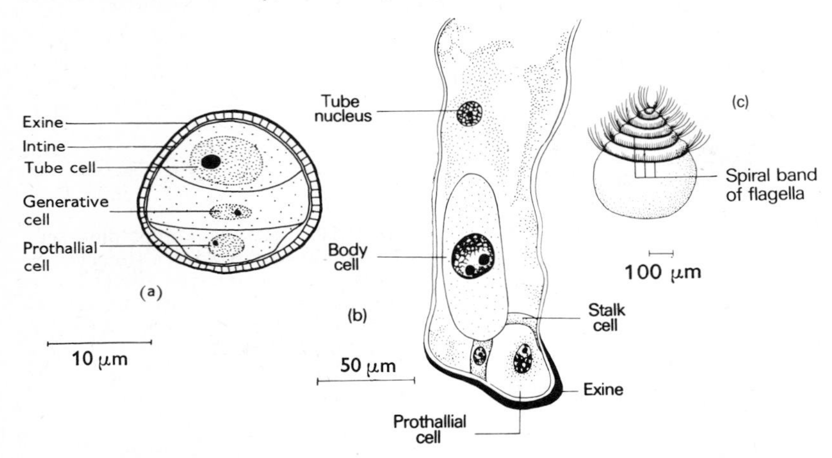

Fig. 7.11 *Cycas* sp. Development of the male gametophyte. (**a**) Vertical section of microspore at liberation. (**b**) Longitudinal section of pollen tube. (**c**) Antherozoid. (**a**, **b** after Swamy, *Am. J. Bot.* **35**, 77 (1948), **c** modified, after Swamy, *Am. J. Bot.* **35**, 77 (1948))

and the proximal part of the male gametophyte bends over the archegonia (Fig. 7.11b). Subsequently the nucleus of the body cell divides to form two coiled, multiflagellate antherozoids (Fig. 7.11c), which in some species may reach 300 μm in diameter. These are finally released from the proximal part of the tube close to the ruptured exine directly into the fluid above the archegonia. One or more eggs become fertilized. The penetrating antherozoid sheds its cytoplasm (including the flagella) in the cytoplasm of the egg, and and its nucleus enters and disperses in the large female nucleus.

This account of the development of the male gametophyte and fertilization is based on the events in *Dioon*, *Zamia* and *Cycas*. Its general validity has yet to be checked, and many features await re-investigation by modern techniques.

EMBRYOGENESIS Although the development of the male and female gametophytes, and the interval between pollination and fertilization is prolonged and may extend over months, the formation of the proembryo proceeds immediately after fertilization. After a period of free nuclear division, in which as many as 256 (2^8) nuclei may be formed, the proembryo becomes cellular. Further growth takes place at the chalazal end, and the embryogeny is evidently endoscopic. At the extreme base of the proembryo is a group of small meristematic cells which develop into the embryo proper, in some species protected on the outside by a layer of cap cells which later degenerate. Above the embryonic cells are a number of elongating cells which form a conspicuous suspensor. The mature suspensor may reach several centimetres in length, but the resistance it meets in driving the young embryo into the nutritive tissue of the female gametophyte causes it to be highly twisted and coiled (Fig. 7.12).

The embryo grows and differentiates at the expense of the female gametophyte, occasionally termed—by analogy with the angiosperms (see p. 320)—a 'pseudo-endosperm'. In the mature seed it has two or several

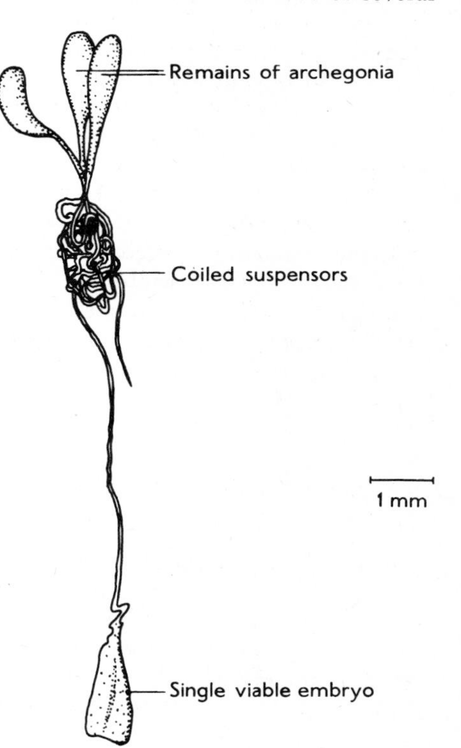

Fig. 7.12 *Cycas* sp. Embryo and suspensor. (After Swamy, *Am. J. Bot.* **35**, 77 (1948))

cotyledons (depending upon the species), directed away from the micropyle and enclosing the stem apex (plumule). Although a short axis is present below the cotyledons (hypocotyl), a root is still lacking at this stage. The fully formed embryo is surrounded by the exhausted remains of the female gametophyte and nucellus, and externally by the integument. Germination occurs as soon as conditions are favourable and the seed has imbibed sufficient water. The hypocotyl pushes its way through the micropylar end of the seed and then begins to develop a strong tap root which persists throughout the life of the plant. The cotyledons remain partially enclosed in the seed, but the plumule emerges and gives rise to mature leaves, the first of which have only a few pinnae.

The fossil history and possible origin of the cycads

The first cycads are found in the late Palaeozoic, both petrified male cones (resembling those of *Stangeria*) and megasporangiophores of the *Cycas* kind having now been discovered in North American rocks.[56a,77a] A cycad of the Mesozoic era is *Bjuvia simplex* from the Rhaetic of Sweden, the remains being sufficient to permit a reconstruction of the plant (Fig. 7.13). There was a general resemblance to *Cycas*, but the leaves were entire instead of pinnate. The megasporangiophores, which showed little development of the distal sterile region, formed a loose terminal cone. *Beania* is a female cycad cone from the Jurassic, resembling that of *Zamia*, but less compact. The same beds yield what are almost certainly male cycad cones and also compressions of leaves, the epidermal features of which are quite similar to those of living cycads.

In their radiospermic seeds and in the complexity of their stelar structure the cycads so closely resemble the later pteridosperms that it seems beyond doubt that they had their origin in some common stock. If so, both leaves and sporangiophores would be in origin lateral branch systems. If we compare the megasporangiophore of *Cycas revoluta*, with its pinnate distal portion, with that of *Zamia*, where the sterile distal portion is lacking, we may see the process by which an ovuliferous megaphyll lost its sterile region and became wholly reproductive. This specialization seems to have been accomplished earlier in the male inflorescences, since in all cycads, even those of Palaeozoic age, the microsporangiophores have little sterile tissue.

Bennettitales (Cycadeoidales) The Bennettitales are wholly fossil, their record extending from the Triassic to the Cretaceous periods. The frequency of their remains is such that they were probably a more conspicuous element of the Mesozoic floras than the Cycadales. In habit there was a general resemblance to the cycads. Some Bennettitales were upright, sparingly branched plants, while others were squat, bearing a crown of leaves near the soil surface.

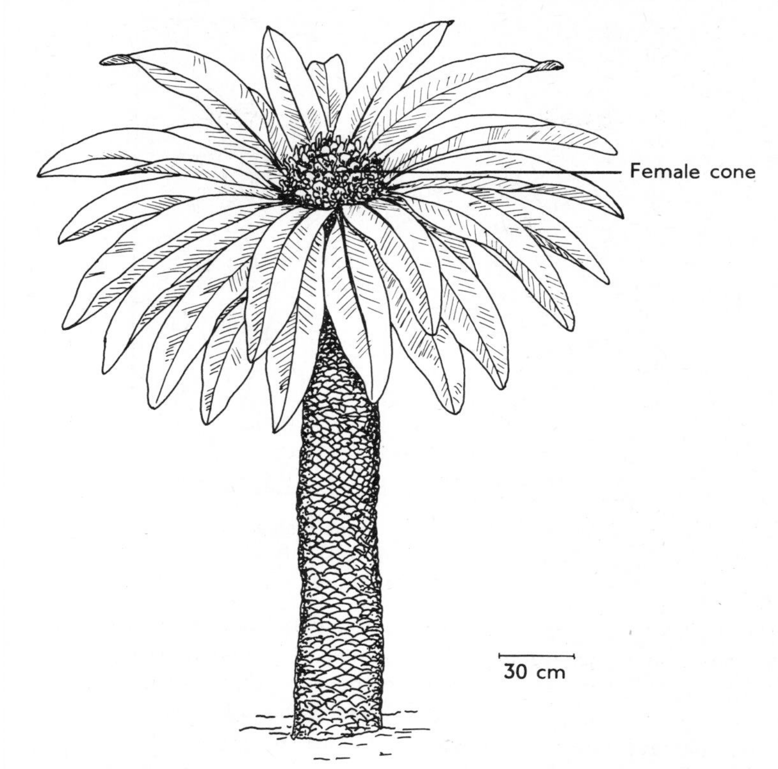

Fig. 7.13 *Bjuvia simplex*. Reconstruction of plant with female cone. (After Florin, from Arnold, *An Introduction to Paleobotany*. McGraw-Hill, New York, 1947)

MORPHOLOGY AND ANATOMY The leaves of the Bennettitales were entire or pinnate, and very similar to those of the cycads. They were not in fact easily distinguishable from those of the cycads until it was discovered that the epidermal features were quite different (Fig. 7.14). In the Bennettitales the walls of the epidermal cells were highly sinuose, and the guard cells and subsidiary cells appear to have had their origin from the same mother cell, giving rise to so-called ***syndetocheilic*** stomata of characteristic form.

The stem structure, so far as it is known, was similar to that of the cycads.

REPRODUCTION The female cone of the Bennettitales consisted of an axis bearing upright ovules interspersed with sterile scales. The male

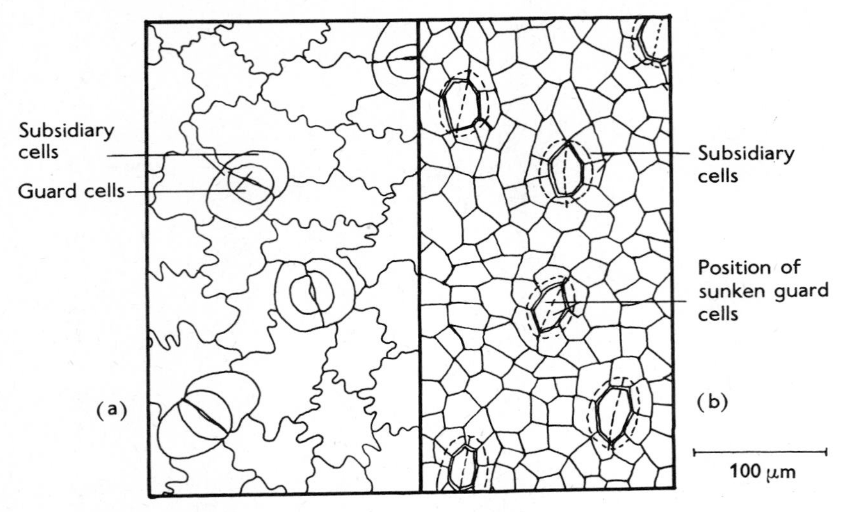

Fig. 7.14 Fossil cuticles of: (**a**) the Bennettitalean type; (**b**) the Cycadalean type. (From a preparation by W. G. Chaloner)

inflorescence was a whorl of microsporangiophores which produced either microsporangia or large complex synangia. The pollen grains were monocolpate, similar to those of the later pteridosperms. In most forms, e.g. *Williamsoniella* (Fig. 7.15) and *Cycadeoidea*, the inflorescence was bisexual, the male portion being produced below the female. The seeds were similar in structure and symmetry to those of the cycads and pteridosperms. In one form, *Bennettites albianus*, embryos can be seen within the seeds, and each has a massive hypocotyl directed towards the micropyle, again resembling the situation in the cycads.

The relationships and origins of the Bennettitales

The general anatomical and reproductive features of the Bennettitales indicate that they also had their origin in some pteridospermous stock, and that they subsequently evolved parallel to the cycads. There is no obvious explanation of why the Bennettitales should have become extinct, whereas the Cycadales were able to persist.

Caytoniales The Caytoniales are based upon Jurassic fossils so far found only in England and Greenland, but nevertheless appearing to be the remains of small plants at least locally abundant. The growth forms of the Caytoniales remain obscure, but the leaves and reproductive organs are now well known. The leaves consisted of a rachis ending in 3–6 leaflets. These were palmately arranged and reticulately veined.

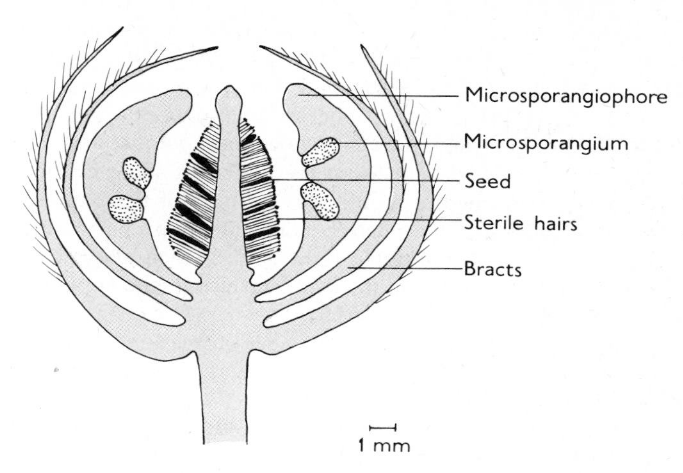

Fig. 7.15 *Williamsoniella coronata*. Longitudinal section of fertile shoot. (After Harris, from Andrews, *Studies in Paleobotany*. Wiley, New York, 1961)

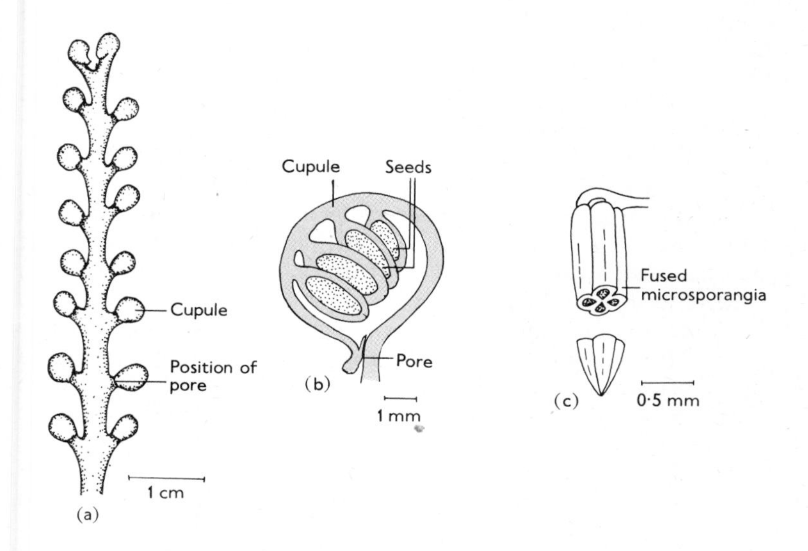

Fig. 7.16 *Caytonia nathorsti*. (**a**) Reconstruction of fertile shoot. (After Thomas, from Andrews, *Studies in Paleobotany*. Wiley, New York, 1961) (**b**) Longitudinal section of cupule. (After Thomas, from Andrews, *Studies in Paleobotany*. Wiley, New York, 1961) (**c**) *Caytonanthus kochi*. Reconstruction. (After Harris, from Andrews, *Studies in Paleobotany*. Wiley, New York, 1961)

THE REPRODUCTIVE ORGANS The female reproductive organ was an axis, 5 cm or more in length, and possibly dorsiventral in symmetry. This axis bore several short pinnae, arranged in more or less opposite pairs, each pinna terminating in a hollow, spherical body about 0·5 cm in diameter (Fig. 7.16a). This sphere contained the ovules, and, because of its analogies with the structures partially enclosing the ovules of some Carboniferous pteridosperms, it is termed a cupule. At the base of the cupule, on the upper side, and concealed by a small flap of tissue, was a pore communicating with the interior. Up to 32 upright ovules, each about 2·5 mm in length, were arranged in lines on the inner surface of the cupule, more or less opposite the basal pore (Fig. 7.16b).

The symmetry of the male reproductive organ was similar to that of the female, but the pinnae terminated in several short irregular branches, each of which bore one or more synangia (Fig. 7.16c). Each synangium was divided longitudinally into four pollen sacs (and consequently, although symmetrical, bore a striking resemblance to the anther of a modern flowering plant), the loculi containing winged pollen grains.

The validity of associating these two fossils rests not only upon their propinquity and structural resemblances, but also upon the occurrence of winged pollen, identical with that in the pollen sacs, near the lip and also inside the ovuliferous cupule. It is not known whether the pollen germinated in these regions, pollen tubes subsequently growing towards the ovules, or whether it was necessary for pollen itself to penetrate the micropyles of the ovules for pollination to be effective.

The possible origin and the significance of the Caytoniales

Although the origin of the Caytoniales is by no means certain, it is again believed to have lain in the pteridosperms. The female inflorescence of *Caytonia* has some resemblance to a reduced branch system, and the ovules are radially symmetrical. Moreover, the male inflorescence is very similar, even to the extent of the winged pollen grains, to certain male reproductive organs from the Triassic of South Africa which are generally accepted as remains of plants of pteridospermous ancestry. It seems unlikely that the Caytoniales left any direct descendants, but they are nevertheless important plants. They show how a closed ovuliferous body, such as we see in the modern flowering plants, could have been evolved by progressive fusion at the orifice of a pteridosperm cupule accompanied by the development of a stigmatic surface.

Cordaitales The Cordaitales, which vegetatively resembled some modern conifers, must have been amongst the most impressive of the seed-bearing plants of the later Palaeozoic. So far as is known, they were all arborescent with columnar trunks, many probably reaching heights of 30 m and diameters of 1 m. The leaves, confined to the upper branches, were spirally arranged and strap-shaped (Fig. 7.17a). In some forms they

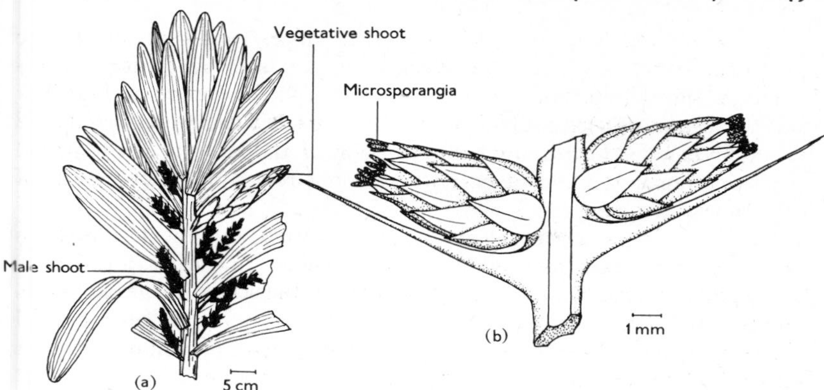

Fig. 7.17 (**a**) Reconstruction of a Cordaitalean shoot. (After Grand'Eury, from Andrews, *Studies in Paleobotany*. Wiley, New York, 1961) (**b**) *Cordaianthus concinnus*. Reconstruction of two male shoots. (After Delevoryas, *Am. J. Bot.* 40, 144 (1953))

were as much as 1 m in length and 15 cm in width. There was regular parallel venation interspersed with longitudinal bands of hypodermal fibres, a structure not dissimilar to that of the leaves of the modern conifer *Araucaria araucana*.

VEGETATIVE FEATURES In general the vascular tissue of the cordaitalean trunks consisted of a large amount of secondary xylem, traversed by narrow parenchymatous rays, surrounding a medullated primary stele. The primary xylem tended to diminish in later forms, leaving a ring of mesarch bundles surrounding an extensive pith, often broken up into lenticular diaphragms. The secondary tracheids showed several series of circular bordered pits on their radial walls, and were closely similar to those of living *Araucaria*. The leaf traces, which passed outwards from the primary xylem, were simple in origin and commonly consisted of two parallel strands. The roots of the Cordaitales are quite well known, since they often became petrified while they were penetrating decaying remains of other plants. They show a triarch stelar structure, and a distinct root cap at the growing tip.

THE REPRODUCTIVE STRUCTURES The reproductive organs of the Cordaitales (known as *Cordaianthus*) were borne on slender branches. Although male and female were separate, they possibly occurred on the same tree. Each reproductive region was basically an axis, from 10 to 30 cm in length (Fig. 7.17a), bearing two rows of bracts in a complanate distichous arrangement. The male and female shoots occurred singly in the axils of the bracts.

The individual male shoots (Fig. 7.17b) were about 1 cm long. Each consisted of a short, stout axis bearing a large number of linear-lanceolate scales, each with a single vein, in a close spiral. The lower scales were sterile, and were acute or obtuse at their apices, but the upper were emarginate and terminated in several (usually six) cylindrical microsporangia. Since both sterile and fertile scales lay in one spiral, they appear to have been morphologically identical. The pollen grains were surrounded by an air bladder formed by the separation of the layers of the wall, the two layers remaining in contact however in one region, possibly the site of liberation of the gametes.

The female shoots were of similar organization, but the fertile scales terminated in ovules instead of microsporangia. In earlier forms (e.g. *Cordaianthus pseudofluitans* (Fig. 7.18a)) the fertile scales (megasporangiophores) projected conspicuously from the shoot, branched, and carried more than one seed. In the later, however, the megasporangiophores were shorter and unbranched, terminating in only one seed concealed amongst the sterile scales (Fig. 7.17a).

The seeds of the Cordaitales were not radially symmetrical, but bilateral, the margin of the seed often being extended as a wing. Because of their characteristic flattened appearance, these seeds are termed platyspermic, and they are readily distinguishable from the radiospermic seeds of the pteridosperms. The integument of a cordaitalean seed (Fig. 7.18b) has the appearance of having been formed by two valves, each containing a vein, coming together and enclosing a megasporangium. This view is supported by the occasional occurrence in female shoots of what are interpreted as abortive megasporangia subtended by two unfused lobes. Each seed would thus have three components, the two outer which form the integument possibly having been derived from sterilized sporangia. It is noteworthy in this connection that the microsporangia were commonly produced in multiples of three.

Platyspermic seeds, similar to those seen in cordaitalean inflorescences, are frequently found detached in Carboniferous deposits and their structure is now well known (Fig. 7.18b). They are about 1 cm in height and only a little less in their major transverse diameter; the minor is of the order of 0.5 cm. The integument was differentiated into one or more layers, at least one of which was sclerenchymatous. The nucellus, except for the basal region, appears to have been separate from the integument. A female gametophyte, surrounded by a distinct 'megaspore membrane', developed within the nucellus, and archegonia were produced on its upper surface, the nucellus above this region becoming differentiated as a pollen chamber. Reticulate markings on the pollen grains lying in this chamber were once interpreted as indicating endosporic germination of the grain, but this is not now generally accepted. The pattern is more probably a relic of the sculpturing of the wall than of an internal cellular structure. Gametes were possibly liberated into fluid above the archegonia. As with the pteridosperms, no seeds have been found containing embryos.

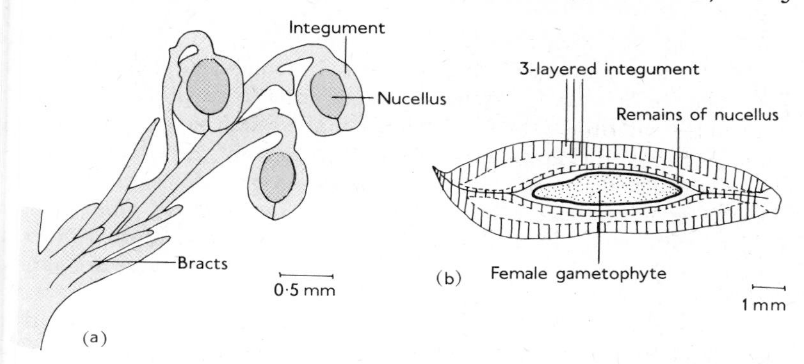

Fig. 7.18 (**a**) *Cordaianthus pseudofluitans*. Reconstruction of female shoot. (After Florin, from Andrews, *Studies in Paleobotany*. Wiley, New York, 1961) (**b**) *Kamaraspermum leeanum*. Transverse section of seed. (After Kern and Andrews, from Andrews, *Studies in Paleobotany*. Wiley, New York, 1961)

The origin and fossil history of the Cordaitales

Little is known of the origin of the Cordaitales, but it possibly lay well before the Carboniferous period. Remains of substantial woody plants, the xylem of which showed araucarian pitting (see p. 254), have been found in rocks as old as the Middle Devonian. In the leaves of some of the Cordaitales of the Lower Carboniferous the nerves branched as they approached the tip, possibly indicating that the leaf originated in a fan-shaped structure. Axes bearing leaves of the kind envisaged are in fact known from the Middle Devonian, but the relationship of these fossils (placed in the genus *Barrandeina*) to the Cordaitales is quite unproven. Nevertheless, the impression is that the Cordaitales were derived from axial heterosporous forms in much the same way as the pteridosperms, but that in the Cordaitales the megaphyll condensed into the characteristic strap-shaped leaf and the integument of the seed evolved in a slightly different way.

The Cordaitales probably persisted into the beginning of the Mesozoic, but then became extinct.

Coniferales The conifers are the most widespread of all the groups of gymnosperms, and they form the climax vegetation at high altitudes and in the colder regions of the temperate zones, particularly the north. They are much less common in the Tropics, and here they are usually confined to mountains and are often mixed with angiospermous trees. Of all the vascular plants discussed so far the conifers are the first of significant economic importance. They are almost all arborescent and the wood is used extensively as timber and as a source of pulp for paper making and related industries.

GROWTH FORMS The growth form of a conifer is frequently pyramidal, the conspicuous main axis being the principal source of the valuable timber. A few conifers of this form attain remarkable sizes and ages. Specimens of *Sequoia*, for example, in California frequently exceed 100 m in height, their trunks reaching diameters of several metres and showing over 2,000 growth rings. *Pseudotsuga* in the forests of the Olympic Peninsula of the Pacific north-west may attain even greater heights (but not girth). The oldest living conifers are probably specimens of *Pinus aristata* at high altitudes on the arid White Mountains of the California–Nevada border. Modern techniques of dating show that some of these are almost 5,000 years old.[30a]

Some conifers (such as the junipers) are bushy, and a few (confined to Australasia) are dwarf, heather-like shrubs of boggy alpine situations. Occasionally the growth form is markedly influenced by the habitat. *Pinus montana*, for example, is a pyramidal tree when growing in acid situations on lower hills, but a straggling shrub with no evident main axis when on limestone at higher altitudes. Most conifers tend to be surface rooted, and many species produce stubby rootlets in the humus layer which are associated with mycorrhizal fungi. *Taxodium distichum*, which grows in swamps in the warmer parts of eastern North America, is outstanding amongst conifers in producing negatively geotropic aerophores which rise above the surface of the water. These specialized roots are, however, rarely produced by specimens planted outside the native habitat.

LEAVES The leaves of the conifers take a variety of forms (Fig. 7.19), but they are nearly always small and simple in shape ranging from needle-like structures several centimetres in length (*Pinus*, Fig. 7.19a) to closely

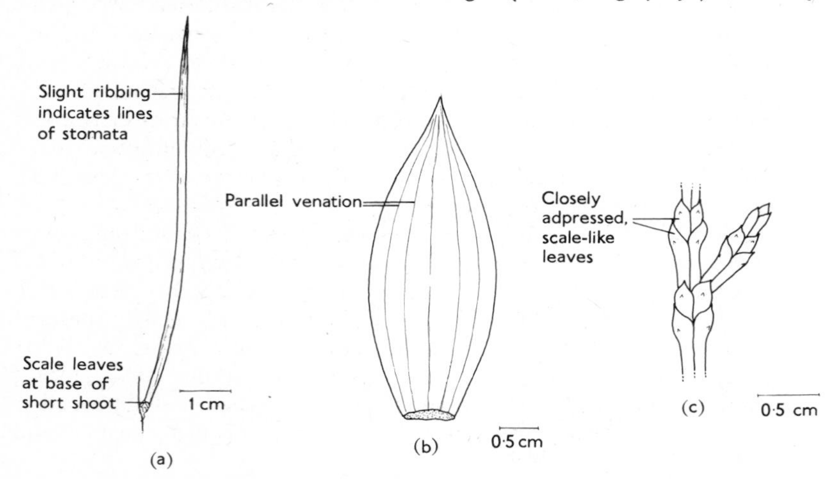

Fig. 7.19 Forms of conifer leaf. (**a**) *Pinus monophylla*. (**b**) *Araucaria araucana* (**c**) *Chamaecyparis obtusa*.

adpressed scales reaching only a few millimetres (as in the Cupressaceae, Fig. 7.19c). *Araucaria araucana* is unusual in having broadly lanceolate leaves 5 cm or more in length (Fig. 7.19b). In the Cupressaceae the plant frequently passes through a juvenile phase in which it produces needle-like leaves. Cuttings or grafts of the juvenile phase sometimes go on producing needle-like leaves indefinitely, and these so-called *Retinospora* forms are common in gardens. The venation of conifer leaves is never reticulate. There are either a number of parallel veins (as in some species of the Araucariaceae, Fig. 7.19b), or a single median vein, often showing a double structure (as in *Pinus*). In some conifers (e.g. *Pinus*) the leaves are borne wholly or principally on short shoots. The leaves of most conifers persist for several seasons; in only a few genera (e.g. *Larix*) are the leaves truly deciduous. In some Taxodiaceae (e.g. *Taxodium*, *Metasequoia*) the leaves are confined to the ultimate branchlets, and the branchlets are shed at the end of the growing season.

THE ANATOMY OF THE STEM AND LEAVES The stems of conifers grow from a group of meristematic cells. In some genera, notably *Araucaria*, the apex is organized into a distinct tunica, in which divisions are regularly anticlinal, and a central corpus, where divisions are in several planes. The latter gives rise to the pith and primary vascular tissue. The mature stems of conifers consist principally of secondary wood, the pith and primary xylem

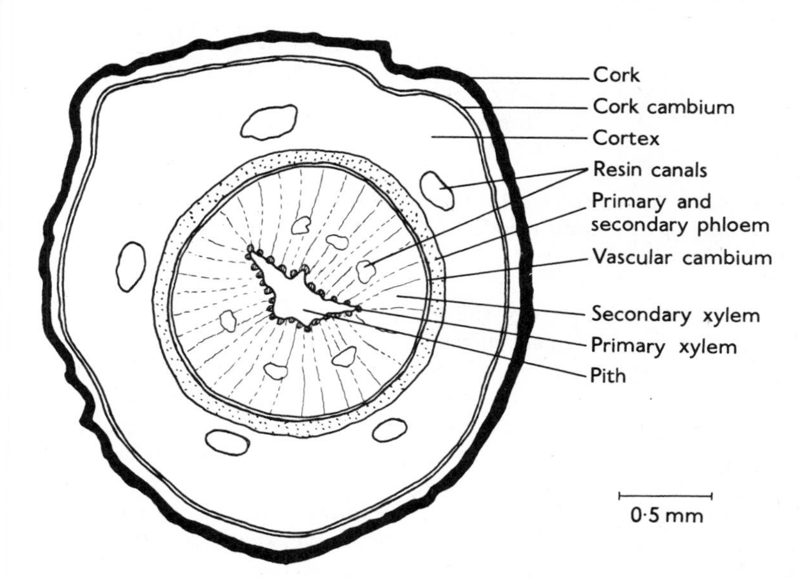

Fig. 7.20 *Pinus sylvestris*. Transverse section of young stem with only one season's secondary vascular tissue.

being relatively inconspicuous (Fig. 7.20). At the outside, the phloem, cortex, and periderm also form a comparatively narrow band. The dense secondary xylem consists of radial files of tracheids, traversed by narrow parenchymatous rays. Wood parenchyma is not conspicuous. In *Pinus* it is confined to the epithelium of the resin canals, and it is entirely lacking in *Taxus*. The tracheids are usually differentiated in distinct annual rings, those formed towards the end of a season's growth being narrower than the preceding (Fig. 7.21a). In *Pinus* the tracheids, which rarely exceed 4 mm in length, bear bordered pits, usually in a single row, on the radial

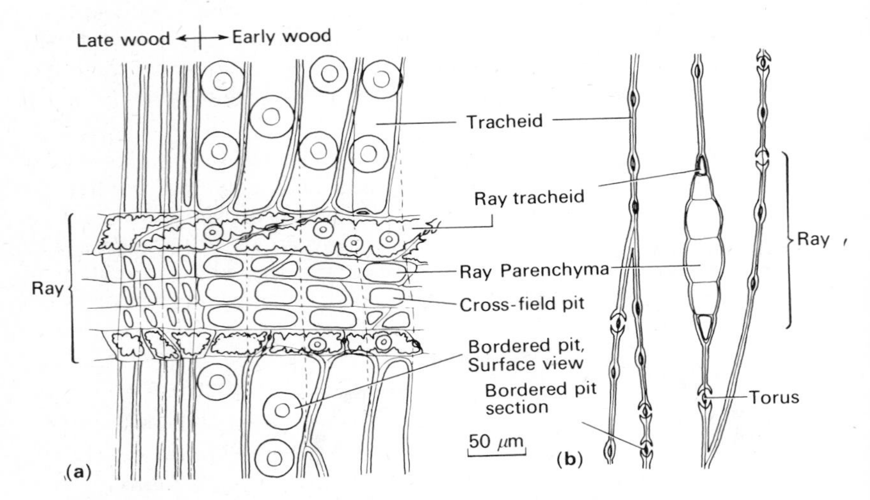

Fig. 7.21 *Pinus sylvestris.* **(a)** Radial longitudinal section of secondary xylem at junction of early and late wood. **(b)** Tangential longitudinal section of a ray similar to that shown in **(a)**.

walls. The central part of the pit membrane is thickened and forms the torus (Fig. 7.21b); thickenings are often present along the margins of the pits, the 'Rims of Sanio'. In *Araucaria* the pits are similar, but in 3–4 rows, the pits of adjacent rows alternating. The tracheids of *Araucaria*, but not of *Pinus*, occasionally have small trabeculae (initially of cellulose, but subsequently lignified) extending across the lumen. These so-called 'Bars of Sanio' also occur in a number of other genera.

Considerable differentiation is sometimes present in the parenchymatous rays of conifer woods. In *Pinus*, for example, the cells of the upper and lower margins in the xylem portion of the ray may form radially orientated tracheids (Fig. 7.21), and cells in a similar position in the phloem closely apply themselves to the sieve cells and become conspicuously rich in cytoplasm. The rays probably provide an important means of transporting materials laterally in the growing stem.

The resin canals of the conifers (Fig. 7.23), which are schizogenous in origin and interconnected, run longitudinally in the xylem and also transversely in some of the larger rays. The resin itself (a complex acidic substance containing oxidized phenols and terpenes) is synthesized in the epithelium of the canals, but the actual site of synthesis in the cells is not yet exactly known. The resin system can be tapped by driving a gutter-shaped steel wedge into the xylem near the base of the tree (Fig. 7.22), and

Fig. 7.22 *Pinus pinaster* (maritime pine). The trunk of this tree (photographed in Portugal) has been tapped for resin. This species of pine is the principal source of natural resin in Europe.

from some species considerable quantities of commercially valuable resin can be collected. Pine resin, for example, is the source of turpentine and colophony, both widely used in the paint and varnish industry.

Many features of anatomical and physiological interest are presented by conifer leaves (Fig. 7.23). The cuticles, for example, are often furnished, especially in the region of the stomata, with distinctive patterns of tubercles and ridges. Palisade and spongy mesophyll are commonly present, and in *Pinus* the walls of the mesophyll cells have ridges projecting into the cell

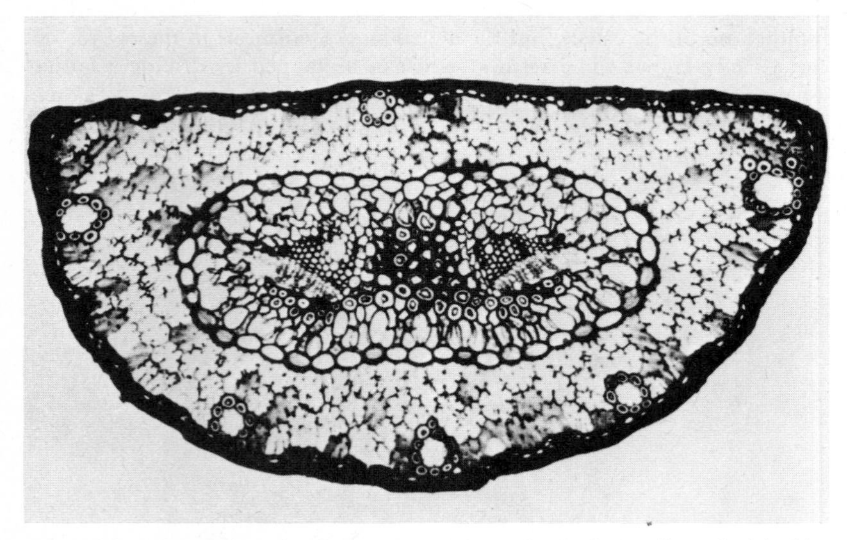

Fig. 7.23 *Pinus sylvestris*. Transverse section of a leaf, the flat adaxial side uppermost, mid-way along its length. The epidermal cells are lignified and the stomata obscured, but the cells of the hypodermis are visible, and adjacent to them the resin canals. Two vascular bundles, separated from the mesophyll by a common endodermis, lie at the centre of the leaf. Thickened and lignified cells are found between these vascular bundles and below the phloem, but the cells elsewhere within the endodermis are largely transfusion tissue. Approx. × 100.

(Fig. 7.23), a character of unknown significance. A well-defined hypodermis, the cells of which may be lignified, is present in many leaves. The cells surrounding the vascular bundles are often intermediate between parenchyma and tracheids, and form a so-called transfusion tissue. Resin canals are frequent, and in some leaves (as of *Thuja*) a prominent gland on the back of the leaf contains a fragrant oil. The leaves of conifers of high altitudes, and of arctic regions, are able to withstand extreme cold, are remarkably resistant to frost damage, and are able to carry out photosynthesis at unusually low temperatures.

The roots of conifers have a simple primary structure, similar to that found in the ferns. The apical meristem is protected by a root cap, and root hairs are produced from a zone immediately behind it. Unlike the ferns, however, secondary vascular tissue begins to be formed at a very early stage, often before the primary tissues are fully differentiated (Fig. 7.24). Resin canals are abundant in the secondary xylem, rays, and cortex.

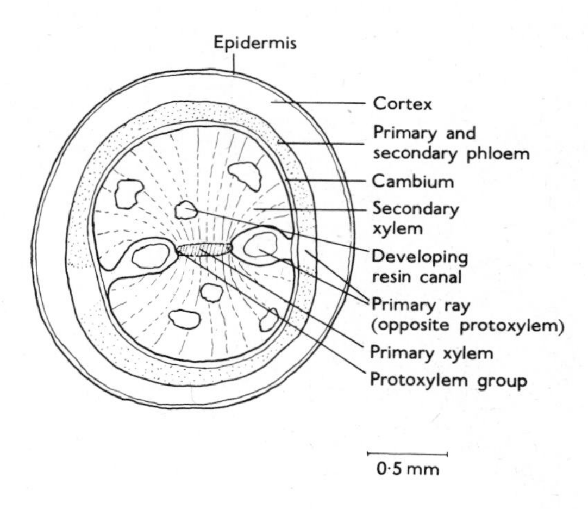

Fig. 7.24 *Pinus sylvestris.* Transverse section of young root.

THE REPRODUCTIVE STRUCTURES As the name implies, the male and female reproductive organs of the conifers are commonly borne in cones (Figs. 7.25 and 7.26). Most conifers are monoecious but diclinous, the male and female cones being produced in different regions. In *Pinus*, for example, the female cones are produced near the apex of the tree and occupy the positions of main lateral buds, while the male cones are produced on the lower branches, usually in groups, each cone occupying the position of a short shoot. A few conifers (e.g. *Taxus* and *Juniperus communis* are dioecious. The reproductive cones are usually compact, but in the Podocarpaceae the female cones are either lax or reduced, and in the Taxaceae the female reproductive region is not cone-like at all. Nevertheless, the general affinities of the Taxaceae are clearly with the conifers.

The male cones are fairly uniform in structure, although they range widely in size. Those of *Taxus*, and of the Podocarpaceae and Cupressaceae are globose, hardly reaching 0.5 cm in diameter, but those of other conifers are commonly elongated, and in *Araucaria* they may exceed 20 cm in length and 3 cm in width. All, however, consist of a central axis bearing

Fig. 7.25 *Larix decidua* (larch). Portion of a shoot showing the short shoots and the female cone in its second year. Note that the female cone is negatively geotropic and that it terminates a lateral axis having a position equivalent to that of a short shoot. × 4/5.

regularly arranged microsporangiophores (Fig. 7.26). These take the form of scales, somewhat peltate in shape, a variable number of pollen sacs being attached to the head and lying parallel to the stalk. The pollen grains of many species are winged and readily identifiable. The grains of *Pinus* (Fig. 7.27), for example, have two asymmetrically placed air bladders (formed by local separation of the layers of the exine) between which the pollen tube emerges. Other grains have characteristic ornamentation; those of *Cryptomeria*, for example, possess a peculiar cuticular hook on one side. The pollen grains often begin to develop internally before being shed. In *Pinus* (Fig. 7.31d) the pollen grain when liberated contains two degener-

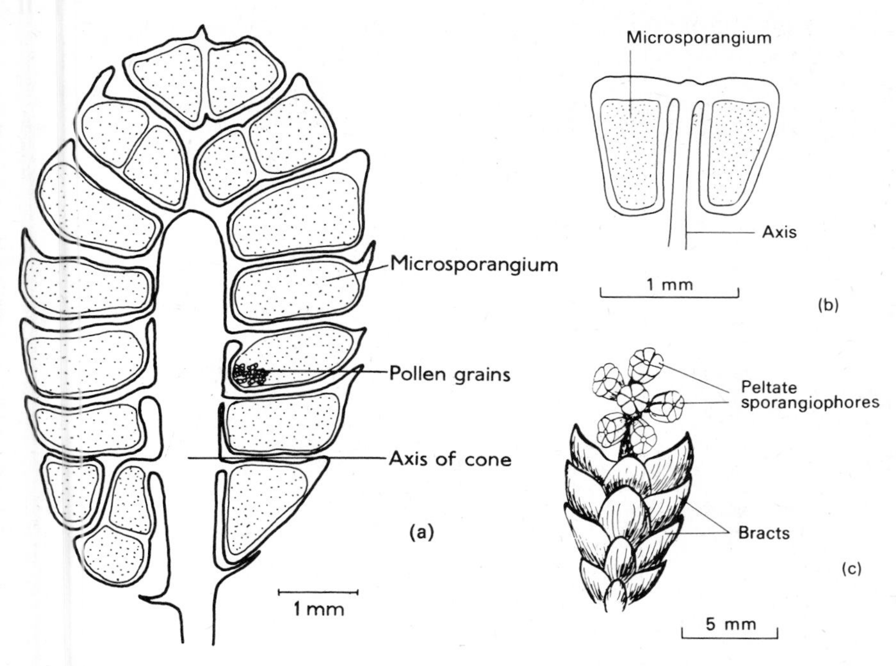

Fig. 7.26 **(a)** *Pinus sylvestris*. Longitudinal section of male cone. Each microsporangiophore bears two pollen sacs. **(b, c)** *Taxus baccata*. Mature male cone, each microsporangiophore bearing 6–8 pollen sacs, and longitudinal section of microsporangiophore.

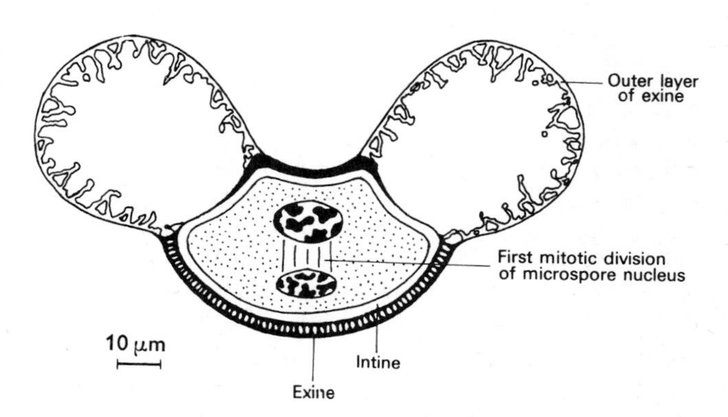

Fig. 7.27. *Pinus banksiana*. Median section of pollen grain showing the first division of the microspore nucleus and the nature of the bladders.

ating prothallial cells, a tube cell and a generative cell (or nucleus). The grains of the Taxaceae, Taxodiaceae and Cupressaceae, however, lack prothallial cells and are uninucleate when shed, whereas the mature grains of the Araucariaceae contain up to 15 prothallial cells. The pollen grains of all conifers germinate distally, i.e. away from centre of the original tetrad.

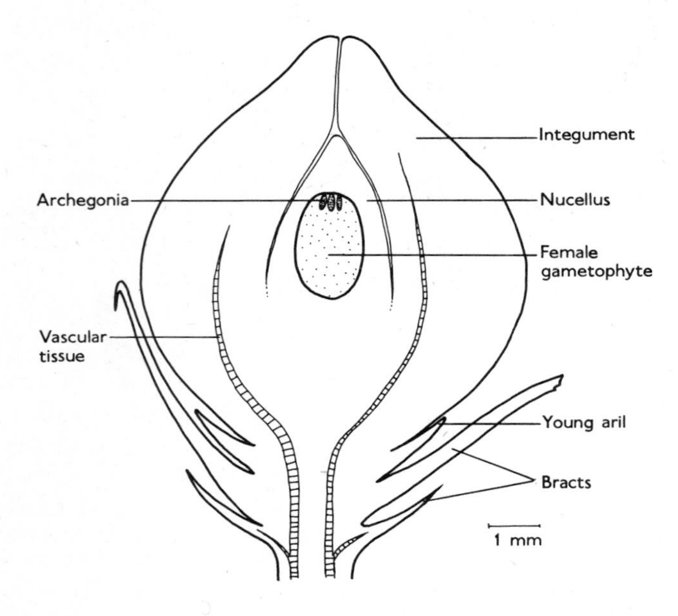

Fig. 7.28 *Taxus baccata*. Longitudinal section of ovule at the time of fertilization.

Of the female reproductive regions, that of *Taxus* is best considered first as it facilitates an understanding of the more complex situation in *Pinus* and other conifers. In *Taxus* the ovule terminates a short shoot bearing three pairs of decussate bracts. The ovule itself is upright and bilaterally symmetrical (Fig. 7.28). The single integument contains a sclerenchymatous layer, and two vascular bundles, diametrically opposed, ascend in the fleshy portion adjacent to the nucellus. A female gametophyte arises in the nucellus as in other gymnosperm ovules, and when mature it bears immersed archegonia in its micropylar surface. The minute short shoot terminating in the ovule is axillary to another short shoot furnished with spirally arranged scale leaves. This whole complex is itself borne in the axil of a normal foliage leaf. Both short shoots of the female reproductive

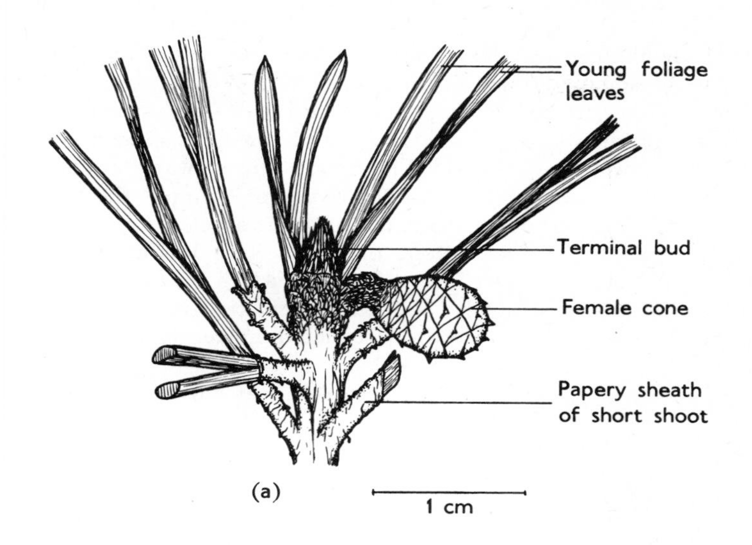

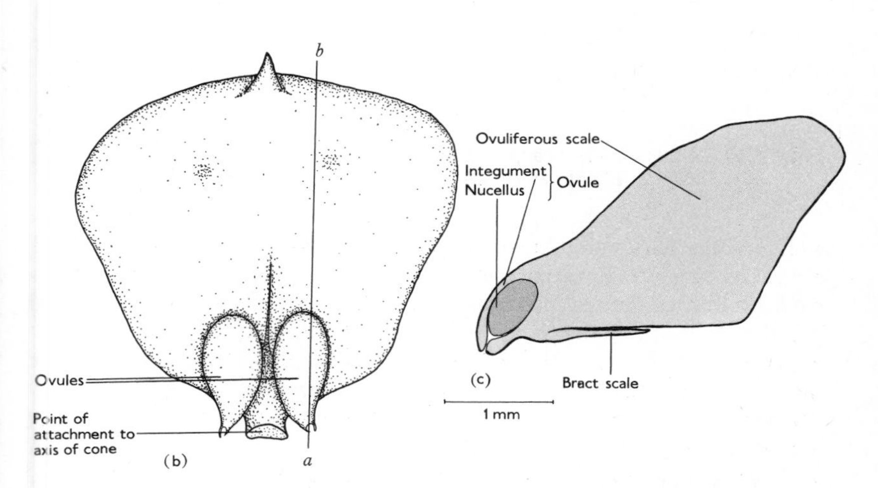

Fig. 7.29 *Pinus sylvestris.* (**a**) Female cone in the early summer of its first year. (**b**) Single scale from cone in (**a**), viewed from above. (**c**) Longitudinal section along line *a–b*.

system in *Taxus* are highly condensed and can be seen only by a careful dissection.

In the female cone of *Pinus* (Fig. 7.29a) we are again concerned with an axis bearing spirally arranged scales in the axils of which are ovuliferous structures (Fig. 7.29b). In *Pinus*, however, the ovuliferous structure is also scale-like, and it is largely fused with and ultimately projects beyond the bract scale in whose axil it arises (Fig. 7.29c). This, however, is not always the situation, even in the Pinaceae. In *Abies*, for example, the bract and ovuliferous scales remain separate, and in some species (e.g. *A. venusta*) the bract scale projects far beyond the ovuliferous. In the Pinaceae the ovuliferous scale bears two inverted ovules near its base, but in other families the number of ovules and their orientation vary. In *Araucaria* (Fig. 7.30) the ovuliferous scale produces and ultimately entirely surrounds

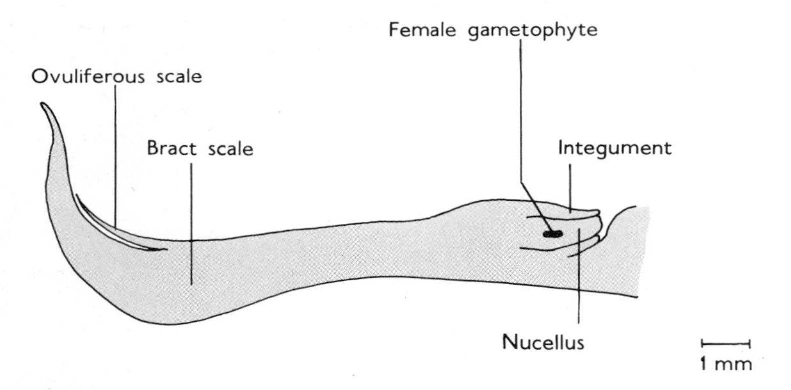

Fig. 7.30 *Araucaria araucana*. Longitudinal section of young ovule. (After Hirmer, *Biblthca bot.* **28**, 27 (1936))

a single inverted ovule. A specialized ovuliferous scale of this kind, also found in some Podocarpaceae, is termed an epimatium. The ovules throughout these more typical conifers are regularly bilaterally symmetrical and the seeds are often winged.

The morphological and anatomical evidence, now supported by the palaeobotanical (see p. 267), points to the ovuliferous scale being a highly modified shoot. The vascular supply to the bract scale, for example, consists of vascular bundles of which the xylem is adaxial, the orientation normal for a leaf trace. The bundles passing to the ovuliferous scale, however, are not only similar in position to those entering an axillary shoot, but the xylem of each is also abaxial, an orientation often seen at the base of a shoot trace.

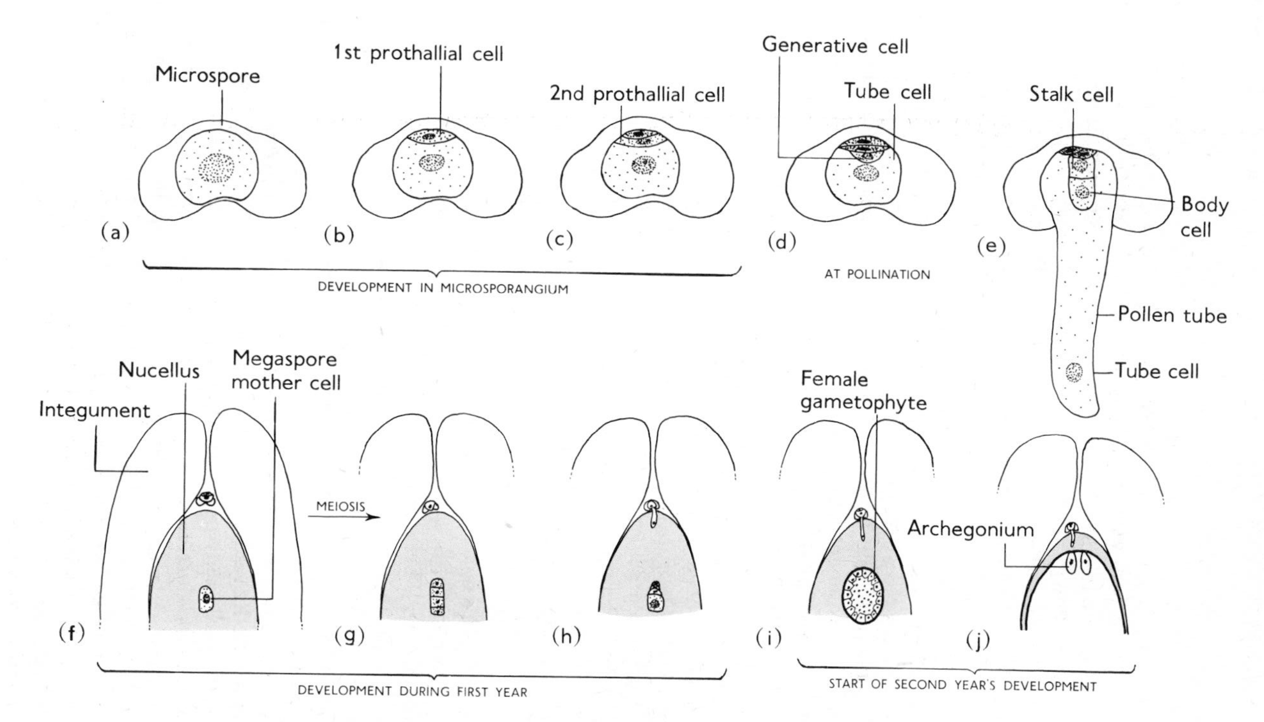

Fig. 7.31 *Pinus sp.* Development of the pollen and ovule. Diagrammatic, not to scale.

POLLINATION AND FERTILIZATION Pollination, which in temperate climates occurs in the spring, involves in most species a pollination-drop mechanism of the usual kind. The development of the female cone is so co-ordinated with that of the male that at the time of release of the pollen the axis of the female cone undergoes general elongation, thus opening the scales and allowing penetration of the pollen. Following pollination, rapid growth of the scales causes them to be tightly packed once again.

In *Pinus sylvestris* the germination and development of the male gametophyte within the female cone are very slow and extend over a whole season (Fig. 7.31), coinciding with meiosis in the nucellus and the initiation of the female gametophyte. A pollen tube emerges from the grain in this first season's growth (Fig. 7.31e), and the generative cell divides into a stalk cell and a body cell. Little further occurs in the winter, but in the following spring development is resumed. After a period of free nuclear division the female gametophyte becomes cellular (Fig. 7.31i) and 1–6 archegonia are formed in its upper surface (Fig. 7.31j), each surrounded by conspicuous jacket cells. The pollen tubes grow towards the archegonia, and when a tube has come to within a short distance of an archegonium the body cell, which has moved into its tip, divides into two male ('sperm') nuclei of unequal size. The tube eventually penetrates the archegonium and the two sperm nuclei are liberated into the egg cytoplasm. The larger sperm nucleus passes into the egg nucleus, whils the smaller sperm nucleus, the stalk cell and the tube nucleus all degenerate. Several archegonia in one ovule may be penetrated by pollen tubes, and this can result in the formation of several zygotes and subsequent simple polyembryony.

EMBRYOGENESIS Free nuclear division occurs in the germination of the zygote of *Pinus*, but it is not extensive, only four nuclei being so formed (Fig. 7.32a). These move to the bottom of the archegonium and form a plate, walls then being laid down between them. These cells divide longitudinally, the cells of each column behaving synchronously. This leads (Fig. 7.32b, c, d) to the formation of a suspensor, tetragonal in section, terminating below in four groups of embryonic initials, each capable of yielding an embryo. This so-called cleavage polyembryony may be further complicated by additional embryos budding off from the basal suspensor cells. Usually only one of these many potential embryos reaches maturity.

The reproductive events in other conifers resemble those in *Pinus*. In *Araucaria* and some other genera, however, the pollen germinates between the scales of the female cone, forming a freely branching, multinucleate, mycelium-like weft, many of the branches penetrating the nucellus. In some genera of the Taxodiaceae the two male gametes are similar in size and are more like distinct cells than in *Pinus*. The female gametophyte in this family may also produce very many archegonia, as many as 60 being found in *Sequoia*. The development of the zygote in other conifers is also different in some details from that seen in *Pinus*. In *Sequoia*, for example, there is no initial free nuclear division, and in the Podocarpaceae the cells of the proembryo pass through a binucleate stage, a feature believed peculiar to this family. In other conifers polyembryony seems less common than in *Pinus*. The embryos of many conifers have several cotyledons; as many as 12 may be present in *Pinus*.

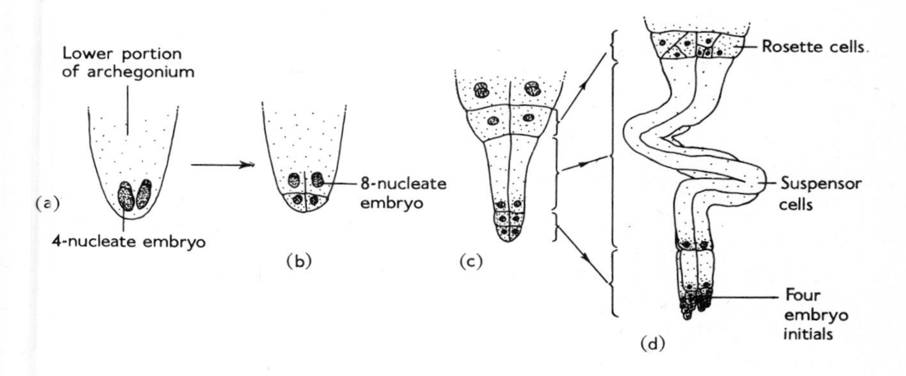

Fig. 7.32 *Pinus sp.* Stages in the development of the embryo. (After Bucholz, from Foster and Gifford, *Comparative Morphology of Vascular Plants*. Freeman, San Francisco, Copyright © 1959)

THE FORMATION AND LIBERATION OF THE SEEDS The mature embryo lies in the remains of the female gametophyte and nucellus, and is surrounded by a hard seed coat formed from the integument (Fig. 7.33b). In some conifers (e.g. *Pinus*) this is expanded as a conspicuous wing assisting the distribution of the seeds by wind (Fig. 7.33a). The female cone often becomes dry and woody during the formation of the seeds and sometimes does not open until a long period after the seeds are mature. In *P. sylvestris* the cone opens and releases the seeds in the second year after pollination (the whole process of reproduction thus extends over three years), but in the 'closed cone' pines of the Pacific coast of North America the cones remain closed indefinitely and the seeds are released by decay of the scales or as a consequence of the singeing of the cones by a forest fire. The cones of some pines are extraordinarily large; that of *P. coulteri*, for example, may reach 40 cm in length and 2 kg in weight. In some Podocarpaceae and in *Juniperus* the ovuliferous scales become fleshy in fruit. In some other podocarps and in *Taxus* the seed becomes surrounded by a succulent aril

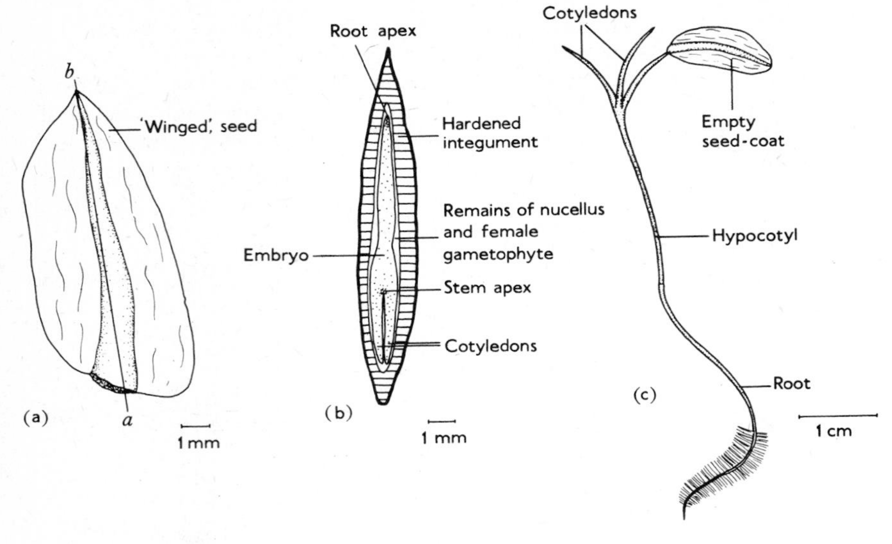

Fig. 7.33 *Sequoia gigantea* (*Sequoiadendron giganteum*). (**a**) Seed. (**b**) Longitudinal section (at *a–b*) of well-soaked seed. (**c**) Young seedling.

which grows up from the base (Fig. 7.34). The bright red aril of *Taxus* is sought after by birds and is probably an aid to dispersal.

GERMINATION In most conifers germination is initiated by the root pole of the embryo elongating and breaking through the seed-coat. The vigorous primary root soon anchors the seedling, and the elongating hypocotyl raises the remains of the seed, from which the several cotyledons are rapidly withdrawn (Fig. 7.33c). All conifer seedlings, so far as is known, are capable of producing chlorophyll in the dark, a remarkable property that distinguishes them from the seedlings of most angiosperms.

The seeds of *Araucaria* often germinate in the cone before it falls apart. In some species the hypocotyl swells to form a tuber, and the seedling is capable of 'resting' in this condition for several months. It was this curious feature that facilitated the transmission of the first specimens of *A. araucana* from Chile to Europe in the eighteenth century.

THE EVOLUTION AND ORIGIN OF THE CONIFERS The current geographical distribution of the conifers presents a number of features of evolutionary significance. Floristically, for example, the conifers of the northern hemisphere are strikingly different from those of the southern, and some families (notably the Pinaceae in the north and the Araucariaceae in the south) hardly cross the equator. Fossils of Quaternary and Tertiary age

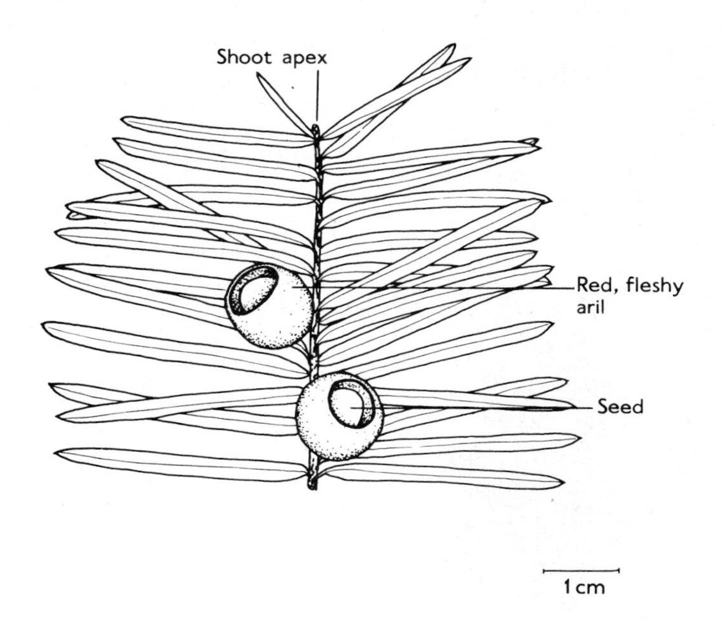

Fig. 7.34 *Taxus baccata*. Shoot bearing ripe fruits.

give no indication that this is a recent segregation, but they do reveal that the distribution of some families was formerly much more extensive than now. *Sequoia*, for example, now confined to the west coast of North America, was once widespread in the northern hemisphere. The distribution of the conifers has clearly contracted with the rise of the angiospermous forests.

The fossil record of the conifers extends back well into the Carboniferous, and the early forms show a close relationship, both vegetative and reproductive, with the Cordaitales.[31] In *Lebachia* (Fig. 7.35), representative of Upper Carboniferous forms, the female cone consisted of an axis, about 8 cm long, bearing spirally arranged, bifid bracts. In the axil of each was a radially symmetrical, but flattened, dwarf shoot terminating in a single erect ovule (Fig. 7.32b). This fertile short shoot corresponds on the one hand with the female flower of the Cordaitales and on the other with the ovuliferous scale of a modern conifer. There is in fact a series of fossils of late Palaeozoic and early Mesozoic age in which the vegetative and fertile parts of the female short shoot become progressively less distinct, leading ultimately to a structure almost identical with an ovuliferous scale. The distinctive form of the female shoot of *Taxus* also appears to have originated at this time. *Palaeotaxus*, vegetatively and reproductively similar to *Taxus*, has been described from the Upper Triassic.

The male cones of the early conifers closely resembled those of the modern, and thus differed sharply from those of the Cordaitales, although the pollen grains of the Cordaitales and early conifers were quite similar.

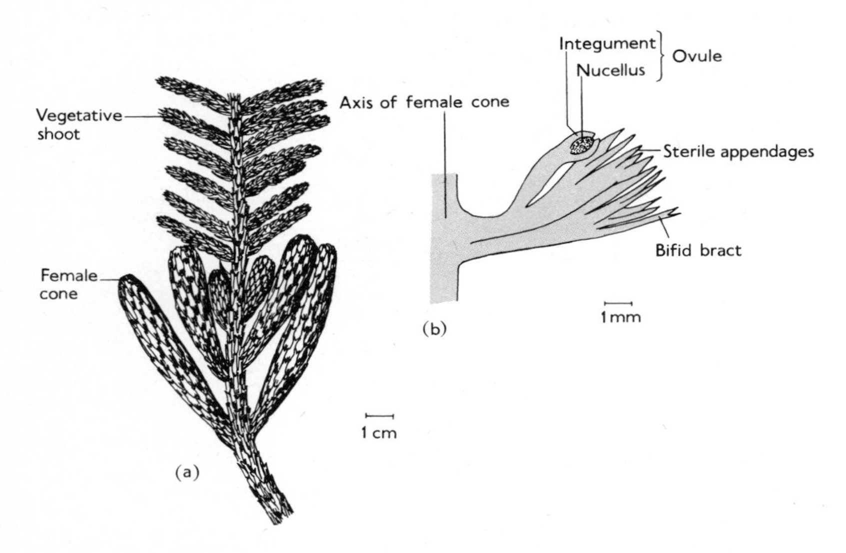

Fig. 7.35 *Lebachia sp.* (**a**) Reconstruction of a shoot bearing upright female cones. (**b**) Longitudinal section of ovuliferous structure. (Both after Florin, from Delevoryas, *Morphology and Evolution of Fossil Plants*. Holt, Rinehart and Winston, New York, 1962)

Nothing is known of fertilization in the early conifers, and antherozoids may have been produced. The arrangement seen in modern conifers, which disposes of the necessity for free fluid at the time of fertilization and possibly reduces the hazards of copulation, may on the other hand have evolved quite early. It was perhaps its advantages which saved the Coniferales from the extinction that befell the Cordaitales.

There are sufficient similarities between the Coniferales and Cordaitales to warrant the assumption of a common origin, the two Orders possibly not diverging until the early Carboniferous.

Ginkgoales This Order is represented today by a single genus and species, *Ginkgo biloba* (maidenhair tree). This remarkable tree, with a striking pagoda-like arrangement of the main branches, was unknown to the Western world until the seventeenth century. It was first discovered in Japan and subsequently in China, but always in cultivation. Suggestions that wild stands of *Ginkgo* may occur in remote parts of China, although not improbable, have never been confirmed. *Ginkgo* is now common in cultivation in all parts of the world.

THE VEGETATIVE FEATURES Fully grown specimens of *Ginkgo* are tall, deciduous trees reaching a height of 30 m or more. The lateral branches bear both long and short shoots (Fig. 7.36), and leaves occur on each. Damage to

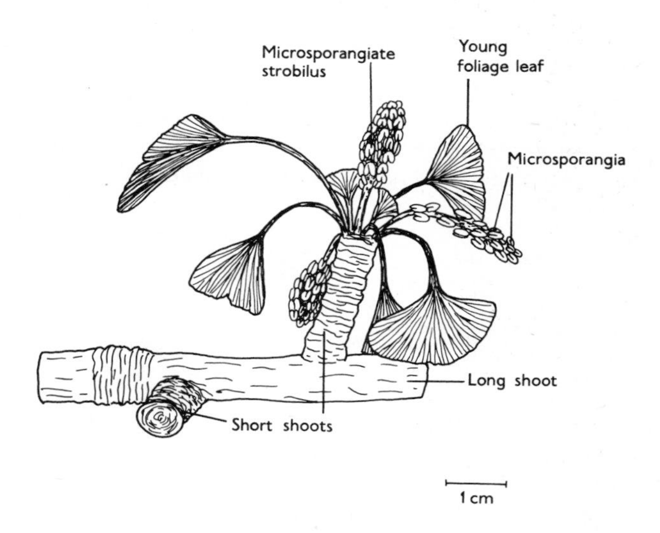

Fig. 7.36 *Ginkgo biloba*. Short shoot bearing male strobili. The leaves have not reached mature size.

a long shoot will cause one or more adjacent short shoots to behave as long shoots, indicating that their manner of growth is not irreversible, and that the maintenance of the dwarf condition probably depends upon the presence of growth-regulating substances produced by the meristem of the long shoot. Anatomically, the apices of the long and short shoots are similar and show well-defined zonation, although no distinct tunica and corpus are present. Growth takes place from a superficial group of apical initial cells. A large proportion of a mature stem consists of secondary xylem, penetrated by narrow parenchymatous rays. The tracheids have bordered pits, usually in a single row, on their radial walls.

The leaves of *Ginkgo* are fan-shaped, usually with a distal notch (hence the specific name). Two vascular bundles ascend the petiole and dichotomize in the lamina, with occasional anastomoses. Short resin ducts may lie between the veins. The distal margin of the leaf is usually irregular, a feature much more marked in juvenile leaves where the distal part of the leaf may even be segmented.

REPRODUCTION *Ginkgo* is dioecious, and sex determination appears to be chromosomal since the male possesses a heteromorphic pair of chromosomes.[52] The male reproductive structures (Fig. 7.36) consist of small strobili, resembling catkins, which arise in the axils of scale leaves of the short shoot. The axis of the strobilus bears a number of microsporangiophores arranged in a loose and irregular spiral. Each microsporangiophore is slightly peltate and the sporangia, usually two, are attached beneath the head. The pollen grains, which have a characteristic furrow in the wall, contain four nuclei when shed, two of the nuclei being associated with rudimentary prothallial cells, and the others identified as the generative and tube nuclei.

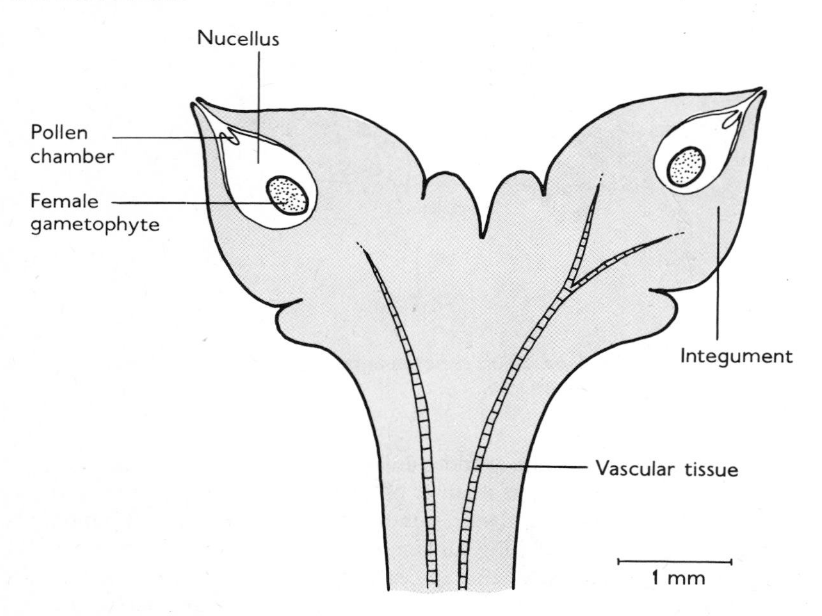

Fig. 7.37 *Ginkgo biloba*. Longitudinal section of female shoot with young ovules.

The ovules are usually borne in pairs, two sessile ovules being symmetrically attached at the end of a stalk-like sporangiophore. Not infrequently, however, the sporangiophore branches irregularly and bears more than two ovules. The sporangiophore itself arises, as in the male, in the axil of a scale or a leaf on a short shoot. The ovules (Fig. 7.37) are about 0.5 cm long and about as broad, and are surrounded at the base by a cushion-like swelling of the end of the sporangiophore. They possess a single integument into which two diametrically opposed bundles ascend. This bilateral symmetry is reflected in the micropyle which is slightly two-lipped at its tip. A megaspore is formed within the nucellus and this yields a female gametophyte bounded by a conspicuous membrane. Two archegonia arise at the micropylar end, and the upper part of the nucellus develops a pollen chamber.

Pollination is assisted by a 'pollination drop' at the micropyle, and the pollen chamber, having received the pollen, then becomes closed above. The cavity in the nucellus progressively deepens, carrying the pollen with it, until it reaches the female gametophyte, the centre of which is extended upwards to form a so-called 'tentpole' (a feature seen also in many fossil seeds). The germinating pollen forms a tube, but, as in *Cycas*, this has a haustorial function. The tube grows backwards into the nucellus, the swollen part of the expanded grain hanging in the chamber above the archegonia. As the archegonia mature the generative cells of the male

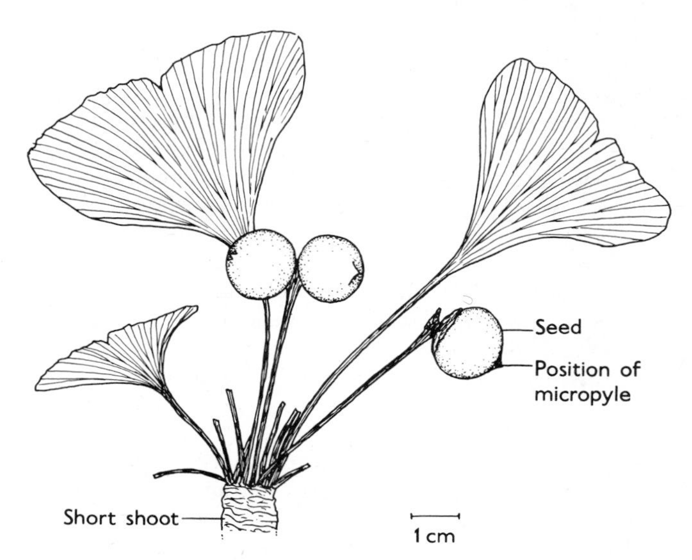

Fig. 7.38 *Ginkgo biloba*. Tip of short shoot bearing ripe seeds.

gametophyte divide. Each yields a stalk and a body cell, and the latter divides again to form two multiflagellate antherozoids whose major diameters are of the order of 100 μm. These antherozoids are released into the fluid above the archegonia and bring about fertilization.

The zygote, which may not be formed until after the ovule has been shed, begins its development by free nuclear division, leading to a proembryo containing about 256 nuclei. Walls then differentiate and a flask-shaped proembryo is formed, the lower part of which becomes the embryo proper. A clearly defined suspensor is thus absent. The mature embryo has two cotyledons. Usually only one of the paired ovules on the sporangiophore develops into a seed. In the mature seed (Fig. 7.38), which reaches a diameter of about 2 cm, the outer layer of the integument becomes fleshy and resinous, and the inner hard. The formation of the seed is completed in one season.

The fossil history and possible origin of the Ginkgoales

There is fossil evidence of the Ginkgoales having existed in the Mesozoic and at that time having been much more widely distributed than today. The leaves of the extinct Ginkgoales were usually much more like the juvenile than the mature leaves of *G. biloba*. The origin of the Ginkgoales is obscure, but it seems likely that they were derived as an offshoot of some cordaitalean stock. The bilateral symmetry of the seed and the dense secondary xylem are indications against affinities with the cycads and pteridosperms.

Gnetales The Gnetales consist of only three genera, *Ephedra*, *Gnetum* and *Welwitschia*. Not only are the Gnetales very different from other gymnosperms, but the genera also differ so markedly amongst themselves that many have considered each to be worthy of independent classificatory rank. They have been studied extensively by morphologists because of certain features which make them appear intermediate between gymnosperms and angiosperms. However, although the Gnetales indicate how certain characteristics of angiosperms may have arisen, they themselves appear to be specialized offshoots from the main evolutionary trends. Unfortunately the fossil record of the Gnetales is very fragmentary and does not extend further back than the Tertiary. For the most part it merely indicates changes in geographical distribution.

Ephedra

Ephedra is widely and somewhat discontinuously distributed. Some 35 species occur in the Mediterranean region, Asia and in the Americas. They are typical 'switch plants', consisting of densely branched axes, the younger of which are green and photosynthetic (Fig. 7.39). The leaves consist of whorls of small scales which soon become scarious. Many species grow in extremely arid situations, such as sand dunes and scree slopes, and these not unexpectedly have an extensive root system. The young twigs of some species have medicinal uses, and the genus is the source of the alkaloid ephedrine.

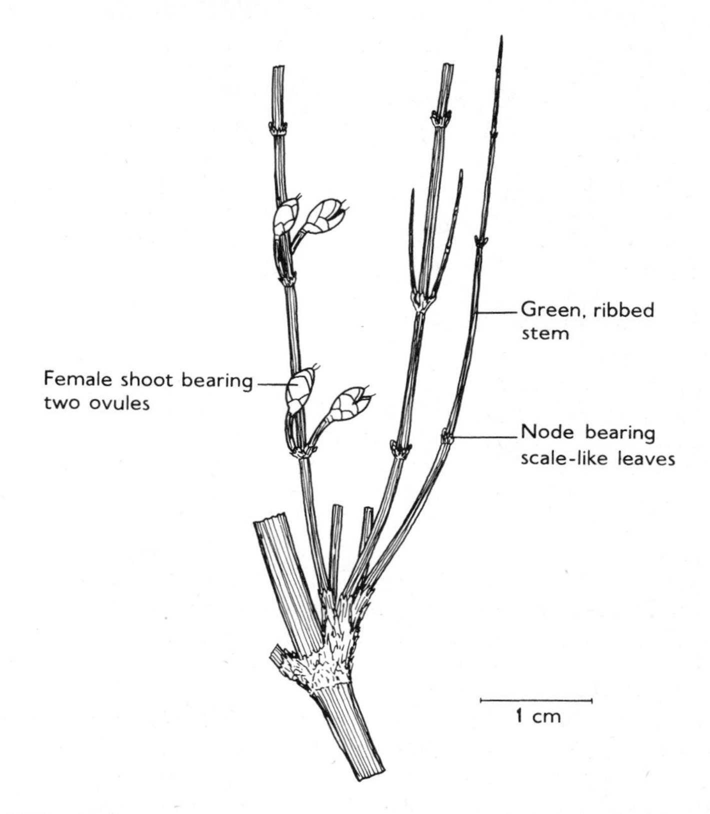

Fig. 7.39 *Ephedra* sp. Portion of shoot system showing the articulated stem, reduced leaves, and female reproductive structures.

THE VEGETATIVE FEATURES The stem of *Ephedra* grows from a group of meristematic cells, and a distinct tunica and corpus are recognizable in the apex. The primary vascular system consists of a number of bundles symmetrically placed around a central pith (Fig. 7.40), the bundles being linked at the nodes by a transverse vascular ring, as in *Equisetum*. The primary xylem becomes surrounded by secondary, traversed by broad parenchymatous rays. The tracheids have bordered pits on their radial walls, well developed tori also being present. Many of the tracheids are arranged in columns, and the end walls are so extensively perforated that they can be legitimately regarded as vessel segments with foraminate perforation plates. The phloem consists of sieve cells and parenchyma, the sieve cells, like those of the conifers, having highly inclined end walls.

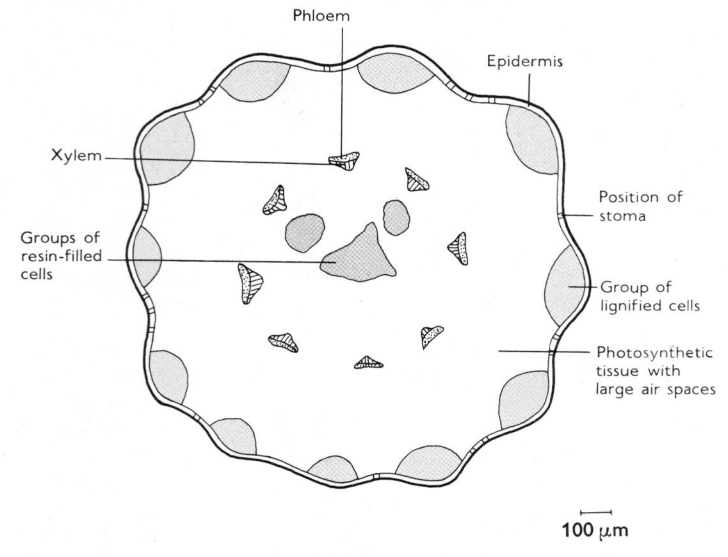

Fig. 7.40 *Ephedra* sp. Transverse section of young stem.

REPRODUCTION *Ephedra* is dioecious. The male reproductive regions are cone-like terminations of short shoots which arise in the axils of the scale leaves. The short shoot bears a number of bracts in decussate pairs, and in the axil of each bract is a male flower (Fig. 7.41). This consists of a microsporangiophore, bearing at its summit 2–8 microsporangia and enclosed in the basal region by 2 medianly placed bracteoles. The pollen grains begin to develop internally immediately after their formation, and when mature contain 4 or 5 nuclei. The sequence of divisions is as in *Pinus* and the first 2 daughter nuclei, which do not again divide, are regarded as prothallial.

The female reproductive organ is similar in structure to the male, but only the uppermost pair of bracts is fertile (Fig. 7.39). Each subtends an upright ovule which is surrounded by a sheath, probably homologous with the two bracteoles of the male flower. The radially symmetrical ovule (Fig. 7.42) is bounded by a papery integument, the apex of which is

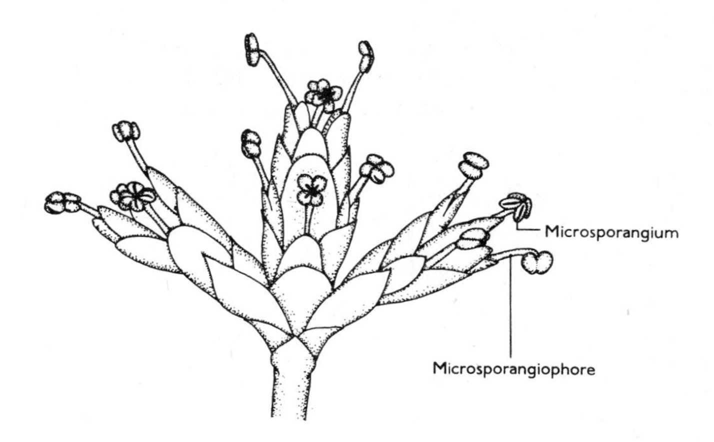

Fig. 7.41 *Ephedra altissima*. Microsporangiate shoot.

prolonged into a micropylar tube, highly cutinized at maturity. The female gametophyte, which is initiated in the nucellus in the usual way, passes through a period of free nuclear division before becoming cellular. Two, rarely more, archegonia are differentiated at the micropylar end, and they are unusual in being quite deeply sunk into the somatic tissue of the gametophyte. As the eggs mature, the upper part of the nucellus breaks down to form a pollen chamber, and complex cytological phenomena, among them the amoeboid migration of nuclei and endomitosis, occur in the upper cells of the gametophyte.

The pollen is distributed by wind and possibly also by insects, and a pollination-drop mechanism is probably present in most species. Germination of the pollen occurs directly on the surface of the female gametophyte and a pollen tube pushes its way into an archegonium. Two sperm nuclei, produced by division of the body cell, enter the egg cell. One fuses with the nucleus of the egg cell and the other, in some species at least, fuses with the ventral canal cell. This 'double fertilization' does not, however, appear to have any special significance, since only the zygote undergoes any further development. The interval between pollination and fertilization, in contrast to the prolonged period in the conifers, may amount to no more than 24 hours in *Ephedra*.

EMBRYOGENESIS The embryology of *Ephedra* is not well known, but it appears established that the zygote first undergoes free nuclear division,

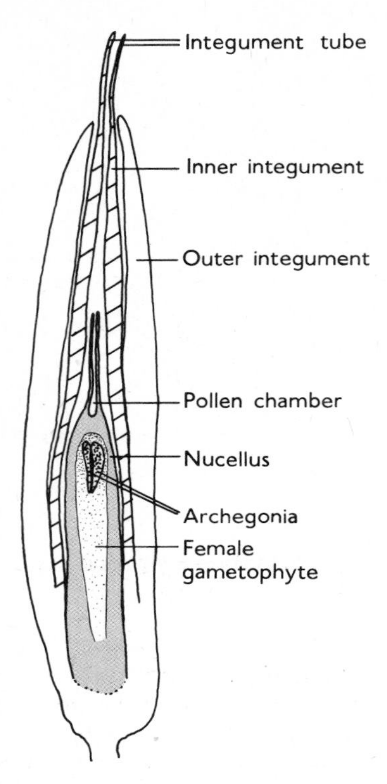

Fig. 7.42 *Ephedra* sp. Longitudinal section of ovule with archegonia. (From a photograph by W. R. Ivimey-Cook in McLean and Ivimey-Cook, *Textbook of Theoretical Botany*. Longmans, London, 1951)

the eight nuclei so formed being the initial cells of proembryos. There is thus potential polyembryony, but only one embryo reaches maturity. A suspensor, of complex, compound origin, drives the embryo into the central region of the female gametophyte, rich in food reserves. The mature embryo has two cotyledons and lies surrounded by the membranous remains of the ovular tissues and the hardened integument. In many species the bracts below the ovules become hard and wing-like in fruit, but in the alpine *E. helvetica* they become fleshy and brightly pigmented.

Gnetum[56]

Gnetum is a tropical genus, occurring in Asia, Africa and South America. Many species are lianes, but others are small trees. In contrast to *Ephedra* the leaves are well developed (Fig. 7.43), and possess broad, oval laminae

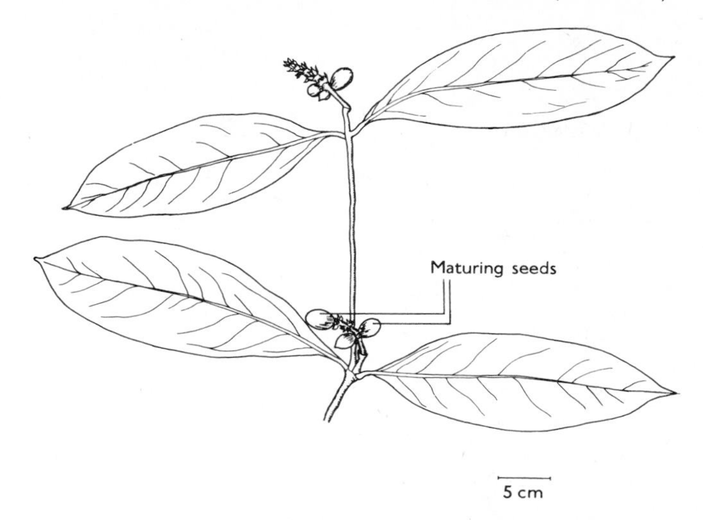

Fig. 7.43 *Gnetum gnemon*. Portion of shoot bearing female strobili. (After Madhulata, from Maheshwari and Vasil, *Gnetum. C.S.I.R., Delhi*, 1961)

with reticulate venation, some of the veins ending blindly in areolae. Vegetatively, therefore, *Gnetum* has very much the appearance of an angiosperm.

THE VEGETATIVE FEATURES The stem of *Gnetum* usually has a small pith, surrounded by a little primary xylem. Most of the xylem is secondary and is interspersed with broad parenchymatous rays. In the climbing forms the stem is eccentric, and successive cambia give rise to a polycyclic stelar structure, the asymmetry in any particular region depending upon its spatial orientation. In general features, therefore, the stem is closer to that of the cycads and pteridosperms than to that of the conifers. A striking difference, however, is the even closer approach in *Gnetum* to the differentiation of authentic vessels in the secondary xylem than in *Ephedra*. Another peculiarity is that in *Gnetum* parenchymatous cells are closely associated with the sieve cells, recalling the companion cells of angiosperms. The cortex adjacent to the phloem is rich in fibres, and the bast of some species is used as cord.

REPRODUCTION The retention of *Gnetum* in the gymnosperms is, however, justified by the nature of the reproduction. Both the male and female reproductive regions are again strobili, usually terminating lateral axes. In the male strobilus (Fig. 7.44a) the axis bears a succession of gallery-like sheaths, usually about eight in number, probably formed from coalesced

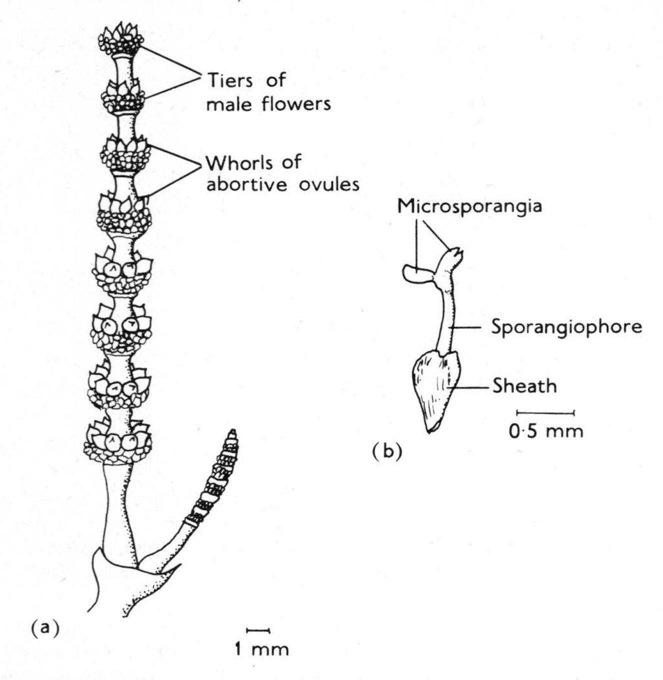

Fig. 7.44 *Gnetum gnemon*. (**a**) Young microsporangiate strobilus with whorls of abortive ovules. (**b**) Dehiscent microsporangia. (Both after Madhulata, from Maheshwari and Vasil, *Gnetum. C.S.I.R., Delhi*, 1961)

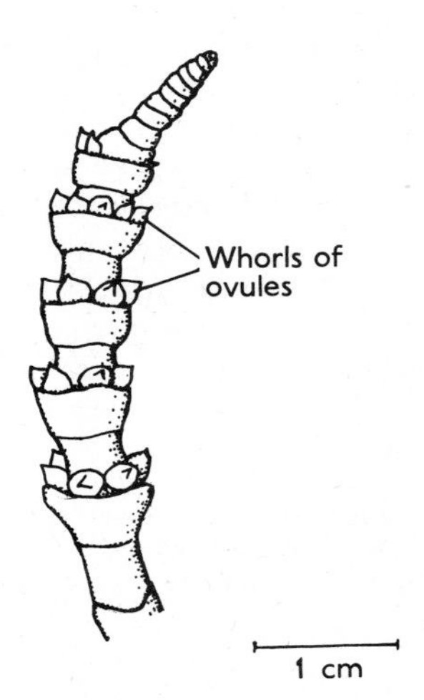

Fig. 7.45 *Gnetum ula*. Megasporangiate strobilus. (After Vasil, from Maheshwari and Vasil, *Gnetum. C.S.I.R., Delhi*, 1961)

bracts. In the axil of each sheath are whorls of male flowers, in some species surmounted by whorls of abortive ovules. The male flower (Fig. 7.44b) consists of a single microsporangiophore, terminating in two microsporangia, surrounded at its base by a delicate membranous sheath. The pollen grains are trinucleate when shed and, although there is some dispute about the identity of these nuclei, it seems clear that if prothallial nuclei are present at all there is never more than one in each grain.

In the female strobilus (Fig. 7.45) the sheaths along the axis each enclose a whorl of female flowers. Each flower consists of a single, radially symmetrical ovule (Fig. 7.46) surrounded by three integuments, the outer of

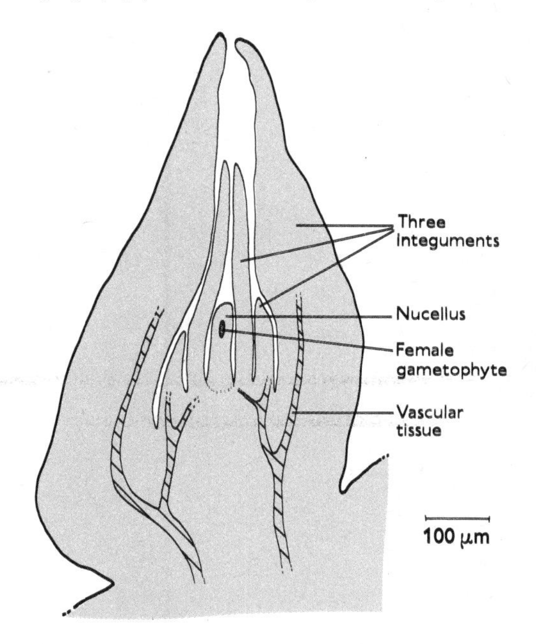

Fig. 7.46 *Gnetum ula*. Longitudinal section of very young ovule. (After Vasil, from Maheshwari and Vasil, *Gnetum. C.S.I.R., Delhi,* 1961)

which is possibly homologous with the basal sheath of the male flower. The inner integument is extended into a cutinized micropyle, and the nucellus beneath becomes transformed into a pollen chamber. One or more of the tetrad of megaspores formed in the nucellus enters into the formation of the acellular female gametophyte.

Pollination, again involving a pollination drop mechanism, initiates renewed growth of the ovule. As the pollen grains pass down into the pollen chamber, the cells lining the micropyle proliferate and occlude the tube. The enclosed grains germinate freely, and this stimulates rapid expansion of the ovule and of the female gametophyte, in which the nuclei divide freely in the growing cytoplasm. The gametophyte, or, as it is more usually called in *Gnetum*, embryo sac, is shaped like an inverted flask (Fig. 7.47). Much of the cytoplasm, in which the nuclei are irregularly scattered, lies at the base of the sac, but the remainder is distributed as a substantial layer around the periphery, a large vacuole occupying the centre. While the embryo sac is completing its development, the pollen tubes, having

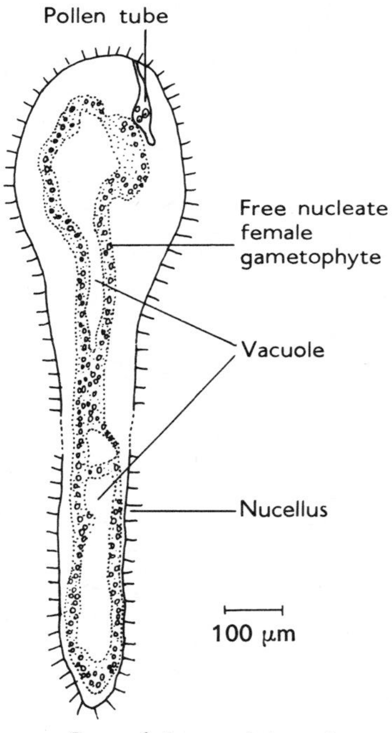

Fig. 7.47 *Gnetum ula.* Longitudinal section of mature female gametophyte (embryo sac) showing entry of pollen tube. (After Vasil, from Maheshwari and Vasil, *Gnetum. C.S.I.R., Delhi,* 1961)

penetrated the nucellus, approach the sac. One of the nuclei in the male gametophyte moves to the tip of the tube and divides into two sperm nuclei. Meanwhile one or more nuclei in the upper part of the embryo sac in the region adjacent to the closest pollen tube become conspicuously large. The pollen tube, having by now made contact with the sac, discharges the two sperm nuclei into it. They immediately migrate to the nearby large nuclei of the sac, which can thus be identified as egg nuclei.

EMBRYOGENESIS The entry of the male nuclei stimulates general division of the somatic nuclei within the embryo sac. The contents of the sac then become cellular, the cells often containing several nuclei, which subsequently fuse. The male nucleus and egg nucleus meanwhile coalesce and form a zygote lying within what can now be regarded as a cellular endosperm. Development of the zygote proceeds at once, but does not involve any free nuclear division. A complicated suspensor, possibly partly haustorial in function, is formed before the embryo proper. Although, since more than one egg nucleus may be present in the embryo sac and several male nuclei may be discharged into it, there is potential polyembryony, only one embryo usually comes to maturity. The embryo has two cotyledons.

The seed of *Gnetum* (Fig. 7.43) has a striking resemblance to that of some of the Bennettitales, but the inflorescence is of course quite different. When ripe, the outer integuments of the *Gnetum* seed become fleshy, and in some species are edible.

Welwitschia

Welwitschia (Fig. 7.48), in respect of habit, is one of the most peculiar plants in existence. The genus is monotypic, and the single species is confined to desert regions of south-west Africa. The stem is short and upright, and mostly below soil level. At the upper end it bears two strap-shaped leaves with indefinite basal growth. Developmentally these are the first pair of leaves after the cotyledons, and growth soon becomes confined to them. A further pair of decussate leaf primordia is formed in the young

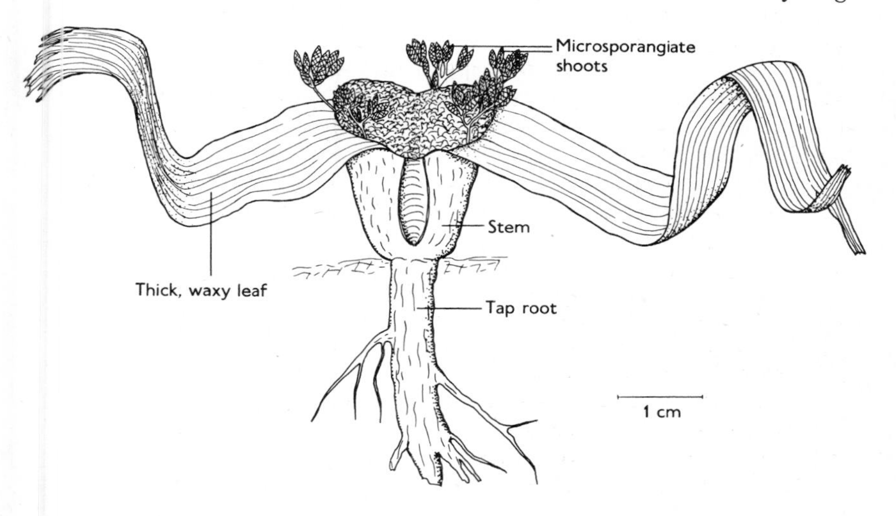

Fig. 7.48 *Welwitschia mirabilis*. Habit. (After Hooker, *Trans. Linn. Soc. Lond.* **24**, 1 (1863))

plant, but these differentiate into horn-like protuberances.[60] Below, the stem passes into a long tap root which gives rise to an extensive root system. The stem contains much secondary tissue, and it shows anatomical peculiarities similar to those of *Gnetum* and *Ephedra*.

REPRODUCTION *Welwitschia* is dioecious, the cone-like inflorescences terminating small branch systems arising in the axils of the leaves. The cones consist of a series of scales, arranged in decussate pairs, in the axils of which are the individual flowers. The male flowers consist of a short axis bearing, first, two decussate pairs of small scales and then a ring of trilocular synangia. These synangia are borne on microsporangiophores which are fused together at the base into a membranous cylinder. In the centre of the flower, and terminating its short axis, is an abortive female flower. The pollen grains are binucleate when shed, but neither nucleus can be regarded as belonging to a prothallial cell.

The female flower consists solely of an ovule surrounded by two integuments, with occasionally two minute lateral outgrowths on its stalk. The outer integument, which is broadly winged tangentially to the cone and is traversed by several vascular bundles, may be homologous with the upper pair of bracteoles of the male flower. The membranous inner integument is extended at its apex into a cutinized micropylar tube. The basic symmetry of the ovule seems to be radial rather than bilateral.

Pollination and the initiation of the embryo sac take place much as in *Gnetum*, but in the later stages, especially at the time of pollination, there are features peculiar to *Welwitschia*. After the initial free nuclear division in the sac, walls are laid down, many of the cells so formed being multinucleate. As the pollen tubes penetrate the nucellus, some of the multinucleate cells in the upper part of the embryo sac give rise to tubular processes. These grow up towards the descending pollen tubes and potential egg nuclei move to their tips. When a pollen tube and process make contact the separating walls dissolve and the sperm and egg nuclei fuse. The zygote then becomes ensheathed in cytoplasm and a cell membrane forms. There is no free nuclear division in the development of the zygote. A suspensor is formed and below it the embryo proper. Only one zygote yields a mature embryo. There are two cotyledons. In fruit the outer integument of the seed forms a broad wing which assists aerial dispersal.

The possible origin of the Gnetales

The origins of the Gnetales remain conjectural. Anatomically, especially in the features of the secondary xylem, they recall the pteridosperms rather than the Cordaitales. This view is strengthened by the symmetry of the seeds, which is primarily radial. Nevertheless, the Gnetales are clearly far from the pteridosperms of the Carboniferous and probably the consequence of quite prolonged evolutionary change.

THE MORPHOLOGICAL SIGNIFICANCE OF GYMNOSPERMY

The diversity of the gymnosperms taken as a whole, in both vegetative and reproductive features, indicates that gymnospermy must be regarded as a grade of evolutionary advance. No close relationship is necessarily to be expected between the Orders which show this kind of reproduction.

If we are correct in believing (pp. 244, 272) that the affinities of the Cycadales are pteridospermous and those of the Ginkgoales cordaitalean, we have two groups of plants whose evolution has probably been independent for many millions of years, but which are at the same level of advancement in respect of the reproductive process. In both, despite their arborescent form, fertilization is still brought about by flagellate antherozoids, liberated from very similar male gametophytes. The amount of fluid required in the fertilization process is of course very much reduced. In the conifers and *Ephedra*, despite the retention of archegonia, antherozoids are eliminated. The pollen tube, at first possibly always a haustorium, as in the cycads and *Ginkgo*, has now taken on the function of delivering the male gametes to the egg.

The highest grade of gymnospermy is clearly shown by the Gnetales. The peculiarities of the female gametophyte in *Gnetum* and *Welwitschia* indicate the kind of developments which, in some early transitional forms, may have led to the angiospermous embryo sac. It is significant that these striking features of the reproduction in the Gnetales should be accompanied by anatomical developments which also foreshadow the angiosperms.

The concept of 'progymnosperms'

The association of gymnosperm-like anatomy with heterosporous reproduction, increasingly encountered in Devonian plants, has given rise to the useful concept of 'progymnospermy'. Although the limits of the progymnosperms, which range from near-psilophytes at one extreme to the immediate antecedents of the earliest seed plants (such as *Archaeosperma*) at the other, are inevitably ill-defined, it is clear that this intermediate level of evolution was of great significance in the phylogeny of vascular plants. The evolutionary tendencies seen in the progymnosperms affected not only the reproductive organs. Some Devonian representatives, for example, showed anatomical features, such as fibres in the cortex and secondary phloem, which are also found in the early gymnosperms. The relationship between the progymnosperms and the gymnosperms was thus in some ways analogous to that now seen between the more advanced gymnosperms (such as the Gnetales) and the angiosperms.

8

The Tracheophyta, IV
(Pteropsida: Angiospermae)

The angiosperms are the most abundant and widely distributed tracheophytes, and of outstanding economic importance. They number some 200,000 species and show remarkable diversity in growth form, morphology and physiology.

Angiospermae

Sporophyte herbaceous or arborescent; branching usually axillary. Leaves various, but regarded as megaphyllous in origin. Secondary vascular tissue commonly present. Vascular system usually consisting of vessels and tracheids, and sieve tubes with distinctive companion cells. Heterospory as in the Gymnospermae, but the ovules borne within a characteristic structure (*carpel*)*, usually closed, the pollen germinating on a specialized region of the exterior* (*stigma*). *Female gametophyte always an embryo sac, lacking archegonia. Fertilization by unspecialized male cells, characteristically double, yielding in each embryo sac a zygote and mostly a triploid endosperm nucleus. Embryogeny endoscopic. Various forms of asexual reproduction not uncommon.*

The growth forms of angiosperms

Examination of an angiosperm flora will usually reveal that it consists of a number of distinct growth forms, ranging from large woody trees to minute herbs. The extent of the representation of these different morphologies in a given vegetation presents numerous ecological problems, the discussion of which is facilitated by Raunkiaer's classification of life forms.[68] This classification rests upon the length of life of the shoots and the position and protection of the resting buds (Fig. 8.1). Those plants in

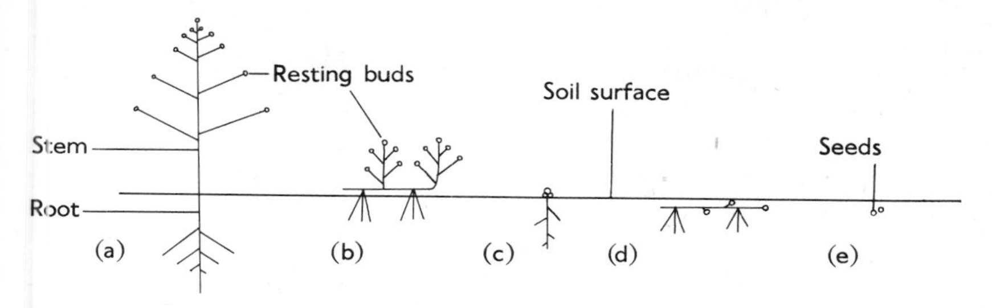

Fig. 8.1 Life forms of angiosperms. The circles indicate the positions of the resting buds. (**a**) Phanerophyte. (**b**) Chamaephyte. (**c**) Hemicryptophyte. (**d**) Cryptophyte. (**e**) Therophyte.

which the shoots are persistent and the buds are carried well above the soil surface are termed ***phanerophytes***, those with resting buds closer to the surface ***chamaephytes***, and those with resting buds at the surface ***hemicryptophytes***. Familiar examples of these three classes are, respectively, the larger woody plants, small bushes such as *Calluna*, and rosette plants such as *Taraxacum*. The classification, except for certain small specialized categories, is completed by the ***cryptophytes*** (geophytes) where the resting buds are below the soil surface, and the ***therophytes***. The latter are those annuals which tide over unfavourable periods as embryos enclosed in seeds. Using this classification it is possible to show in a precise statistical manner that, for example, the vegetation of the humid tropics consists predominantly of phanerophytes, and that in the northern hemisphere the percentage of hemicryptophytes in general increases with latitude (Table 8.1).

Table 8.1 The relationship between life form and latitude. (Data from Raunkiaer.[68] Succulents, water plants and specialized epiphytes are omitted)

Flora	Approx. latitude	Percentage representation Ph	Ch	H	Cr	Th
Amazonian rain forest	0°	95	1	3	1	0
St. Thomas and St. Jan (West Indies)	18°N	58	12	9	3	14
Denmark	57°N	7	3	50	22	18
Iceland	65°N	2	13	54	20	11
Spitzbergen	77°N	1	22	60	15	1

Ph: phanerophyte Ch: chamaephyte H: hemicryptophyte
Cr: cryptophyte Th: therophyte

For convenience of description we can regard the plant body of an angiosperm as consisting of three morphological categories, stem, leaf and root, although, as we shall see later, there are good reasons for regarding the leaf as a megaphyll and hence of axial origin. The main stem is usually upright, displaying negative geotropism and positive phototropism. Branching occurs in the axils of leaves. The final orientation of these laterals is probably determined by a combination of complex geotropic (termed plagiogeotropic) and phototropic responses. Although such branch systems are usually aerial, they may be subterranean (when, of course, light will no longer affect morphogenesis). The genus *Parinarium*, for example, is represented by normal trees in tropical Africa, but in *P. capense* (Fig. 8.2), a species of the colder South Africa, the stem, although

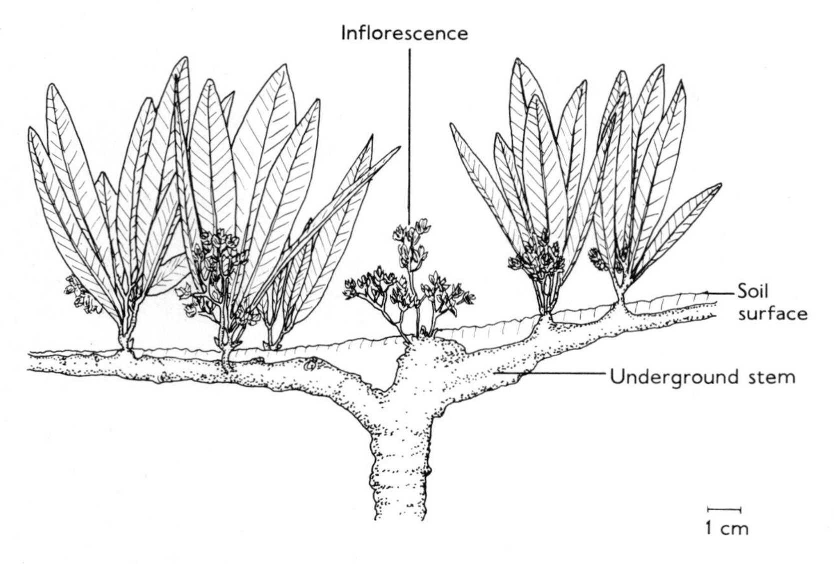

Fig. 8.2 *Parinarium capense*. The suffrutescent habit. (After Davy, *J. Ecol.* **10**, 211 (1922))

of similar woodiness and ramification, is below soil level, only small shoots appearing at the surface. This so-called ***suffrutescent*** habit is also encountered, although less strikingly, in alpine willows.[24]

In the development of the branch system the main axis may originate in two ways. Either the apical bud continues its vegetative growth indefinitely, or it is extinguished at the end of each season, when either it gives rise to a reproductive system of limited growth, or it aborts. Where the apical bud remains active (Fig. 8.3a), the main axis is of simple origin, and growth is said to be ***monopodial***. Where it is extinguished (Fig. 8.3b), growth is continued in the following season by the uppermost lateral and the positional readjustments are often such that seasonal discontinuities are hardly visible in the mature axis. The axis is, nevertheless, of compound origin, and growth is said to be ***sympodial***. These two kinds of growth are represented amongst plants of all life forms. Amongst phanerophytes, for example, the growth of the ash (*Fraxinus*) is monopodial and that of the lime (*Tilia*) is sympodial. The factors which cause abortion of terminal buds in plants of sympodial growth are complex, but in trees day length is often of paramount importance. The shortening days of late summer and autumn stimulate the synthesis of a growth inhibitor in the leaves. This is

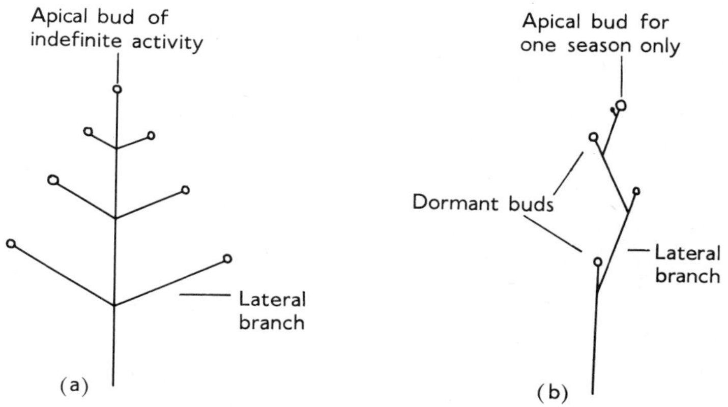

Apical bud of indefinite activity

Lateral branch

(a)

Apical bud for one season only

Dormant buds

Lateral branch

(b)

Fig. 8.3 The basic types of branching. (**a**) Monopodial. (**b**) Sympodial.

in turn transmitted to the terminal buds, causing either dormancy or in some species abortion. In long days the inhibitor is lacking so that in experimental conditions plants of this kind can be made to assume a quite different manner of growth.

In some angiosperms, as in some Filicales, the main axis is in the form of a horizontal rhizome. In *Aegopodium podagraria* (goutweed), and probably in other species, the rhizome is able, by adjusting the orientation of its growing region, to maintain itself at an almost constant depth beneath the surface of the soil.[9] The physiological mechanism underlying this remarkable behaviour is still not wholly known. Other highly modified stems are seen in corms, bulbs and some tubers, in which the stem, which is adapted for the storage of food, either in itself (corms, stem tubers) or in the associated swollen leaf bases or scale leaves (bulbs), shows very little

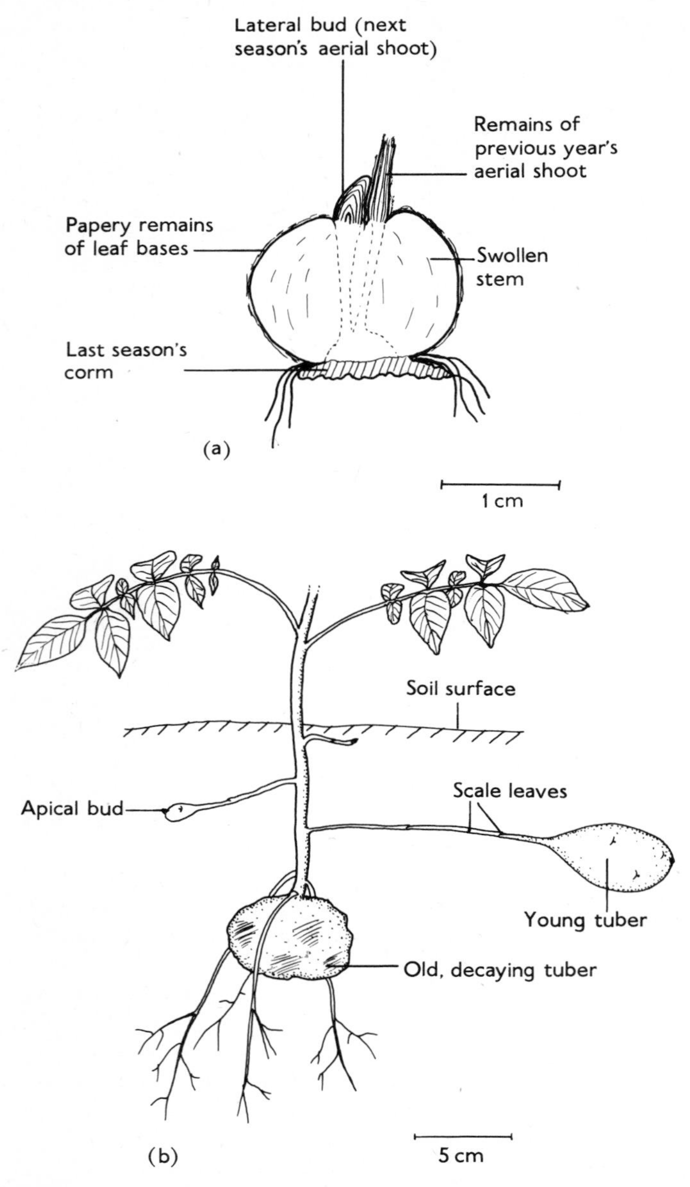

Fig. 8.4 Stems modified for storage. (**a**) Longitudinal section of the corm of *Crocus* (winter condition). (**b**) Stem tuber formation in *Solanum tuberosum*.

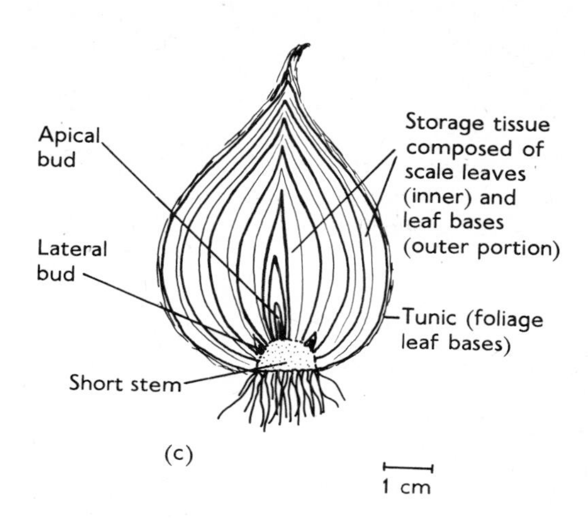

Fig. 8.4 Continued, (**c**) Longitudinal section of the bulb of *Allium cepa*.

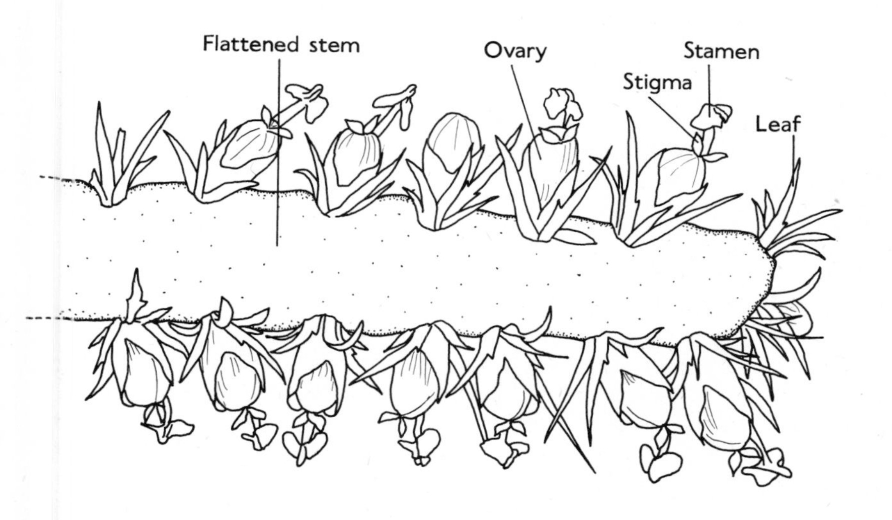

Fig. 8.5 *Butumia marginalis* of the Podostemaceae. Flowers are produced at the margin of the flattened, thalloid stem. (After Taylor, *Bull. Br. Mus. nat. Hist. Bot.* **1**, 53 (1953))

elongation (Fig. 8.4). A very peculiar stem, resembling a thallose liverwort, is found in the Podostemaceae (Fig. 8.5), a family of small plants growing on rocks by tropical streams.

The angiospermous stem

GROWTH AND ANATOMY The stem of the angiosperm grows from a group of meristematic cells lying near the summit of the apex. The apex itself (Fig. 8.6) is usually organized into two distinct zones, recognizable by their geometry and the directions of division of the cells.[21] In the

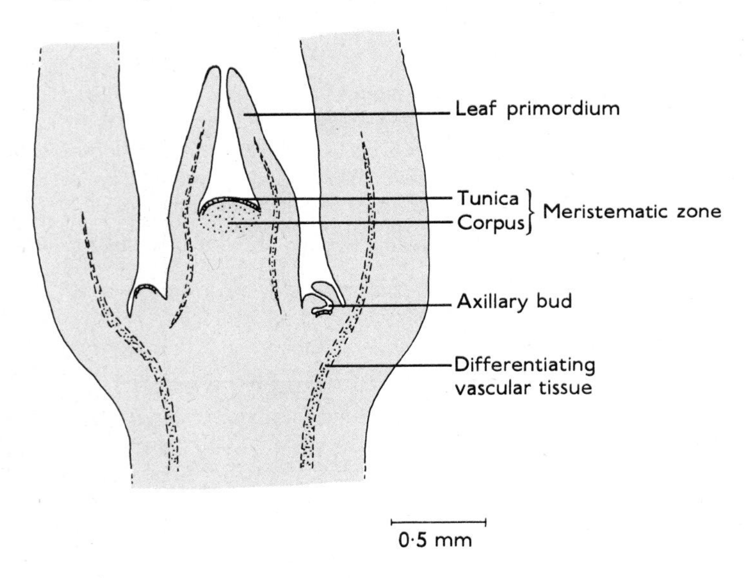

Fig. 8.6 *Syringa*. Longitudinal section of the stem apex.

centre is the corpus, where divisions are in various directions, thus adding to the width of the apex as well as to its length. The corpus is covered by the tunica, a layer of cells in which divisions are principally anticlinal, thus increasing the surface of the apex. The tunica is usually two cells thick, but in some plants the thickness may amount to only a single cell and in others to four or five. That these two zones are to some extent independent is shown by the existence of chimaeras in which the cells of the tunica acquire (as a consequence, for example, of treatment with colchicine) aneuploid or polypoid nuclei and retain them through all subsequent divisions, while the nuclei of the cells of the corpus remain diploid. The tunica yields the leaf primordia and ultimately the cortex of the mature stem, and the corpus the vascular tissue and associated parenchyma. Although this kind of apical organization is foreshadowed in the conifers (particularly in *Araucaria*, see p. 253), it is nowhere so distinct as in the angiosperms.

The angiosperms fall into two major groups, referred to as the monocotyledons and dicotyledons, in which the embryos are commonly furnished with one and two cotyledons respectively. These groups also differ in many other features, among them the form of the primary vascular tissue. In the dicotyledons, which are the more numerous, the primary vascular tissue usually consists of a ring of collateral bundles in which the differentiation of the xylem is centrifugal and of the phloem centripetal (Fig. 8.7). The

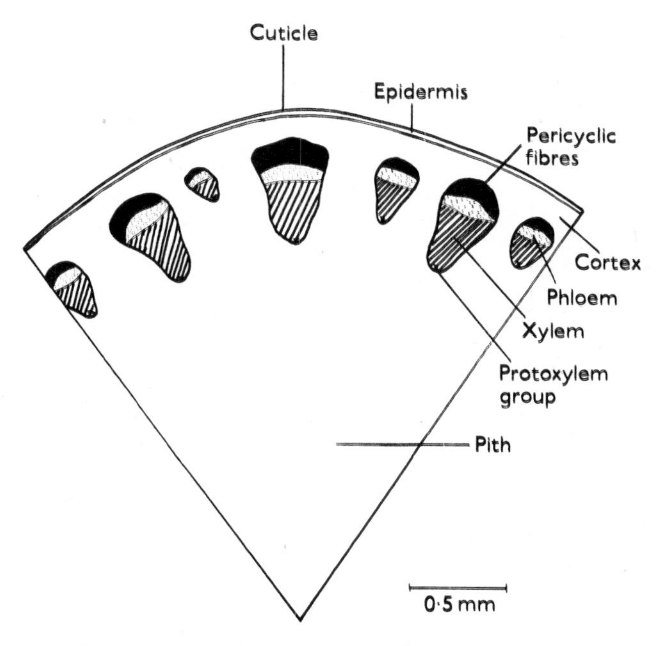

Fig. 8.7 *Dahlia*. Transverse section of a segment of a young stem.

xylem and the phloem usually remain separated by a thin layer of undifferentiated cells, later recognizable as the ***intrafascicular*** cambium. In the monocotyledons, however, the bundles, although often collateral and mostly orientated with the xylem adaxial, are not in one ring but are irregularly scattered in a parenchymatous matrix, referred to as the ground parenchyma (Fig. 8.8). The bundles of the monocotyledons also lack potentially meristematic tissue between the xylem and the phloem, and are consequently said to be closed.

Usually secondary vascular tissue soon appears in a dicotyledonous stem. The undifferentiated layer in the primary bundle becomes meristematic and continuous with a similar layer between the bundles (the ***inter-***

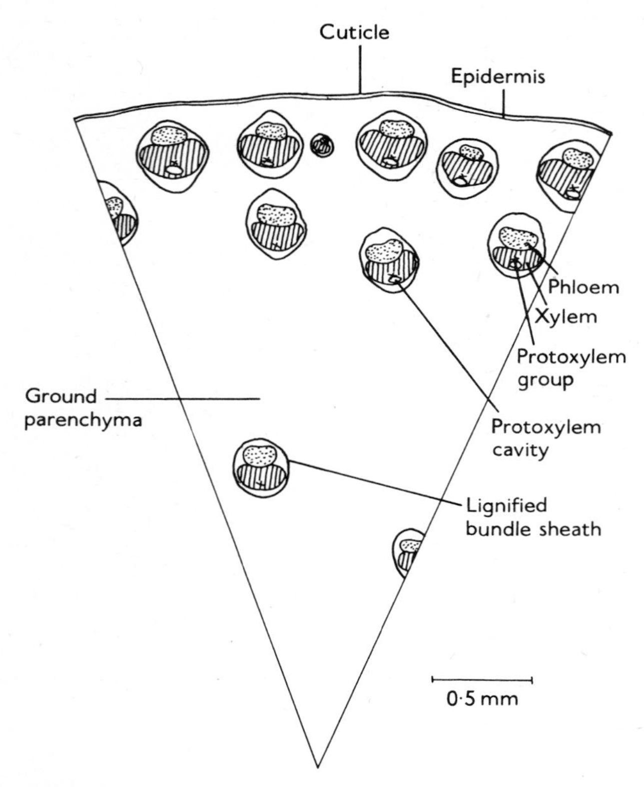

Fig. 8.8 *Zea mays*. Transverse section of a segment of stem.

fascicular cambium), and subsequent activity of the cambial ring resembles that already seen in the gymnosperms (Fig. 8.9). The first cambium does not always continue to function indefinitely; other cambia may arise outside the vascular cylinder, recalling the situation in *Cycas* (p. 237). In a few arborescent monocotyledons a meristematic zone at the periphery of the stem gives rise to additional collateral vascular bundles as the girth of the stem increases. This activity is, however, limited and in most arborescent monocotyledons, such as the palms, the plant remains for several years as a widening bud bearing a rosette of leaves only a little above soil level. When the bud approaches its mature diameter elongation of the stem takes place very rapidly. For this reason, coconut palms, for example, are rarely seen with their trunks half extended. The manner of growth of the

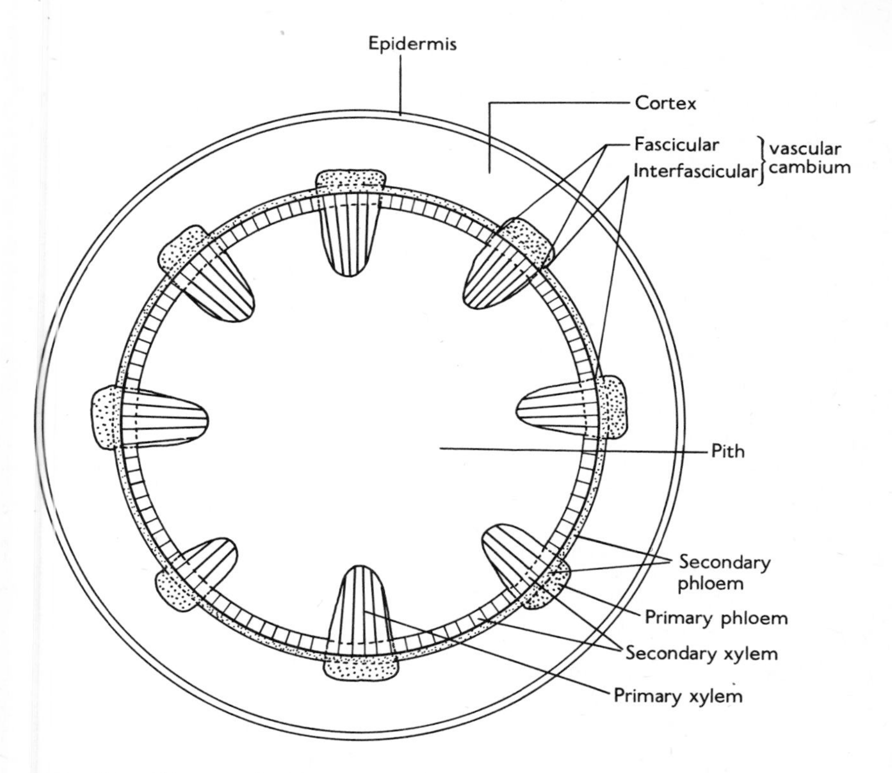

Fig. 8.9 Diagram showing the method of production of secondary vascular tissue in a dicotyledonous stem.

monocotyledonous trees is thus quite different from that of dicotyledonous forms.

The vascular tissue of angiosperms is distinguished from that of almost all gymnosperms by the presence of vessels in the xylem and of companion cells in the phloem. Vessels are long tubes, commonly about 10 cm long, but in some plants, such as lianes, reaching or exceeding a length of 5 m. These tubes are composed of segments (Fig. 8.10), each of which is derived from a single cell and is equivalent to a single tracheid. During differentiation of a vessel, however, the end walls of the segments disintegrate, leaving merely rims marking their position, or highly perforated diaphragms, referred to as end plates. Vessels are often accompanied by tracheids, and there are a few angiosperms in which the conducting elements of the xylem remain wholly tracheidal.

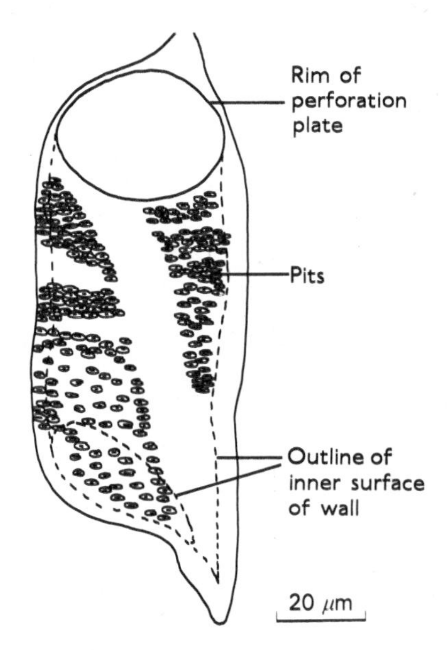

Fig. 8.10 *Fraxinus*. A single vessel element from a macerate of the secondary wood.

The sieve cells of the angiosperms are arranged in longitudinal rows, the whole column being referred to as a sieve tube (Fig. 8.11). The end walls of the sieve cells (or 'sieve tube elements') are usually inclined and bear one or more sieve plates. The contents of the sieve cells remain organized, although the cytology is peculiar and the organelles, including the nucleus, are partly degenerate. Deposits of an amorphous polysaccharide, callose, often appear on the sieve plates at the end of the growing season. The companion cell or cells, closely applied to the sieve cell, may be essential to its physiology. The companion cell has a dense cytoplasm and a prominent nucleus; it arises from the same mother cell as the adjacent sieve cell.

ECONOMIC PRODUCTS The stems of angiosperms frequently contain abundant fibres, and the formation of periderm in the outer cortex or even closer to the vascular tissue is often extensive. Some trees generate cork very freely, that of *Quercus suber* (cork oak) in the Mediterranean region being periodically harvested and finding many uses in commerce. A large part of the economic value of angiosperms in fact lies in the stems. Quite apart from the often very valuable timber, stems yield materials as diverse as starch (e.g. sago from the palm *Metroxylon* and arrowroot from the rhizome of South American *Maranta*), sugar (from sugar cane and in the form of syrup from the stem of *Acer saccharinum*, the sugar maple), fibres,

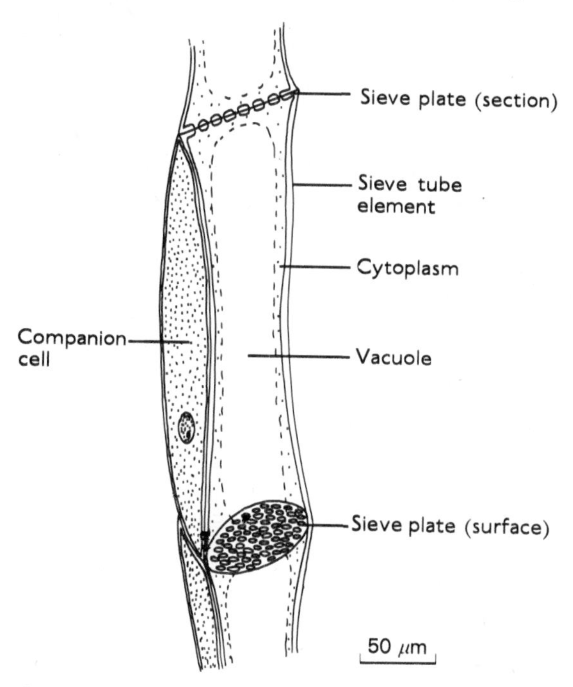

Fig. 8.11 *Cucumis*. Longitudinal section of a sieve tube and companion cell.

spices (e.g. cinnamon, the bark of *Cinnamomum zeylanicum*), rubber (formed from the latex of *Hevea*, and less importantly from that of certain other species) and drugs (e.g. quinine from the bark of *Cinchona*).

Leaves

THE FORMS OF LEAVES The leaves of angiosperms show a wide range of size and shape. A petiole and lamina are usually distinguishable, and two small lateral outgrowths at the base of the petiole, termed stipules are often present. In many species the leaves are pinnately branched, sometimes even as far as a third order, causing the whole to resemble a lateral branch system.[6] This resemblance is occasionally enhanced (e.g. in *Thalictrum aquilegifolium* (Fig. 8.12)) by the presence of small outgrowths, recalling stipules, at the points of branching. At the other extreme leaves may be little more than scales, photosynthesis being carried on principally if not

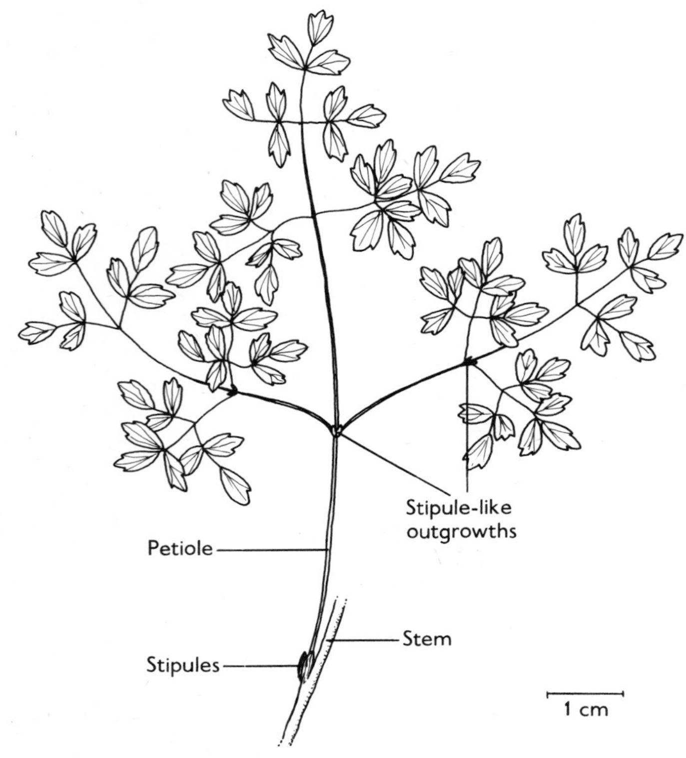

Fig. 8.12 *Thalictrum aquilegifolium.* Compound leaf structure.

entirely in the stem. In *Casuarina*, a tree of the tropics and sub-tropics of the Old World, with leaves of this kind, the leaves are arranged in whorls so that the whole shoot comes to bear a striking external likeness to that of *Equisetum*. In a few plants (e.g. *Ruscus aculeatus*, butcher's broom) the reduction of the leaves to scarious scales has been accompanied by the transformation of lateral shoots of limited growth into flattened, leaf-like structures called ***phylloclades*** or ***cladodes*** (Fig. 8.13).

An intermediate condition is where photosynthesis is carried on in what morphologically are broadened and flattened petioles, the laminae being absent or rudimentary. These structures, called ***phyllodes***, are found in many species of *Acacia*.

A classification has been developed for leaves, depending upon the area of the mature lamina, and clear relationships have emerged between

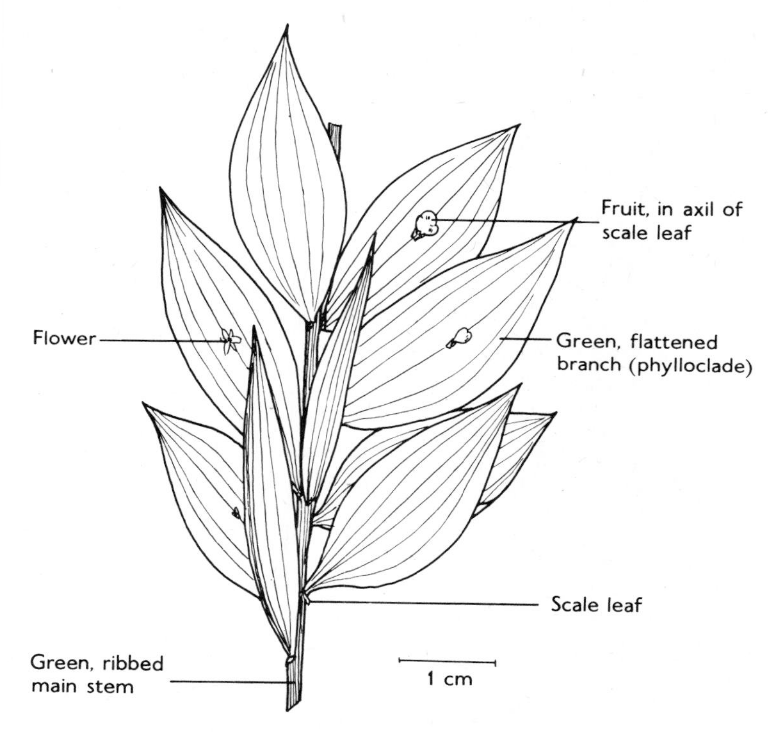

Fig. 8.13 *Ruscus aculeatus.* Habit of shoot.

climate and leaf size. In the tropical rain forest, for example, the leaves of many species tend to be of similar area, but in the dry, scrubby vegetation of the Mediterranean region the leaves by contrast show a much wider range of considerably smaller areas. Amongst other relationships, mostly of unknown significance, is the rarity of leaves with toothed margins in tropical vegetation, and their abundance in temperate. Rather leathery leaves, with their apices extended into 'drip tips', are also characteristic of vegetation in regions of high rainfall.[70]

Although the leaves of most species are differentiated into a lamina and a petiole, in some, the grasses providing familiar examples, a petiole is absent and the leaves are consequently termed sessile. Most leaves are orientated with the plane of the lamina more or less horizontal, and the insertion into the stem transverse. There are, however, some plants in which the plane of the leaf is vertical and the base of the leaf clasps the stem as a rider a horse, an insertion consequently termed equitant. An abscission mechanism is often present at the base of the petiole, and in some species motor cells, usually aggregated into a distinct ***pulvinus***, are able to alter the orientation of the leaf in a striking fashion. In the tropical rain tree (*Pithecolobium saman*), for example, the leaves appear to collapse with the diminishing light of the afternoon (a so-called 'sleep movement'), and in *Mimosa pudica* (sensitive plant) similar movements take place if

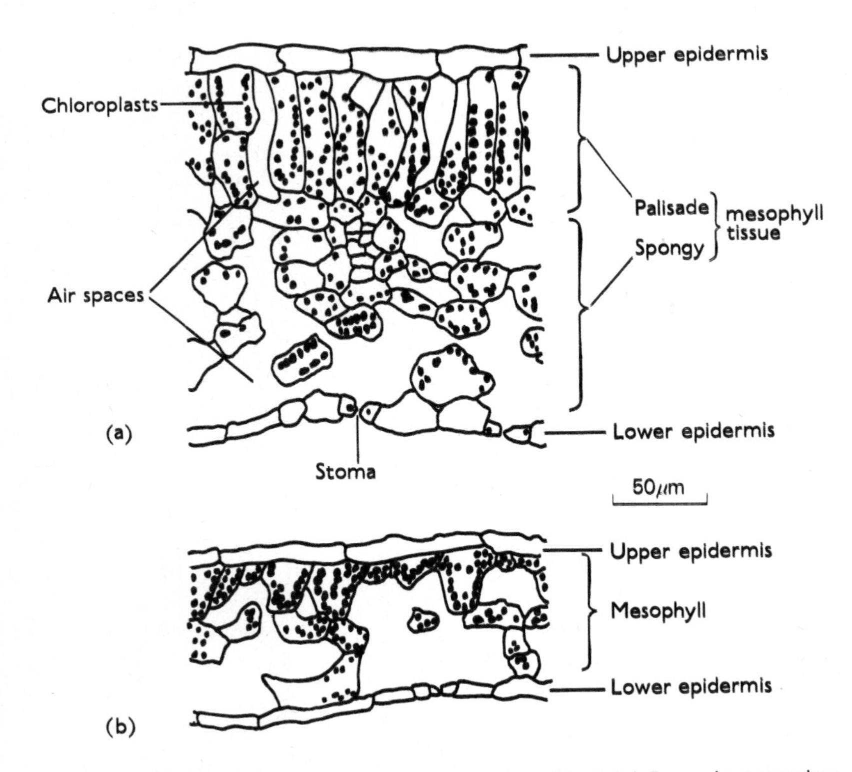

Fig. 8.14 *Impatiens parviflora*. Transverse section of leaf. (**a**) From plant growing in full sunlight. (**b**) From plant growing in 7% sunlight. (Both after Hughes, *J. Linn. Soc. Bot.* **56**, 161 (1959))

the leaves are mechanically disturbed. The leaves of the 'compass plants', of which *Lactuca scariola* (a wild lettuce) is a notable example, move in relation to the sun so that only one edge is fully insolated at any one time. The motive effects of pulvini depend upon rapid changes in turgor and the ability of the cytoplasm in certain conditions to move water actively in and out of vacuoles.

THE ANATOMY AND DEVELOPMENT OF ANGIOSPERM LEAVES Although the leaves of angiosperms are structurally similar to those of gymnosperms, there are a number of new features. The palisade tissue, for example, is frequently sharply differentiated from the rest of the mesophyll and in some plants, especially in those that are shade tolerant (e.g. *Impatiens parviflora*, balsam), the form of the palisade is markedly influenced by light intensity (Fig. 8.14). Equitant leaves are usual bifacial, stomata and palisade being symmetrically placed on each side. Amongst other structures found in leaves are oil glands, often giving the leaf a fragrance when crushed, and, usually at the margins or on the petiole, nectaries (as, for example, in cherry). The venation of the lamina is commonly reticulate, patterns of extreme intricacy often being generated by the minor veins, many of which end blindly in the areolae (Fig. 8.15). The symmetry of the vascular supply ascending the petiole is usually clearly dorsiventral, but it becomes almost perfectly radial in the petioles of peltate leaves (e.g. *Tropaeolum majus*, the garden nasturtium).

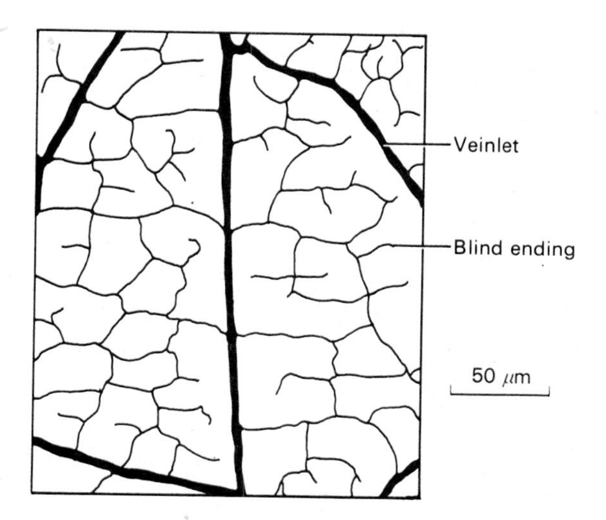

Fig. 8.15 *Impatiens parviflora*. The pattern of leaf venation. (After Hughes, *J. Linn. Soc. Bot.* **56**, 161 (1959))

The leaf primordium grows from initial cells at its margin, and at first cell division predominates over expansion. The pattern of the main veins soon, however, becomes established, and also in dicotyledons of any branching of the leaf. This is brought about by the marginal meristem becoming discontinuous and its activity confined to definite areas of the periphery. Segmentation in the leaves of monocotyledons (e.g. palms) is a more complicated process and involves the degeneration of tracts of tissue between areas which will subsequently become pinnae.

As growth of a leaf primordium proceeds, cell expansion comes to predominate over division, one of the last products of cell division being the guard cells of the stomata. It is evident that the surface growth of the lamina is closely co-ordinated with the extent of the vascular framework, and it is possible to disturb this co-ordination by experimental means and so to produce deformed leaves. These experiments have the additional interest of revealing that the metabolic factors influencing the growth of the vascular skeleton are different from those influencing the expansion of the lamina. Standing apart from this general scheme of development are again the leaves of some monocotyledons. The leaf of a grass, for example, grows from a meristem at the base of the lamina, above which there is continuous basipetal differentiation until the leaf reaches its mature length.

JUVENILE AND MATURE FORMS OF LEAVES As in the conifers, the leaves

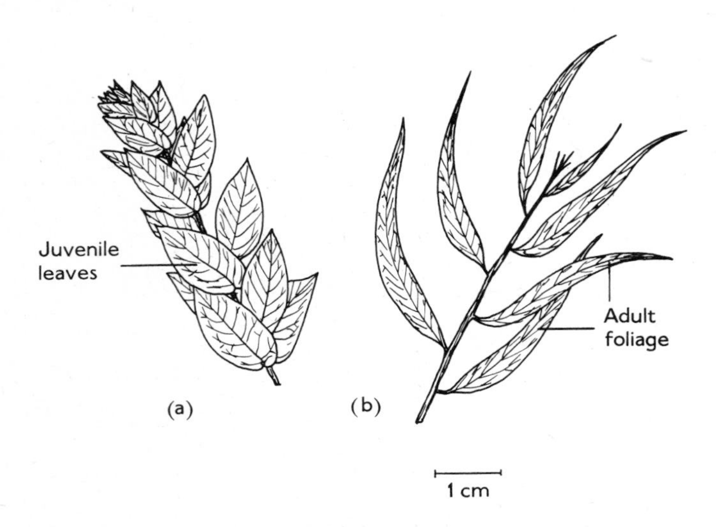

Fig. 8.16 *Eucalyptus globulus.* (**a**) Shoot with juvenile leaves. (**b**) Shoot with adult leaves. (Both after Niedenzu in Engler and Prantl, *Die natürlichen Pflanzenfamilien*, III, 7. Engelmann, Leipzig, 1898)

of young plants sometimes differ from those of the mature in both form and arrangement. *Eucalyptus globulus* (Fig. 8.16) provides a striking example of this phenomenon. The juvenile leaves are ovate, decussately inserted, and bifacial in structure, whereas the mature are falcate, spirally inserted and possess normal dorsiventral structure. A young tree has a cone of juvenile foliage within a crown of mature foliage, showing that the change takes place more or less simultaneously at all the apices when a tree reaches a certain maturity. Sometimes, as in *Hedera helix* (ivy), mature foliage does not appear until the approach of reproduction. Once the production of mature foliage has begun, the system is remarkably stable. Reproductive branches of *Hedera*, for example, can be struck as cuttings, and these yield small bushes quite different in appearance and growth form from the scandent vegetative plant. It is very difficult, even by experimental means, to cause the apices of mature plants to revert to juvenile growth. It appears likely that maturity is a consequence of complex and co-ordinated changes in the ribonucleic acids and proteins in the meristematic cells, and that the original system is only readily re-created during sexual reproduction.

THE ECONOMIC AND ECOLOGICAL IMPORTANCE OF ANGIOSPERM LEAVES

Angiospermous leaves are a rich source of foodstuffs and raw materials. Fodder crops often consist largely of leaves, and the leaves of many species are prized as pot-herbs because of the aromatic oils they contain. The leaves of a wide range of species find medicinal uses, and they are the commercial source of a number of important drugs, among them atropine and hyoscyamine (from *Atropa* and other members of the Solanaceae) and cocaine (from *Erythroxylon*). The leaves of many species yield valuable fibres, and in tropical regions palm leaves provide a ready and efficient material for thatching. Apart from such direct utilization, angiospermous leaves play a large part in maintaining the fertility of the soil. Since they contain a higher proportion of nitrogenous substances and fewer antiseptic materials, such as tannins and phlobaphene, the leaves of angiosperms decay more rapidly than those of gymnosperms. Angiospermous litter is thus quickly reduced to humus, the organic matter that is the basis of soil fertility. The productivity of much natural vegetation in tropical regions depends upon the steady deposition of litter. Thoughtless removal of this natural cover and destruction of the ecosystem can result in areas of persistent and intractable barrenness.

Roots

The roots of the angiosperms are in general organized similarly to those of the gymnosperms. Those of some species are able to develop chlorophyll when illuminated. In *Taeniophyllum* (Fig. 8.17), a peculiar Malaysian epiphytic orchid, photosynthesis is in fact confined to band-like aerial roots, the stem remaining little more than a bud until the production of the inflorescence. A number of angiosperms, especially those of temperate regions that are biennial in habit, develop a swollen tap root which overwinters. These are often used as vegetables, carrots and parsnips being familiar examples. Some swamp plants, notably the mangroves

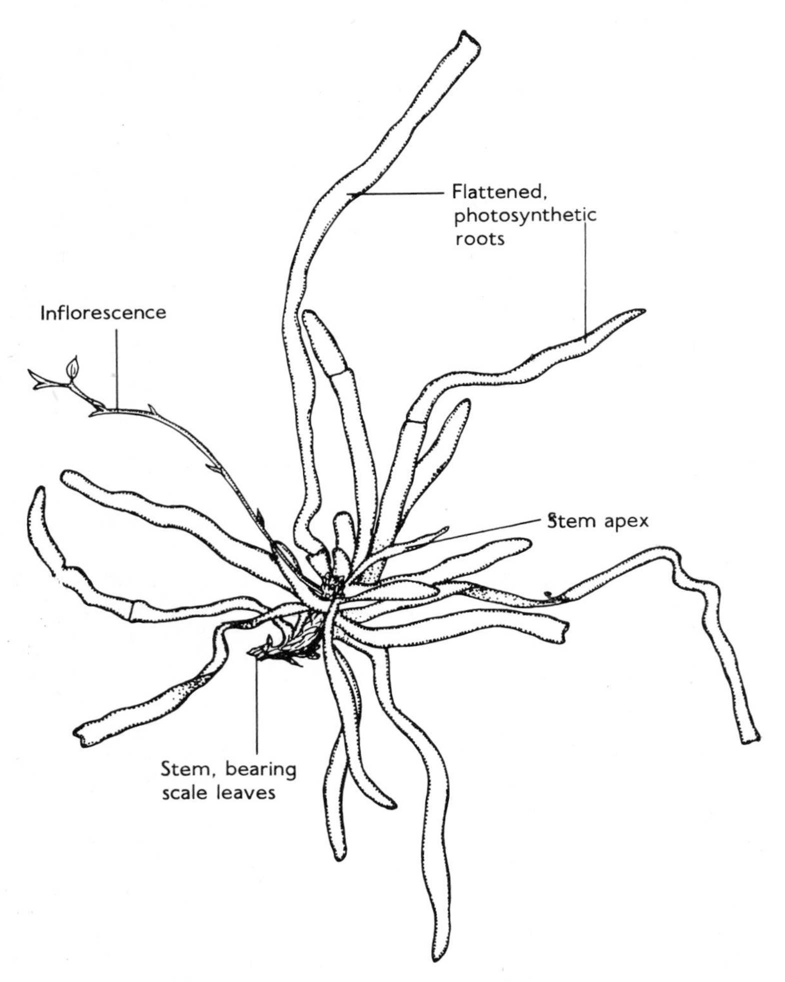

Fig. 8.17 *Taeniophyllum* sp. Habit showing flattened, photosynthetic roots. (After Goebel, *Organographie der Pflanzen*. Fischer, Jena, 1933)

(which colonize brackish estuaries in the Tropics), produce negatively geotropic aerophores from the root system, a development foreshadowed in the conifer, *Taxodium* (see p. 252). A few tropical trees, notably figs, show 'prop roots'. These leave the trunk at about a metre above ground level and descend obliquely towards the substratum. They soon become secondarily thickened and the tree comes to be supported by a cone of such outgrowths around its base. In a number of dicotyledonous trees markedly asymmetrical thickening of the principal roots around the foot of the trunk leads to the production of remarkable buttresses.

Associations between roots and micro-organisms are not infrequent. In forest trees the upper rootlets, usually found proliferating in litter, often possess mycorrhizal fungi. The roots of orchids are also mycorrhizal, but it has been discovered that these plants, which are all herbaceous, can be grown in pure culture in the absence of the fungus provided suitable nitrogenous substances are supplied in the medium. In the Leguminosae (the family containing the peas, beans and several important fodder plants) there is a regular association, involving anatomical modifications, between the roots and certain soil bacteria capable of fixing atmospheric nitrogen. Similar associations with other nitrogen-fixing micro-organisms have been confirmed in *Alnus* (alder), *Hippophaë* (sea buckthorn) and *Myrica* (myrtle), and probably exist in many other plants.

The manner of growth of the angiosperm root has been studied in some detail, since it is not complicated by the presence of leaf primordia. Apart from the root cap, which is maintained by a meristem adjacent to the surface of the apex proper, meristematic activity lies principally in the summit of the apex (Fig. 8.18). There is, however, convincing evidence, supported by experiments with radioactive isotopes, that a group of cells at the centre of the apex experiences few if any divisions. These cells, forming a so-called 'quiescent centre', can be stimulated into division by wounding and radiation damage,[21] so they are not deficient in essentials. Their normal inactivity must therefore depend upon the physiological organization of the intact apex, but the precise factors involved are still unknown. Behind the meristematic area the cells elongate, and measurements have shown that there is considerable synthesis of ribonucleic acid and protein in the cells concerned at this time. In some cells, often those concerned with the production of root hairs, the nuclei undergo a curious form of mitosis without associated division (endomitosis), leading to a polyploid condition.

A transverse section of a young root (Fig. 8.19) shows a central core of xylem with radiating arms of protoxylem, between which lies the phloem. In dicotyledons the number of protoxylem groups is usually small, but in the monocotyledons the number is much greater, usually in the region of 15 or 20. An endodermis, separated from the vascular tissue by a zone of

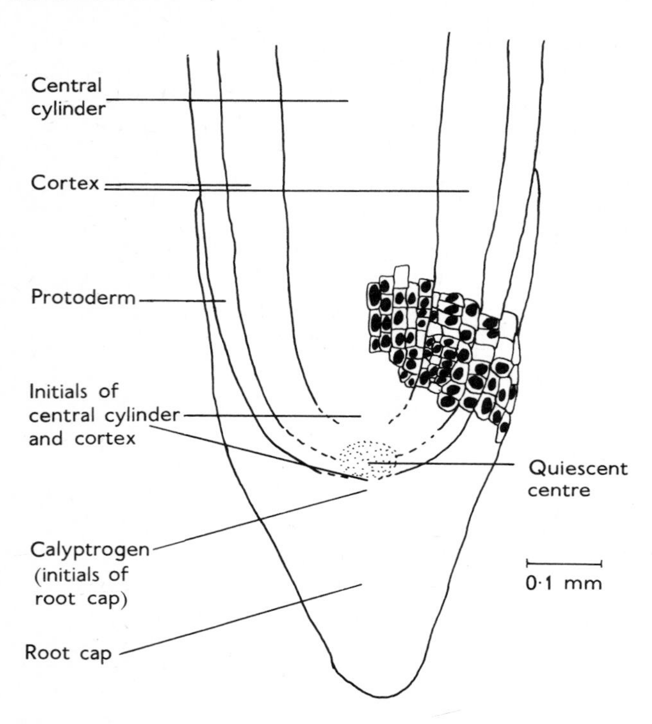

Fig. 8.18 *Trillium* sp. Longitudinal section of root apex.

parenchyma (pericycle), is usually present. The root is the only region in the plant body of an angiosperm where this layer of peculiar cells occurs with regularity. The meristems of branch roots arise in the pericycle, an origin termed endogenous, and very different from the more superficial (exogenous) origin of branches of the stem. Where cork is produced, the cork cambium arises in the pericycle, thus causing the whole of the primary tissue external to it ultimately to die and slough off. In herbaceous plants the parenchymatous tissues of the roots sometimes become locally distended with food materials, occasionally forming distinct tubers (e.g. *Dahlia*, Fig. 8.20). These serve as organs of perennation.

Roots, especially those that are tuberous, are frequent sources of drugs and folk medicines. Aconitine, obtained from the roots of *Aconitum napellus*, frequent in alpine meadows and seen occasionally by shaded streams in Britain, is a well-known example. Derris, a powder made by grinding the roots of *Derris elliptica*, a climbing shrub of Asia, is a powerful insecticide harmless to man.

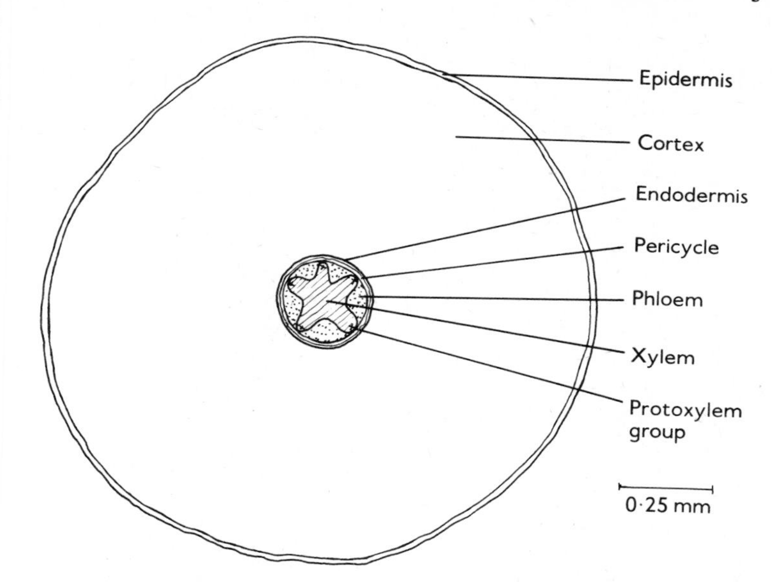

Fig. 8.19 *Ranunculus* sp. Transverse section of young root.

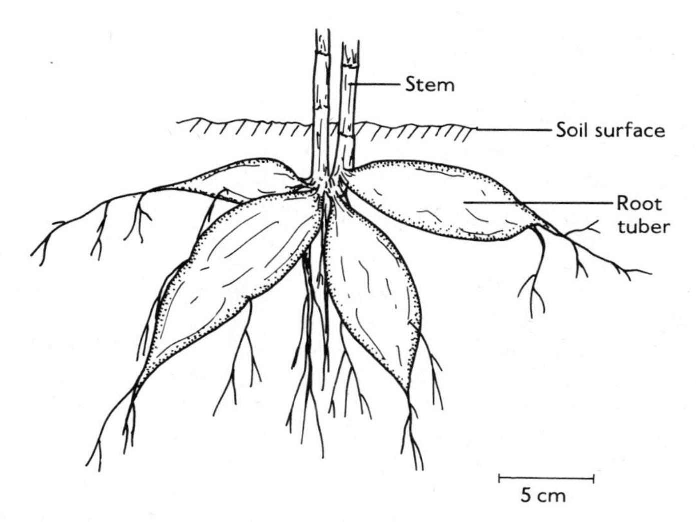

Fig. 8.20 *Dahlia* sp. The production of root tubers.

Reproduction

In the angiosperms the onset of the reproductive phase is frequently (but not always) dependent on the length of day ('photoperiodism'), both short-day and long-day plants being clearly recognizable. The reproductive axes are ordinarily short and of limited growth. The reproductive axis and its attendant structures, which are often conspicuously coloured or shaped, is termed a flower. Flowers may be borne either singly (as in *Anemone nemorosa*) or severally, the branching system bearing the flowers then being called an inflorescence. Sometimes the inflorescence is contracted and superficially resembles a single flower (as in many Compositae), or even a strobilus (familiarly termed a catkin) or cone (as in many temperate trees). The parts of a plant giving rise to flowers or inflorescences are not normally very different from those that are vegetative, but distinct functional separation occurs in some tropical trees. In *Couroupita guianensis* (cannon ball tree), for example, the uppermost branches are densely leafy and solely vegetative. The reproductive branches, which are almost leafless and bear numerous flowers, hang down from the crown around the upper part of the trunk. In some other tropical trees, flowers are produced only from special regions in the lower part of the trunk, a phenomenon known as ***cauliflory***.

In annuals and certain longer-lived plants of warmer regions (which may even be trees) the production of flowers heralds the end of the life-span. Such plants are said to be ***monocarpic*** or ***hapaxanthic***.

Reproduction in angiosperms may be either monoecious or dioecious. Monoecious species are further divisible into the monoclinous, where the male and female sporangiophores are together in the same flower, and the diclinous, where they are in separate flowers (as in the cucumber, *Cucumis sativa*). Functional separation of the sexes in bisexual flowers is often achieved by the organs of the two sexes maturing at different times, a phenomenon termed ***protandry*** if the male precede the female, and ***protogyny*** if the female precede the male. Some orchids show an extreme form of protandry in which the ovules are not formed unless the flower is pollinated.

THE MALE REPRODUCTIVE STRUCTURES The male sporangiophore of the angiosperms is termed a stamen (Fig. 8.21). Typically this consists of a stalk (filament) terminating in four pollen sacs, the sacs being in two pairs and these lying side by side and joined by the connective. The whole of this region is called the anther, and seen in transverse section (Fig. 8.22) resembles the male synangia of some pteridosperms, and of *Caytonia* (see p. 248). Apart from congenital fusion of stamens, discussed later, there is some variation in the form of the individual sporangiophore. In *Degeneria*, for example, and in some members of the Magnoliaceae, the 'filament' is in fact a broadly ovate scale, about 5·0 mm long and 2·0 mm

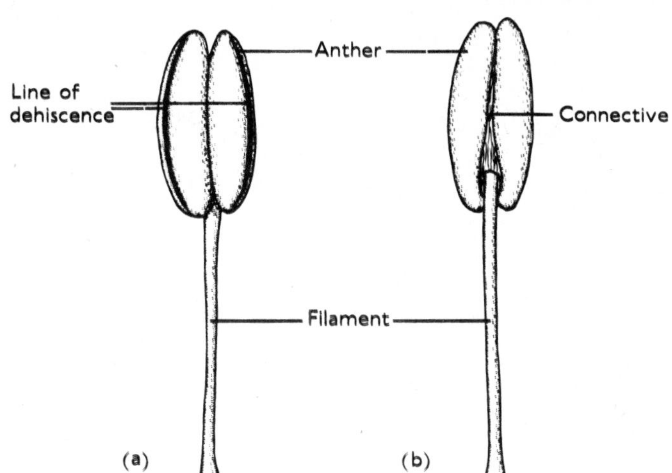

Fig. 8.21 Diagram showing the features of a typical angiosperm stamen. (**a**) Viewed from the front. (**b**) Viewed from the back. In this anther the pollen sacs face the floral axis (introrse), but the reverse position is also found (extrorse).

wide, with four pollen sacs partially embedded on the abaxial surface (Fig. 8.23). At the other extreme the filament may be lacking, one or two

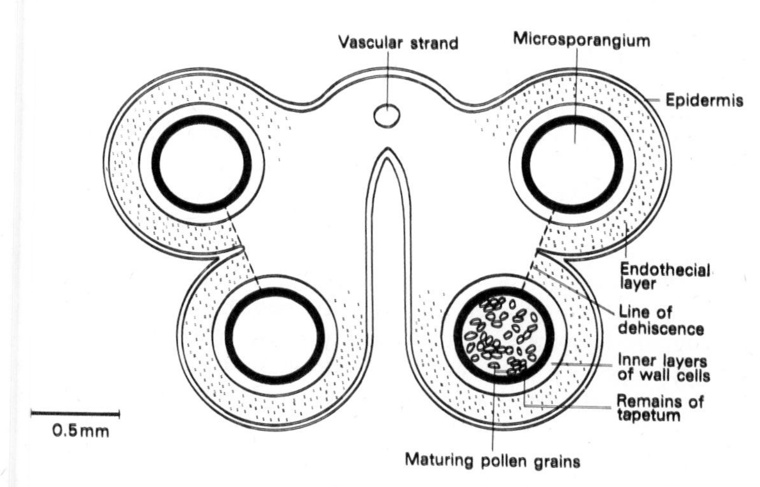

Fig. 8.22 *Lilium longiflorum*. Transverse section of almost mature anther.

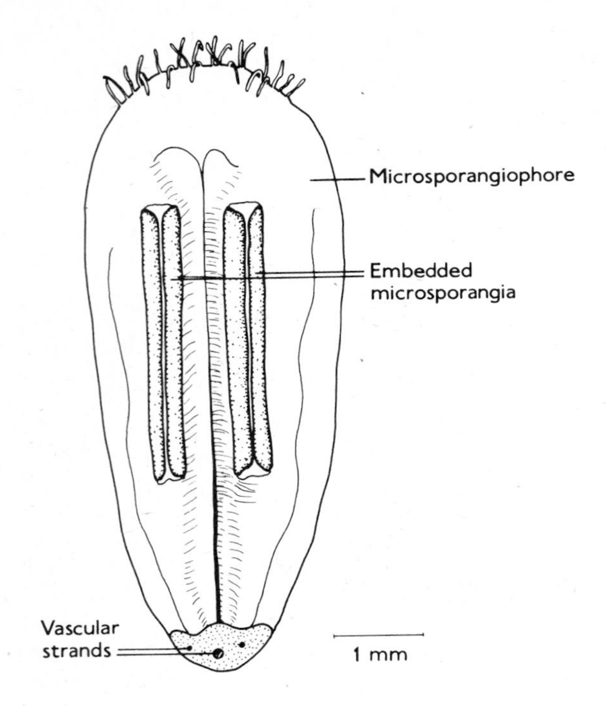

Fig. 8.23 *Degeneria vitiensis*. Adaxial surface of stamen. (After Canright, *Am. J. Bot.* **39**, 484 (1952))

anthers being attached directly to a modified floral axis, as in the complex flower of the orchids. Besides simple dehiscence along a longitudinal stomium, other methods of opening, such as the development of pores or the differentiation of distinct valves, are encountered in some species.

The pollen of angiosperms develops in much the same way as in gymnosperms, but more details are known of the cytology. A peculiarity in the life of the pollen mother cells is a phase in which portions of cytoplasm move from one cell into another, pushing their way through pores in the walls. The significance of this phenomenon, referred to as ***cytomixis*** is unknown. Shortly after the formation of the microspores they become coated with sporopollenin (probably a highly polymerized mixture of esters of complex fatty acids), a material extremely resistant to decay. The pattern with which the sporopollenin is deposited is characteristic of a species (Fig. 8.24), a feature which makes isolated pollen grains identifiable, even those partially fossilized in peats and lake muds. Most of the sporopollenin (or its immediate precursors) is synthesized towards the end of the life of the tapetum,[46] but the manner in which the materials are added to the wall is unknown. In *Lilium*,[25a] and probably elsewhere, the pattern of the wall is established before the grain is released from the tetrad, and the subsequent growth of the wall may be by simple accretion.

Development of the microspores usually begins in the anther, and they contain two nuclei (in a few families three because of division of the generative nucleus while still in the grain) when shed (Fig. 8.32a). One of these,

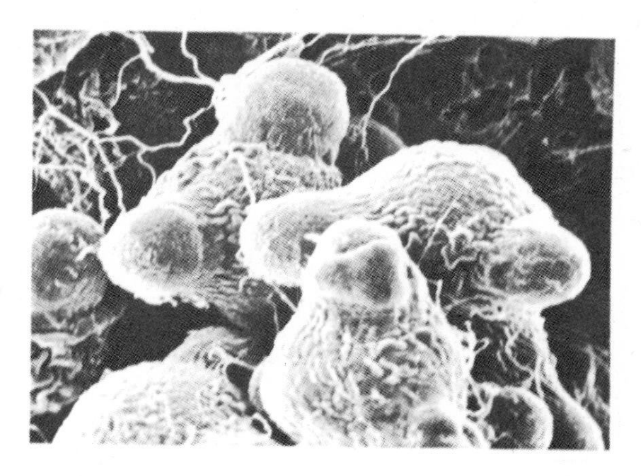

Fig. 8.24 *Oenothera organensis.* Pollen grains (each with three projecting colpi), enmeshed in threads of viscin (a polymer produced by the tapetum), lying on the stigma. (Photomicrograph by H. Dickinson). ×3,000

the vegetative nucleus, becomes large and diffusely staining, and the other, the generative nucleus, becomes elongate and dense. These nuclei lie on a radius of the original tetrad, and there is evidence that their conspicuously different behaviour depends upon gradients set up in the cytoplasm of the cleaving pollen mother cell and persisting in that of the grain.[72]

THE FEMALE REPRODUCTIVE STRUCTURES The megasporangiophore is the distinguishing feature of the angiosperms for it is normally a closed body (termed a carpel), furnished with a distinct stigma (often elevated on a style) on which the pollen germinates (Fig. 8.25). In only a few genera (e.g. *Reseda*) are the carpels open at maturity. A carpel is dorsiventral in symmetry and the fertile region is adaxial. One or several ovules may be present, and if the latter they are commonly borne in two series along the so-called 'ventral suture' of the carpel (Fig. 8.25).

The ovules are much smaller than those of most gymnosperms, but they

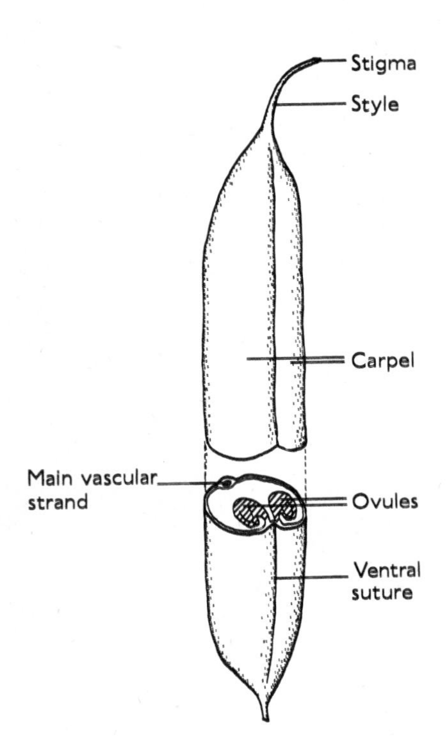

Fig. 8.25 The simple carpel characteristic of the family Leguminosae. This arrangement of the ovules is referred to as 'marginal placentation'.

are structurally similar. They may be either upright, inverted or occasionally more or less horizontal, these three orientations being termed orthotropous, anatropous and campylotropous respectively (Fig. 8.26). One integument may be present, but in many families there are two, and in some as many as four. The last formed integument, irrespective of the total number, sometimes becomes transformed into a fleshy aril in the fruit (as in the durian, *Durio*). The integuments normally develop uniformly, indicating the basic radial symmetry of the angiospermous ovule. A slender vascular trace enters the stalk of the ovule, but in only a few families does this extend into an integumentary vascular system.

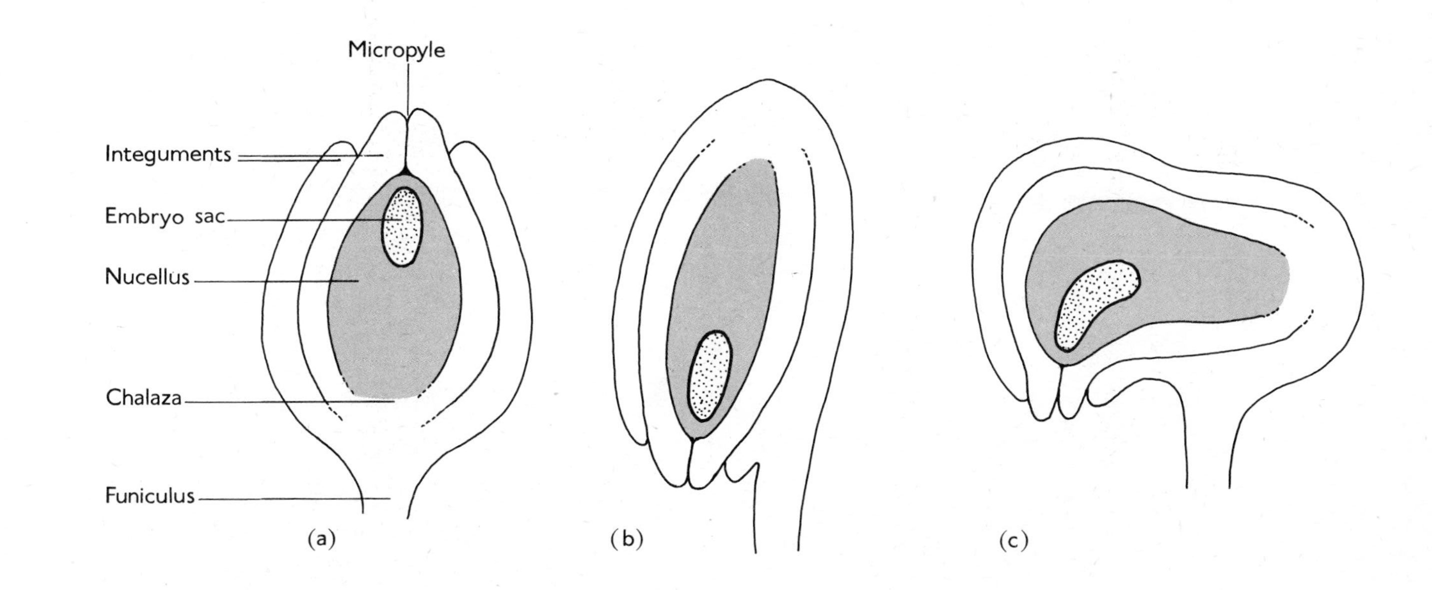

Fig. 8.26 Forms of ovule orientation (diagrammatic). (**a**) Orthotropous. (**b**) Anatropous. (**c**) Campylotropous.

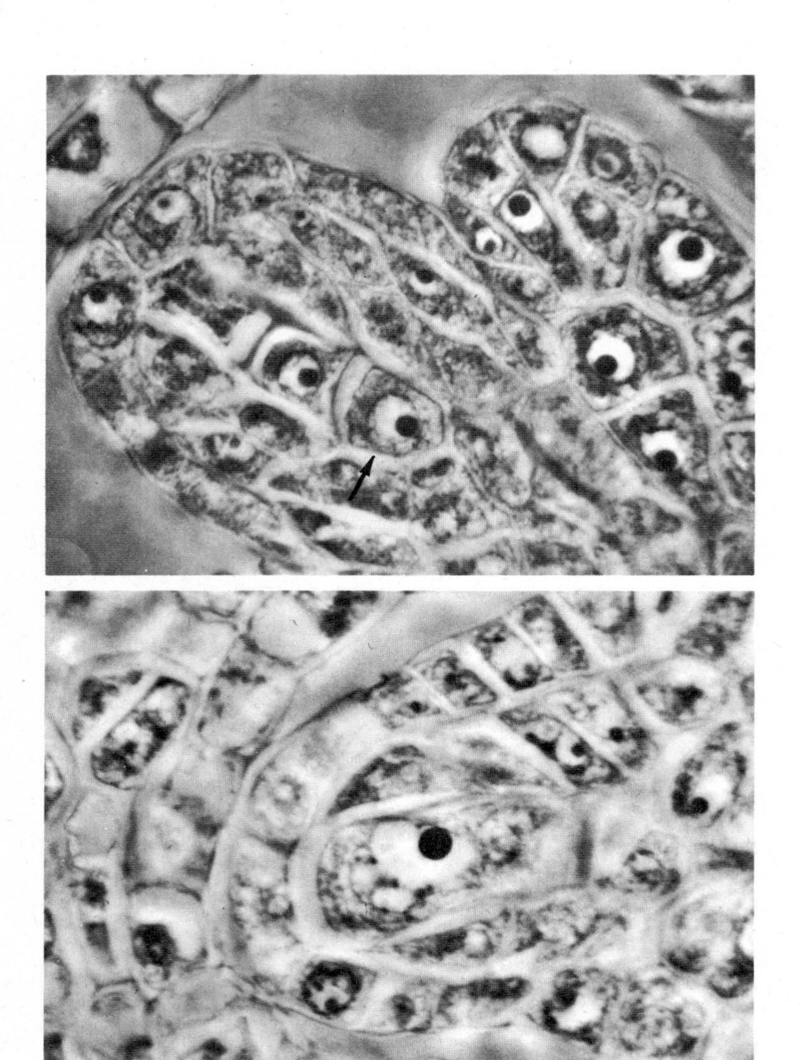
(b)
(a)

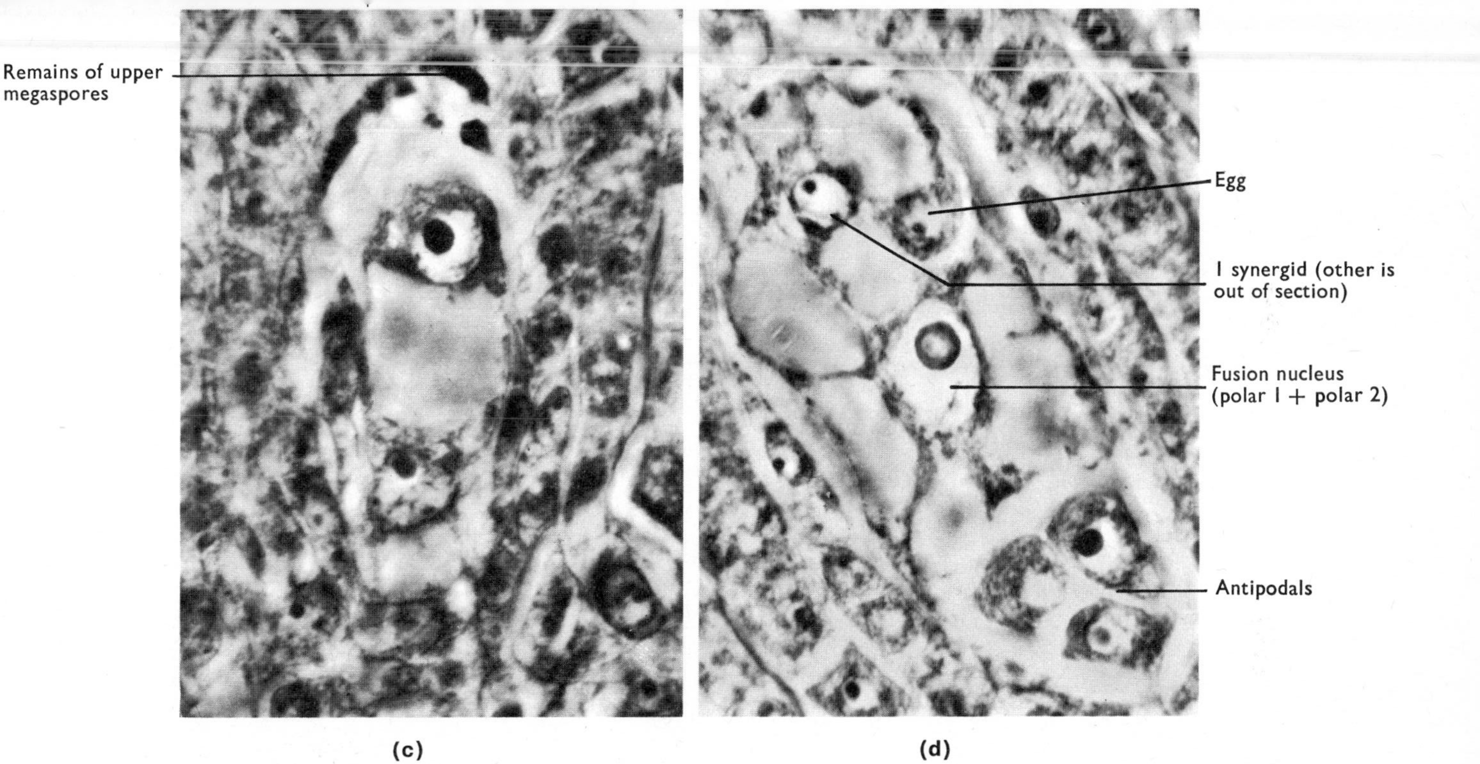

Fig. 8.27 *Myosurus minimus* (mousetail). Stages in the development of the female gametophyte. (**a**) Megaspore mother cell. (**b**) The tetrad of spores resulting from meiosis. The innermost (indicated by the arrow) becomes the megaspore. (**c**) The two-nucleate embryo sac, with the remains of the three upper megaspores still visible above the sac. (**d**) The mature, eight-celled embryo sac. (**a**) and (**b**) ×1,350. (**c**) ×1,500. (**d**) ×1,200.

THE FEMALE GAMETOPHYTE[55] The development of the female gametophyte (Fig. 8.27), in the angiosperms termed uniformly the embryo sac, begins in a familiar fashion (Fig. 8.28). A cell in the upper part of the nucellus, immediately below the layer of cells at its surface, becomes conspicuously large (Fig. 8.28a). Meiosis then leads to a linear tetrad of megaspores (Fig. 8.28b). Although the term 'megaspore', because of the evident homology of this cell with the megaspore of the gymnosperms, is retained, the megaspore in the angiosperms is frequently smaller than the

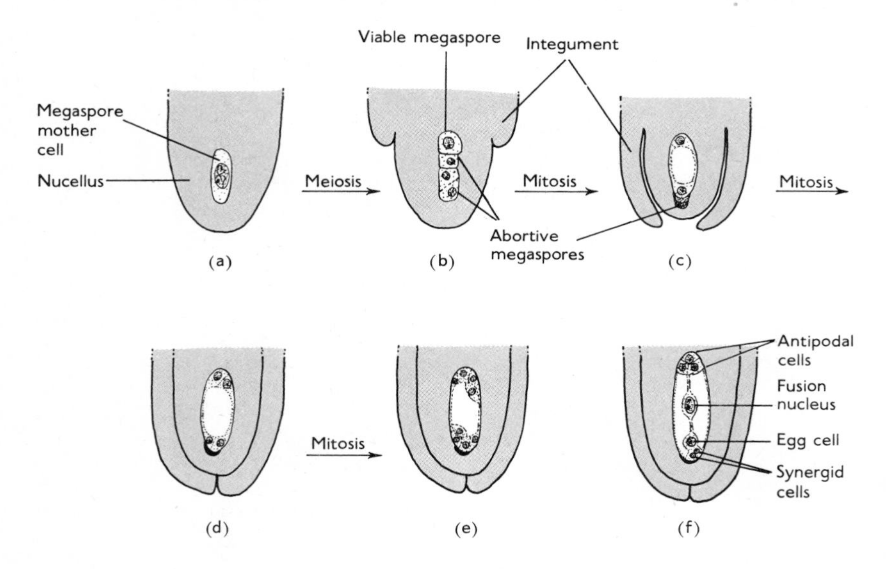

Fig. 8.28 The sequence of divisions leading up to the production of a monosporic, eight-nucleate embryo sac. From flowers of *Myosurus minimus*. Diagrammatic; not to scale.

microspore. Normally only the inner of the tetrad of megaspores undergoes further development, and the female gametophyte is consequently said to be monosporic in origin. The nucleus of the megaspore enters a succession of mitoses, and, in step with the expansion of the ovule as a whole, the multinucleate, acellular embryo sac is formed (Fig. 8.28e). In a frequent type of embryo sac there are eight nuclei, produced by three successive mitoses, which associate themselves with cytoplasm and arrange themselves in a definite pattern (Fig. 8.28f). Eight cells are usually identifiable at this stage, each bounded by a delicate membrane. At the micropylar end of the sac is a group of three cells, of which one (frequently with a weakly staining nucleus) is the egg cell. The cells accompanying the egg are termed synergid cells. At the other end of the sac (the chalazal end) is another group of three cells, forming the so-called antipodals. The nuclei of the remaining two cells (termed polar nuclei) come together at the centre of the sac and eventually fuse, thus giving rise to a central diploid fusion nucleus (Fig. 8.27d).

It should be clearly realized that the foregoing is only one of the several kinds of development of the female gametophyte found in the angiosperms. In some families, notably the Onagraceae, the embryo sac develops from the outer and not from the inner megaspore, a feature of unknown cytological and physiological significance. Sometimes, as in *Allium* (onion), meiosis in the formation of the megaspores is incomplete, a diad of spores being produced instead of a tetrad. The embryo sac, which develops directly from the inner spore, is then said to be bisporic in origin. Finally, in another kind of development, termed tetrasporic, the nucleus of the megaspore mother cell divides reductionally, but the cell itself does not divide at all. Instead, mitotic divisions follow and the cell expands without interruption to form the sac. Bisporic and tetrasporic embryo sacs occur in a wide range of families (and are present in *Gnetum*), and there is no justification for regarding this kind of origin as abnormal. Another kind of variation lies in the number of nuclei in the mature sac. Although this is commonly eight, irrespective of the mode of origin of the sac, others are known containing four or sixteen. The cytology of development is also variable. Although where sexual reproduction is normal the egg nucleus remains haploid, the chalazal polar nucleus may contain more than one set of chromosomes so that the central fusion nucleus becomes triploid or even, as in *Fritillaria*, tetraploid.

THE ARRANGEMENT OF THE FERTILE REGIONS IN THE FLOWER The arrangement of the sporangiophores in the flower often is of considerable complexity, and floral morphology is a specialized field with its own terminology. Here we shall use only as much of this terminology as is necessary for general discussion, and for further details the reader must consult specialist works. To facilitate illustration of the principles involved in the structure of the flower, we shall first consider a bisexual flower in which the components remain separate (Fig. 8.29), a situation seen, for example, in *Ranunculus*. The flower terminates an axis, and the transition from the vegetative to the reproductive region is marked by the occurrence of one or more whorls of sterile structures resembling rudimentary leaves. These form collectively the ***perianth***. Where two whorls are present the segments of the outer whorl are frequently green (and termed sepals), whereas those of the inner are brightly coloured (and termed petals), and the two parts of the perianth are termed calyx and corolla respectively.

Following closely upon the perianth are the stamens, forming collectively the androecium. The stamens may be either indefinite in number and spirally arranged (as in *Ranunculus*), or definite and arranged in whorls, the positions of individual stamens often bearing a clear relation to each other and to those of the preceding perianth segments. This relationship is often one of alternation, but sometimes, especially in the Caryophyllaceae, the stamens of the outer whorl may stand opposite the petals.

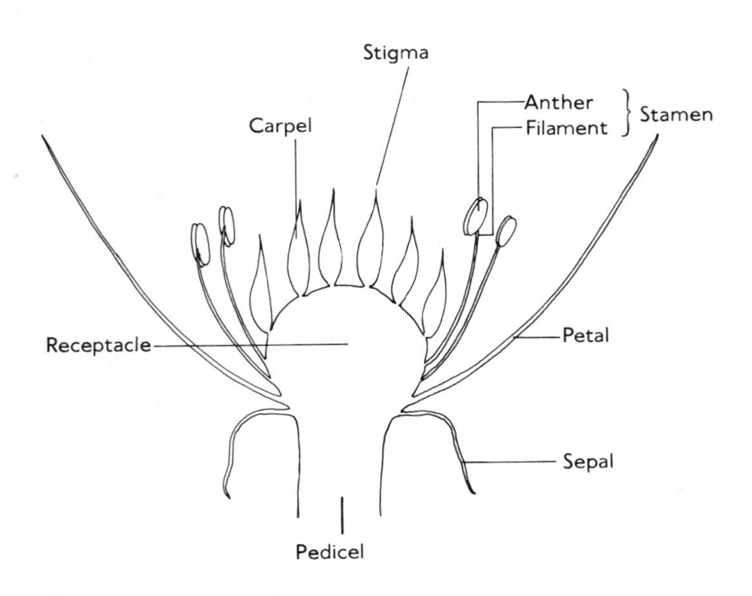

Fig. 8.29 Longitudinal section of a flower with separate components. Diagrammatic.

The termination of the reproductive axis, often termed the receptacle, bears the carpels, this female region being called the gynaecium or ovary. The carpels may again be spirally arranged, as in *Ranunculus*, or in a single whorl. In the latter event the carpels usually alternate in position with the stamens of the uppermost whorl.

In unisexual flowers the sporangiophores of one sex are absent or non-functional, although the bisexual condition can sometimes be brought about by treating the very young flower with growth substances such as gibberellic acid. Unisexual flowers may approach the limits of reduction. In *Euphorbia*, for example, the male flower consists of a single stamen (Fig. 8.30), a small joint at its base indicating the transition from axis to flower. Congenital fusion of parts of similar nature is widespread in flowers. Both the calyx and corolla independently may become tubular, the composite nature of the tube being indicated by lobes or teeth at its mouth. Filaments of whorled stamens may also fuse for most or only part of their lengths, leading to a cylindrical androecium. Fusion of carpels leads to a syncarpous ovary, in which the individual carpels are often represented by

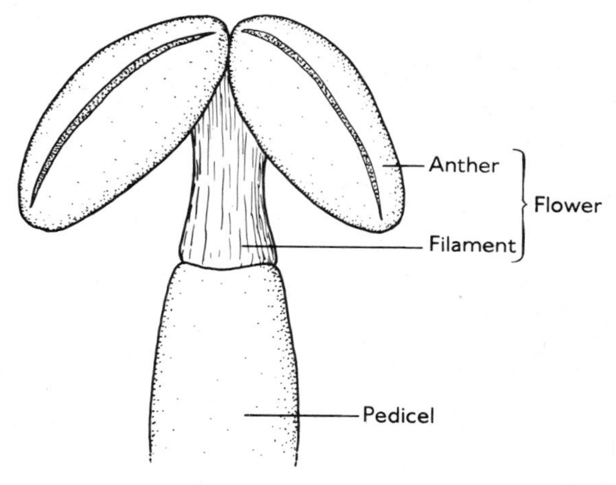

Fig. 8.30 *Euphorbia canariensis*. The highly reduced male flower. (After Kerner von Marilaun and Oliver, *The Natural History of Plants*. Blackie, London, 1902)

compartments (or loculi), as in the Liliaceae. Externally the composite nature of the ovary is often indicated by lobing, or by the style, which rises from the centre of the ovary, breaking up into branches equivalent in number to the constituent carpels. The ovary may sometimes be sunken in the receptacle (as in *Rosa*) or otherwise surmounted by the perianth and androecium, leading to clearly epigynous flowers. Sterilization of one or more stamens may occur, giving rise to ***staminodes***, which may either persist as rudimentary structures (as in *Scrophularia*) or occasionally become enlarged and brightly coloured, and play a conspicuous part in the organization of the flower (as in *Canna* and *Iris*). Both sterilization and congenital fusion of parts are involved in the complex flowers of the orchids.

THE ONTOGENY OF THE FLOWER Although in the gymnosperms the apices giving rise to the reproductive and vegetative axes differ little in organization, in angiosperms they are usually easily distinguishable. In becoming reproductive an apex flattens considerably and may even become concave. The perianth whorls and sporangiophores are often produced in acropetal sequence, but this is not always so. The direction of initiation may even become locally reversed, and primordia appear between the whorls of those already laid down. This phenomenon affects particularly the androecium, with conspicuous results in the symmetry of the mature flower.[30] In some Caryophyllaceae, for example, a whorl of five stamen primordia appears, each stamen alternating in position with the preceding petals. A second whorl of stamens is then initiated basipetally, the stamens alternating with those of the first whorl, and hence standing opposite the petals. In this way an ***obdiplostemonous*** flower is formed.

POLLINATION AND THE GROWTH OF THE MALE GAMETOPHYTE Pollination in angiosperms is occasionally dependent upon aerial dispersal of the grains (as in catkin-bearing trees), but frequently involves the participation of insects as carriers. Various mechanisms have been evolved which ensure that insects visiting flowers in search of nectar (usually secreted by glands situated towards the interior of the flower) become dusted with pollen, which then becomes transported to stigmatic surfaces in other flowers.[51] A familiar example of such a pollination mechanism is provided by *Salvia pratensis* (sage), a species pollinated by the bumble-bee. The stamens have short filaments, but a greatly elongated connective (Fig. 8.31a). The longer upper portion of the connective terminates in an anther, but the shorter lower part in a plate blocking the approach to the nectary. A bee attempting to reach the nectar displaces the plate. Since the connective is hinged about the filament, the anther is in this way forced on to the insect's back, coating it with pollen (Fig. 8.31b). A bizarre form of pollination is found in some orchids (e.g. *Ophrys speculum* of the western Mediterranean) where the pattern and conformation of the lip recall the female of the insect concerned. The male insect, in attempting to copulate with the lip of the flower, detaches the pollen (which coheres in masses called pollinia) and carries it to another flower.[77] A few tropical flowers are pollinated by birds and bats. The appropriate authorities must be consulted for more detailed accounts of pollination mechanisms.

The pollen, having been brought into contact with the stigma, germinates freely. A pollen tube breaks through a thinner region in the exine of each grain (Fig. 8.32) and penetrates the style. The subsequent growth of the tube towards the ovules often follows a tract of specialized, thin-walled, transmitting tissue, the direction of growth probably being determined by metabolic gradients of a quite simple kind.[61] The pollen tubes eventually

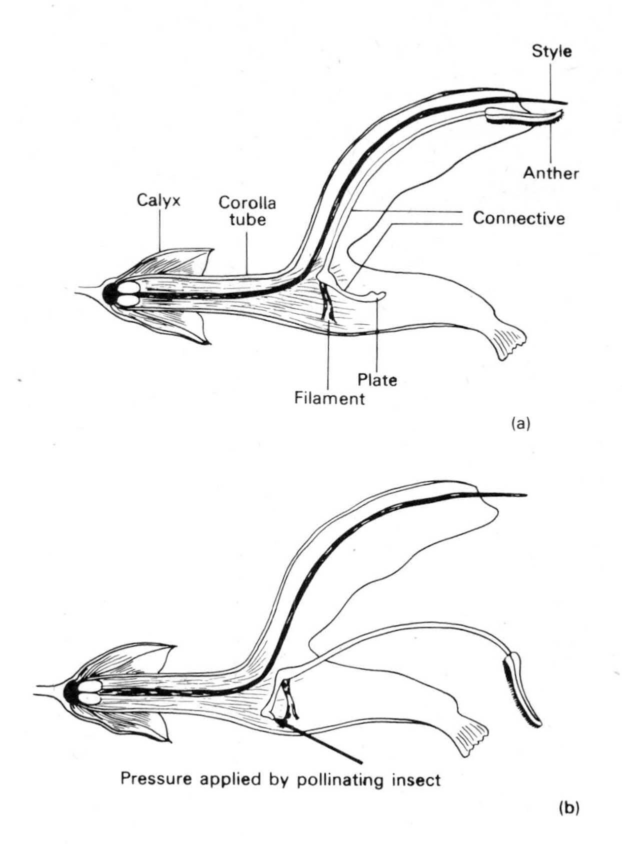

Fig. 8.31 *Salvia pratensis.* (**a**) Half flower showing the normal position of the stamen. (**b**) The same after the plate has been depressed. (After Kerner von Marilaun and Oliver, *The Natural History of Plants*. Blackie, London, 1902)

reach the embryo sacs, penetrating the ovules either by way of the micropyle or by growth through the chalazal end. Meanwhile the generative nucleus in each tube has divided into two sperm nuclei and these, each associated with a little cytoplasm, move towards the tip (Fig. **8.32c**).

In some plants some of the flowers (often produced late in the flowering season) do not open, but are nevertheless fertile. Such flowers are termed ***cleistogamous*** (and the normal ***chasmogamous***). A familiar example is provided by *Viola canina.* The normal flowers of spring produce little seed, but the bud-like cleistogamous flowers of summer are fully fertile. They contain only minute petals and all but the two abaxial stamens are abortive. The pollen of these, however, germinates in the anther and the pollen tubes grow through the wall into the stigma.

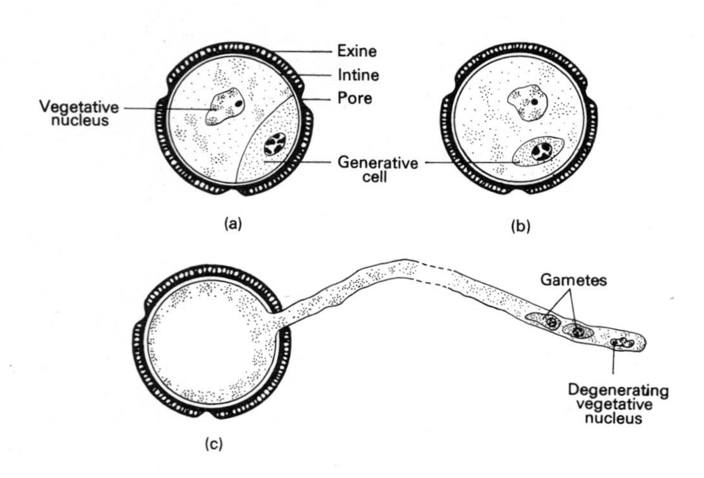

Fig. 8.32 Development of the pollen grain. (**a**) Condition in the anther during final stages of the growth of the wall. (**b**) Condition during liberation and pollination. (**c**) The final nuclear division during the growth of the pollen tube.

FERTILIZATION A pollen tube usually enters the embryo sac itself at the micropylar end, and a region near the tip of the tube, possibly in consequence of enzymes produced by the synergid cells, soon breaks open. The contents of the tube are discharged into the sac, and the 'double fertilization', characteristic of the angiosperms, then ensues. One male nucleus penetrates the egg; the other wanders through the sac and fuses with the central fusion nucleus, forming the polyploid primary endosperm nucleus. Almost simultaneously the other nuclei in the sac degenerate, the zygote and the endosperm nucleus alone remaining sites of further growth.

THE FEATURES OF ENDOSPERM The formation of the endosperm usually begins before the division of the zygote. The primary nucleus and its daughters undergo successive mitoses, giving rise to an endosperm that is either cellular from the first, or initially acellular (Fig. 8.33a) and only later partly or wholly cellular. Cytologically the endosperm is a remarkable tissue, with unique cytoplasmic and nuclear properties, significantly different from the haploid 'pseudo-endosperm' of the gymnosperms (see p. 242). It is a rich source of growth-regulating substances or their immediate precursors, many of them not yet chemically identified. The successful artificial culture of a number of plant tissues, for example, is still impossible in fully defined media, but is facilitated by the addition of coconut milk (the liquid endosperm of the coconut seed). Endosperm itself can sometimes be obtained in pure culture and, since the cell walls of young endosperm often have little thickening, such cultures have been used to study the details of mitosis *in vivo*. In some endosperms the nuclei

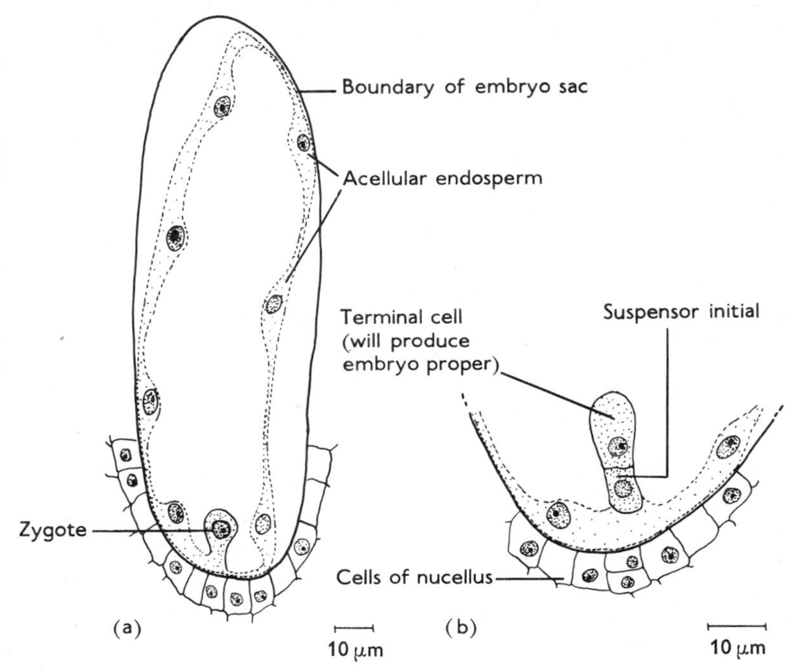

Fig. 8.33 *Myosurus minimus*. Longitudinal section of embryo sac following fertilization. (**a**) After several divisions of the primary endosperm nucleus. (**b**) After the first division of the zygote. The zygote is at the micropylar end of the sac.

reach high and irregular levels of endopolyploidy, and amitotic nuclear division has been reported in a number of instances.

In keeping with the function of providing for the nutrition of the embryo and young plant, the cells of the mature endosperm are often filled with food materials. Carbohydrates, fats and proteins are present in various labile forms. The occurrence of these materials is sometimes so abundant that the seeds concerned acquire vast economic importance. The cereal grains, where the endosperm yields starch and protein in a form readily palatable to humans, and the oil seeds, such as *Ricinus* (castor bean) and *Linum* (linseed), are familiar examples. Polymers of mannose (mannans) also occur as reserve products, a peculiar example being provided by *Phytelephas*, a palm. The large endosperm of this species becomes so heavily indurated with a mannan that it enters commerce under the name of 'vegetable ivory', once used for the manufacture of buttons and billiard balls. In a few plants, notably the orchids, the endosperm undergoes only trifling development and in others it is remarkably aggressive. In *Pedicularis*, for example, the endosperm produces haustoria which invade the integumentary tissue, leading eventually to its complete resorption.

EMBRYOGENESIS[79] As in other seed plants, the first dividing wall of the zygote lies transverse to the longitudinal axis of the ovule (Fig. 8.33b), and

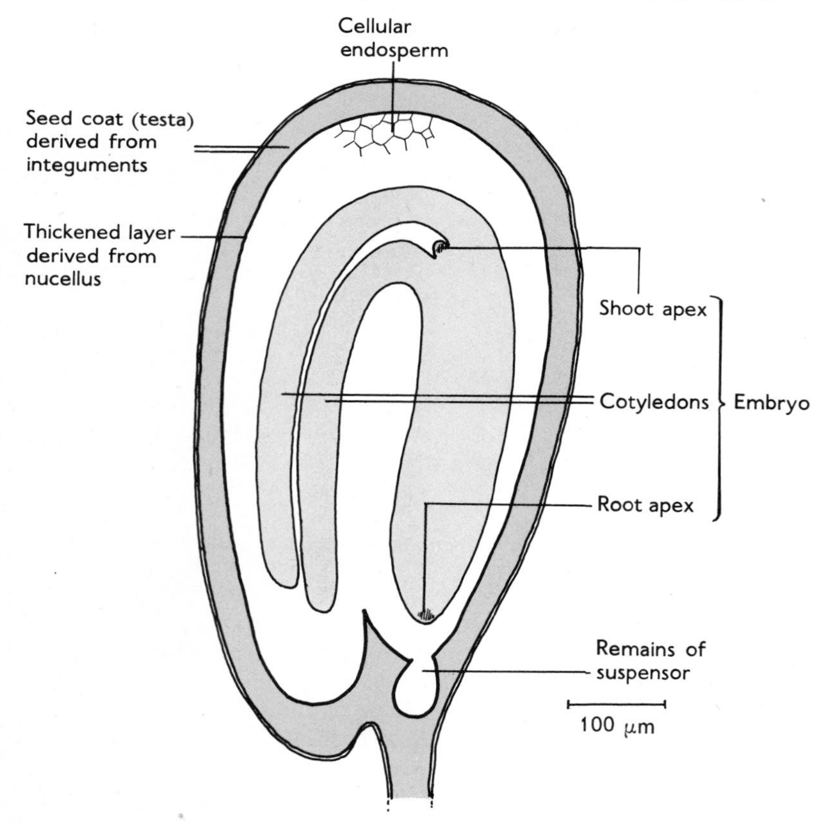

Fig. 8.34 *Capsella bursa-pastoris*. Longitudinal section of a young seed containing a well-developed embryo.

the subsequent embryogeny is endoscopic. Free nuclear division is not, however, a characteristic of the development of the proembryo, as it is in many gymnosperms. Instead, embryogenesis begins with a series of rather precise cell divisions, showing little variation between species, and meristematic activity does not pause until the embryo differentiates and the seed is formed. The mature embryo (except in the orchids where it remains undifferentiated) possesses a stem apex (plumule), one or two cotyledons, and a root apex (radicle). Even though the mature embryo may be folded (as in *Capsella bursa-pastoris* (shepherd's purse), Fig. 8.34), the radicle is always directed towards the micropyle. The extent to which the fully formed embryo has drawn upon the food reserves of the endosperm varies widely with the species. In 'endospermous' seeds this reserve remains considerable, but in the 'non-endospermous' little remains, and much of the food material in these seeds often becomes transferred to the cotyledons (Fig. 8.34). Examples of such seeds are those of the legumes where the swollen cotyledons entirely replace the endosperm in both space and function.

THE MATURATION OF THE SEED AND FRUIT Associated with the maturation of the embryo are considerable changes in the nucellus and integuments. In some seeds food reserves are laid down in the nucellus, and in the mature seed this then forms a distinct tissue known as the ***perisperm***. Frequently, however, only a little of the nucellus remains, and this together with the integument becomes transformed into the outer covering of the seed. This often involves the deposition of much cutin and lignin, and the resulting seed coat (***testa***) is consequently often remarkably impervious. The wall of the carpel (***pericarp***), together in some instances with the receptacle, also undergoes changes as the seed matures, leading to the production of a fruit. Where the seeds are distributed while still enclosed in the carpel, the latter may develop appendages, such as hooks, wings, and parachute-like fringes of hairs, which assist dispersal by wind, animals or other agents. Other angiosperms have evolved mechanisms whereby tensions are set up in the carpel wall as it dries, causing eventually an explosive dehiscence which scatters the seeds far and wide. Frequently, instead of becoming dry and brittle, the maturing carpel develops a thick fleshy wall, the palatability of which appeals to humans and other animals. In these instances the seeds commonly have a testa sufficiently resistant to survive the passage through the digestive tract. This brief summary of a wealth of morphological detail shows how the angiosperms have evolved a great variety of ways which ensure their wide dispersal,[70a] a feature in which they differ sharply from the gymnosperms, and which has undoubtedly contributed to their present success.

DORMANCY AND GERMINATION[23, 61a] The cause of the cessation of growth of the embryo as the seed matures is not altogether clear, but the partial dehydration of the interior of the seed at this time is probably an important factor. Certainly the imbibition of water is essential for renewed growth, but this by itself is often insufficient for the production of viable seedlings. A period of dormancy ('after ripening') must often ensue before successful germination will occur. The seeds of some species also need to be chilled, a treatment which apparently activates the enzymes which make available to the embryo the food reserves of the endosperm. This cold requirement also provides a biological advantage, since the seeds will not germinate until after the winter, and the seedlings thus avoid the rigours of this unfavourable season. Other seeds, e.g. those of lettuce, germinate only after illumination, and the phytochrome system (which depends upon a protein sensitive to red light) is here involved in the renewal of growth. The seeds of many tropical plants, on the other hand, germinate immediately in moist conditions, and soon lose their viability if stored. The tropical mangroves, plants of coasts and estuaries, are outstanding in that the embryo continues to develop in the ovule and a swollen radicle protrudes from the withered flower. The young plant eventually falls, with the radicle correctly orientated for penetration of the mud beneath, and so establishes itself.

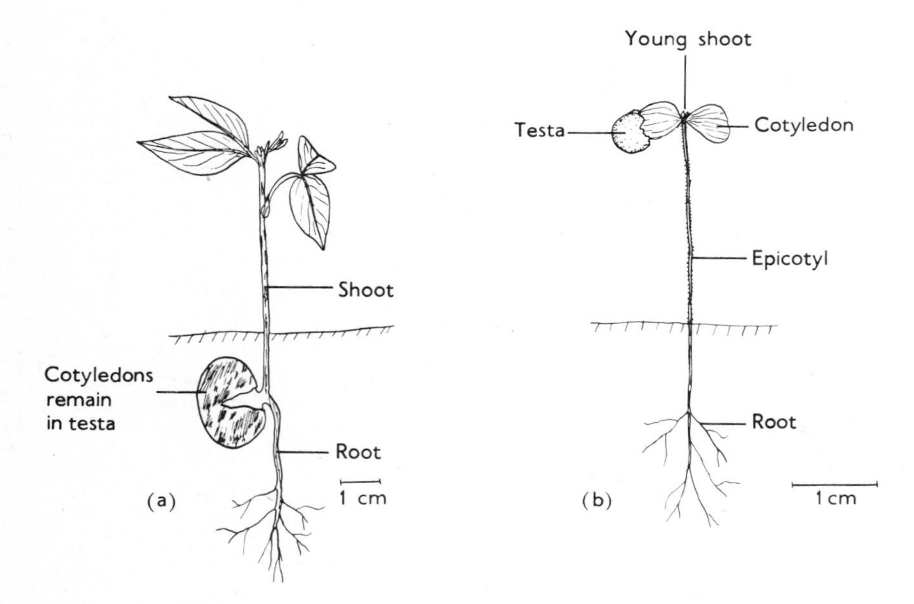

Fig. 8.35 (**a**) *Vicia faba*. Hypogeal germination. (**b**) *Sinapis* sp. Epigeal germination.

As a germinating seed imbibes water there is usually considerable swelling and eventually rupture of the testa. The radicle is the first organ to emerge, followed closely by the plumule. The cotyledons may remain within the seed at or below soil level (hypogeal germination), as in *Vicia faba* (broad bean) Fig. 8.35a, or become elevated (epigeal germination) and photosynthetic, as in *Sinapis* (mustard) Fig. 8.35b. Germination is basically similar in monocotyledons and dicotyledons, but in the grasses the plumule and radicle are first enclosed in sheaths called the coleoptile and coleorhiza respectively.

Asexual reproduction

In addition to sexual reproduction, the angiosperms show many forms of asexual reproduction. One form of such reproduction, termed apomictic, superficially resembles sexual reproduction. Although its presence can be readily inferred from the genetics of the species concerned, the elucidation of the accompanying cytology has sometimes demanded very careful investigation. In one kind of apomixis, parthenogenesis, the egg develops without fertilization. Where this occurs regularly, the nuclei involved in the formation of the embryo sac and the nucleus of the egg contain an unreduced number of chromosomes. Such embryo sacs may result from a modification of the meiosis which yields the megaspores (as in *Taraxacum*, dandelion), or from the spatial and functional replacement of the megaspore by an adjacent nucellar cell (as in *Hieracium*, hawkweed). Although the formation of the embryo occurs without fertilization, the stimulus of pollination is often needed before the process will begin (a phenomenon known as pseudogamy), and an endosperm may even be formed in the normal manner. The pollen, although without genetic effect in the species producing it, is sometimes able to form hybrids with related species which reproduce sexually. This parallels the situation in apogamous ferns, such as *Dryopteris borreri*, where the male gametes, although normally functionless, are not impotent (see p. 220).

Another form of apomixis involves the formation of embryos, often in addition to the normal zygotic embryo, directly from the cells of the nucellus. A classical example of this phenomenon, known as adventive embryony, is provided by *Citrus* (Fig. 8.36), where the mature seeds may contain several viable embryos. Although in some species adventive embryony is independent of pollination, there are others (probably including *Citrus*) in which the embryos do not mature unless it occurs. In these instances the germinating pollen probably provides a chemical factor essential for continued growth. It is significant in this connection that in one orchid, *Zygopetalum machayi*, adventive embryony, although dependent upon pollination, is quite as effectively stimulated by pollen from another genus as by that from the species itself.

The only remaining form of apomixis is more conspicuous since it involves the formation of bulbils or dwarf shoots in place of flowers. *Saxifraga cernua* and *Festuca vivipara* provide examples. Some species (e.g. *S. cernua*, Fig. 8.37) producing these structures are also able to reproduce in a normal sexual manner, bulbils being borne in the lower part of the inflorescence and flowers in the upper. In some grasses, and probably in other plants displaying this phenomenon, the way the plant reproduces can be influenced by the length of day in which it is grown.

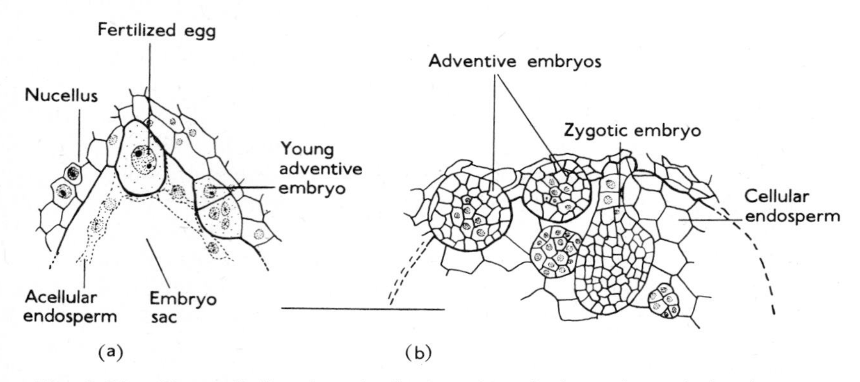

Fig. 8.36 *Citrus trifoliata*. Longitudinal section of micropylar end of embryo sac in which adventive embryony is occurring. (**a**) Immediately after the fertilization of the egg. (**b**) After the endosperm has become cellular. Only the true embryo has a suspensor. (Both after Osawa, from Maheshwari, *An Introduction to the Embryology of Angiosperms*. McGraw-Hill, New York, 1950)

Other forms of asexual reproduction are purely vegetative. Stems may arch over and root themselves, as in *Rubus*, and leaves of a number of species produce plantlets either at the summit of the petiole, as in *Tolmiea* (a frequent house plant), or marginally, as in *Kalanchoë*. Species which grow from bulbs usually reproduce themselves by the production of axillary buds at the base of the axis which develop into daughter bulbs. Corms multiply themselves by a similar process. Fleshy roots which are able to give rise to buds largely account for the success with which plants such as *Convolvulus arvensis* (bindweed) and *Lepidium draba* (hoary cress) are able to multiply themselves. In addition to these various forms of vegetative reproduction occurring in nature, layering, budding and grafting are frequently used in horticulture, and are the only means of propagating many valuable varieties whose sexual reproduction is defective.

Most remarkable of all, perhaps, is that in experimental conditions isolated cells and even pollen grains will give rise to whole plants of normal growth and form.[76a]

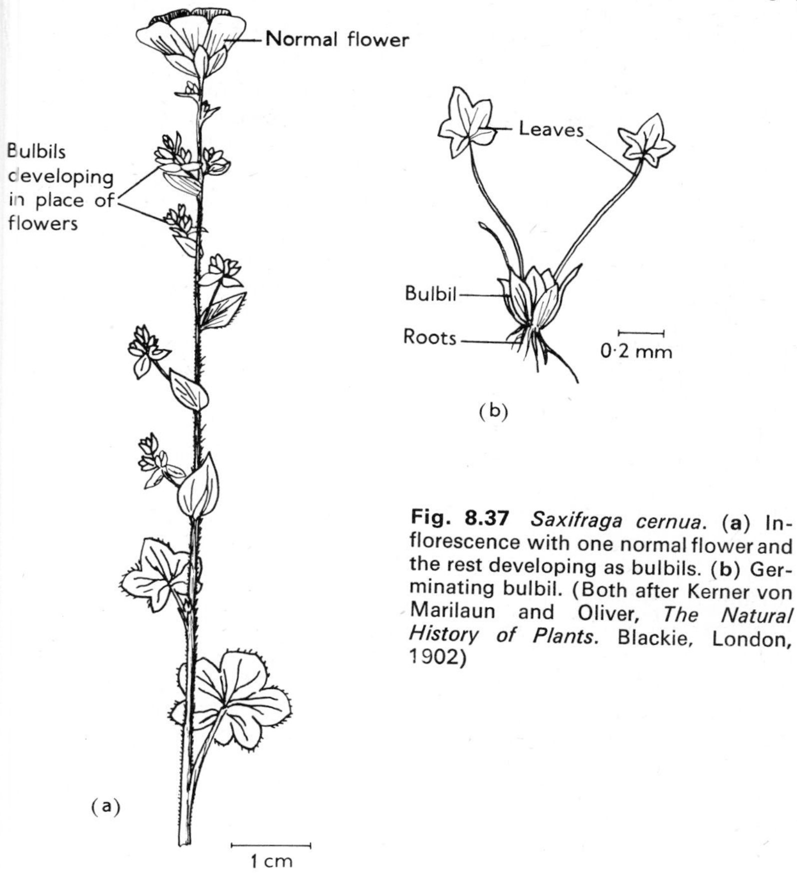

Fig. 8.37 *Saxifraga cernua.* (**a**) Inflorescence with one normal flower and the rest developing as bulbils. (**b**) Germinating bulbil. (Both after Kerner von Marilaun and Oliver, *The Natural History of Plants.* Blackie, London, 1902)

CORRELATION WITHIN THE PLANT BODY The angiosperms, more than any other class of plants, have been investigated with object of discovering those factors responsible for co-ordinating the growth of the plant as a whole. In the presence of an apical bud, for example, lateral buds commonly remain inactive, an instance of correlative inhibition. Correlations of this kind are clearly of great importance in determining the morphology of the mature plant.

Correlation is an example of cell interaction, but the interacting cells are often separated by others not directly concerned. These interactions at a distance are brought about by growth-regulating substances which are able to move through the plant in a manner not yet wholly understood. Research has now revealed a whole series of such substances, which for present purposes can be classified as auxins, kinins, abscissins, gibberellins and ethylene.[80a] Auxins (of which indole-3-acetic acid is one of the best known) are those substances which will cause curvature resembling that produced by unilateral illumination if placed asymmetrically on the tip of a decapitated oat coleoptile. They are usually produced by meristematic cells and are involved (amongst other effects) in maintaining apical dominance and initiating differentiation. Kinins (cytokinins, phytokinins) are substances which promote cell division. They are found in fruits and seeds, but may also be generated in damaged cells and cause proliferation around wounds and their subsequent repair. Abscissins, probably produced in leaves, are effective in causing leaf abscission, and dormancy in apical buds, common phenomena in trees of temperate regions. Gibberellins were first extracted from the fungus *Gibberella* which, when infecting rice, causes the plant to be excessively tall. They are now known to exist in higher plants, and early experiments showed that when administered to dwarf mutants as, for example, of maize (corn), these could grow to their normal stature. Additionally, they are probably involved in photoperiodic responses and the change from vegetative to reproductive growth. Ethylene has long been known to be produced by ripening fruits, and conversely the process of ripening in stored immature fruits can be controlled by regulation of the amount of ethylene in the atmosphere.

Although detailed investigation of the growth-regulating substances (which include many others less well known in addition to those mentioned here) is the province of the physiologist, no morphologist enquiring into why a given species always grows in its characteristic and immediately recognizable way can ignore them. It was inevitable that the flowering plants, because of their ubiquity, familiarity, and economic importance, should be chosen for the bulk of this kind of work, but the study of simpler systems in the algae and archegoniate plants might at this stage be more rewarding from a scientific point of view.

EVOLUTION WITHIN THE ANGIOSPERMS

The evolution of the angiosperms has many different aspects. Of particular interest to systematists is the evolution of individual species, genera and families. Another approach, of more appeal to morphologists, is to consider the evolution of certain morphological or anatomical features —such as the form of leaves and the nature of vascular tissue—in the angiosperms as a whole. Unfortunately the fossil record of the angiosperms has little information to offer on any of these points. As soon as the angiosperms become well represented in the fossil floras of the Cretaceous, they are largely referable to modern families and even genera.

Recent evolution in angiosperms

Knowledge of recent evolution in the angiosperms has come principally, as with the ferns, from studies of chromosome pairing in hybrids, often produced under experimental conditions. Hybridization of diploid species, followed by allopolyploidy, appears to have been the origin of a number of species, among them *Spartina townsendii* (cord-grass) and *Galeopsis tetrahit* (hemp nettle). Observations of chromosome size may assist in assessing evolutionary relationships of a wider character. The family Commelinaceae, for example, includes some genera in which the chromosomes are small, others in which they are strikingly large, and yet others in which the nuclei contain chromosomes of both sizes. A few fossil series of recent age are known, mostly of angiospermous seeds, but they indicate little other than minor changes in structure and shape. Seeds of the water plant *Stratiotes*, for example, are found in successive strata throughout the Tertiary period. There was evidently a slight, but progressive, lengthening and narrowing of the seed during this period, together with minor changes in the relative development of the parts.

Another source of evidence relating to recent evolution is comparative biochemistry. In *Aesculus*, for example, an investigation of some curious and distinctive amino acids produced as metabolites has given a number of clues to interspecific relationships. Comparison of extractable proteins by serological methods is another promising technique for detecting affinities. Qualitative and quantitative comparisons of the occurrence of a number of different substances in extracts at one and the same time is made possible by chromatography. Two species which yield chromatograms which are closely similar are reasonably regarded as having diverged in relatively recent time.

The growth forms of the earliest angiosperms[10,76]

A point of considerable interest is the habit of the earliest angiosperms. It is generally held that they were woody plants, probably phanerophytes (see p. 285), growing in tropical and warm temperate regions. This view derives partly from the fact that about three-quarters of the families of living angiosperms are primarily adapted to such environments (where woody plants are the predominant terrestrial vegetation), and partly from the apparent absence of herbaceous forms amongst the pteridosperms, the most likely ancestors of the angiosperms. It is also true that one would expect woody forms to be more primitive in general than herbaceous, since the life cycles of herbaceous forms are usually shorter. Many more generations are therefore possible in a given time, and hence there is a greater opportunity for genetic mutation and recombination than with the slower breeding woody plants. Herbaceous forms could, therefore, reasonably be expected to evolve at the greater rate, and primitive features consequently to persist principally in the woody forms.

It is therefore of considerable significance that the forests of warm and tropical regions contain a small number of woody angiosperms in which the xylem lacks vessels and contains only tracheids. These tracheids are similar in size to those of *Lyginopteris* (p. 231), and are consequently longer than the segments of vessels of many angiosperms. Such evidence as we have points to the earliest angiosperms having been bushes or small trees with anatomical features of this kind.

The evolution of xylem

Consideration of the evolution of morphological and anatomical features within the angiosperms as a whole is facilitated by the general structural relationship between the angiosperms and the gymnosperms. There are, of course, many features which are peculiar to the angiosperms, but these features are not present in all species. Since the fossil evidence, as will be discussed later (p. 337), indicates that the angiosperms were derived from some gymnospermous source, the more gymnosperm-like an anatomical or morphological character in an angiosperm, the more likely it is to portray the state of that character in the earliest representatives of the Class.

This concept can be usefully employed in considering the evolution of xylem, especially of the woody plants. Examination of large numbers of dicotyledonous woods has shown that the lengths of the vessel segments, although varying little within a species, show wide variation between species. There is also variation in the form of end-plate. Various combinations of length and form of end-plate are found, indicating that these two features can vary independently, but certain combinations occur more frequently than others. Long vessel segments (1·3–2·0 mm) tend to be associated with obliquely placed end-plates with scalariform perforations, and short segments with transverse end plates with single large pores (see Fig. 8.10). Long vessel segments are also associated with a number of other features, such as the angularity of the outline in transverse section, small cross-sectional area and uniform thickening of the walls. These associations, which are based upon the examination of large numbers of woods, are statistically sound and consequently require a biological explanation.

The most satisfactory interpretation of these data makes use of the 'principle of correlation', which maintains that in allied organisms undergoing evolution primitive features will remain associated with each other more frequently than would be expected were they associated solely by chance. Consequently we can identify long vessel segments and the features most frequently associated with them, since they are all suggestive of tracheids, as indicating the most primitive state of xylem vessels. Conversely, the features associated together at the other end of the range, namely shortness of vessel segments ($\leqslant$0·3 mm), circularity of outline in transverse section, and transversely placed and simply porous end plates, are those of the most recent kind of vessel.[39] Using this principle of correlation it is possible to identify the primitive and advanced states of other features of the xylem, such as the arrangement of the vessels in relation to each other and the nature of the parenchymatous rays.

The evolution of leaves

Although little can be said with certainty concerning the evolution of leaves, there are many indications of stem-like properties in the leaves of angiosperms, indicating their affinity with megaphylls (p. 187) rather than microphylls (p. 150). Apart from the stem-like branching of some leaves (p. 295) and the readiness of others to yield shoot buds (p. 326), many pinnate leaves are hardly distinguishable from lateral shoots of limited growth, bearing leaves in a single plane. In *Chisocheton*, a tree of south-east Asia with a large pinnate leaf, the apex of the main rachis even remains active for several seasons, producing a pair of pinnae annually.[22]

Comparatively recent evolution in the form of leaves has evidently been concerned in some instances with modification for a particular function. The insect traps of *Drosera* (sundew) Fig. 8.38, and of *Nepenthes* (pitcher plant) Fig. 8.39, are examples of this kind of development (see also Fig. 8.40). Sometimes evolution has been in relation to, and perhaps coupled with, that of a particular insect, usually a species of ant, which lives in and

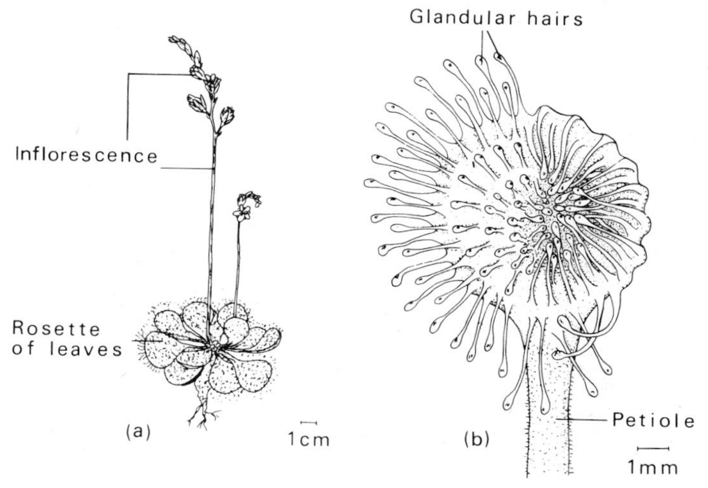

Fig. 8.38 *Drosera rotundifolia*. (**a**) Habit. (**b**) Single leaf in which some of the glandular hairs have closed on an entrapped insect. (After Kerner von Marilaun and Oliver, *The Natural History of Plants*. Blackie, London, 1902)

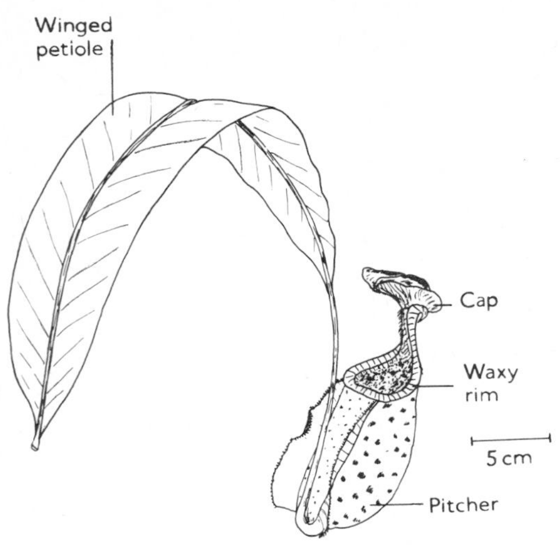

Fig. 8.39 *Nepenthes.* Leaf modified to form a pitcher-like insect-trap. (After *Bot. Mag.*, from Wettstein, *Handbuch der systematischen Botanik*. Deuticke, Leipzig, 1935.) The inner surface of the pitcher is covered with tile-like plates of wax making it impossible for the insect to climb out.

Fig. 8.40 *Dionaea muscipula* (Venus fly trap) : an insectivorous plant in which the leaf is highly modified. The terminal lamina of the leaf is in two halves which rapidly close together when the sensitive hairs on the surface are touched, the marginal teeth interlocking. Entrapped insects are digested by proteolytic enzymes coming from glandular hairs on the inner (adaxial) surface of the lamina. × 7.5.

is nourished by the plant concerned. In several tropical species of *Acacia*, for example, the long, thorn-like stipules are hollow and provide shelter, while the leaflets terminate in small globules of parenchyma (Belt's corpuscles) which are devoured by the ants (Fig. 8.41). Succulence of leaves (the anatomical and physiological modification of the mesophyll so that it is able to retain large quantities of water) also reaches its highest development in the angiosperms and is a feature of many desert and maritime species.

The evolution of flowers

The evolution of flowers presents numerous complex problems, in many instances evidently bound up with the evolution of particular insects or other organisms upon which the species have become dependent for pollination. The first problem, however, is to identify the most primitive kind of flower. Here we can again make use of the principle of correlation (p. 330), and the significant fact emerges that in those woody species in which the xylem consists wholly of tracheids the members of each region of the flower are separate from each other. Congenital fusion of parts (which involves, of course, no actual fusion, but the interpolation of intercalary growth in development) can with fair certainty be looked upon as a later feature in the evolution of the flower.

Quite apart from that towards the congenital fusion of parts within the flower, another tendency, clearly evident in some alliances, has been towards the evolution of compact inflorescences. The ultimate form of such

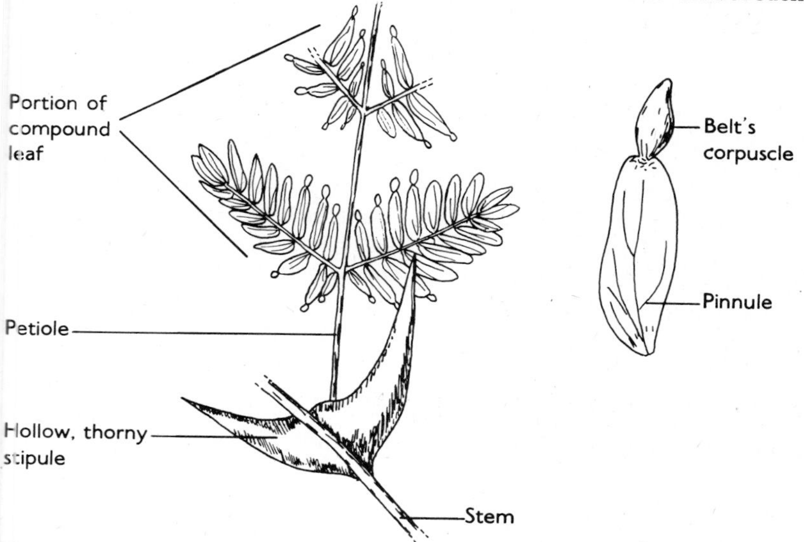

Fig. 8.41 *Acacia sphaerocephala*. Portion of leaf showing the hollow stipule, and Belt's corpuscles.

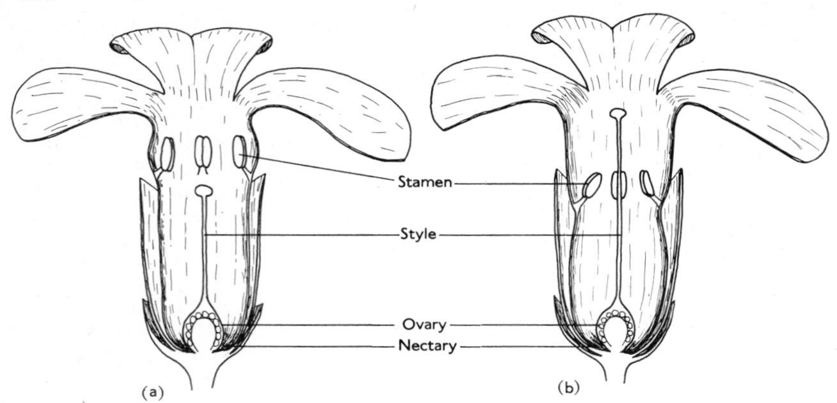

Fig. 8.42 *Primula* sp. Half flowers showing (**a**) thrum arrangement; (**b**) pin arrangement.

an inflorescence is the dense capitulum, occurring in a number of families, but characteristic of the Compositae. In many instances, owing to the differentiation of disc and ray florets, capitula have come to resemble superficially simple flowers, a feature well shown by the common daisy, *Bellis perennis*. In *Syncephalantha*, a Central American composite, we even find a racemose aggregate of capitula which, taken as a whole, also has a flower-like form owing to the asymmetry of the marginal capitula.[5] The sporadic occurrence of these flower-like inflorescences in a number of families indicates that the radiate form of the simple flower has some biological significance. Natural selection would account for the repeated emergence of this pattern with the increasing morphological complexity of the reproductive region.

The evolution of spurred and otherwise zygomorphic perianths, distinctive scents, nectaries and peculiar stamens (see p. 318), and also of striking pigmentation of the corolla, must clearly be considered in relation to pollination. This specialized field will not be considered further here, and reference must be made to the appropriate texts. Less conspicuous, but no less important, has been the emergence of mechanisms which prevent self-pollination and its possibly deleterious genetic effects. In *Primula*, for example, two kinds of flowers are present, 'thrum-eyed' with a short style and elevated stamens, and 'pin-eyed' with a long style and lower stamens (Fig. 8.42). An insect searching for nectar at the base of the corolla tube will tend to pass pollen from the stamens of the 'pins' to the styles of the 'thrums', and *vice versa*. This morphological device is accompanied by a physiological difference which causes the imperfect growth of 'pin' and 'thrum' pollen on their own styles. The mechanism, which is genetically controlled, depends upon the presence of antibodies in the style which precipitate the enzymes in the pollen tubes of pollen from the same or like flowers. The pollen of the thrum flowers is some 50 per cent greater in diameter than that of the pin, and this is correlated with a greater rate of extension of the pollen tube of the thrum pollen.

Other physiological systems are known which militate against self-pollination. Thus, in *Linum* successful growth of the pollen tube can take place only if the ratio of the osmotic pressure of the style to that of the pollen is of the order of 1:4, a relationship never present between pollen and style of the same flower. Continued outbreeding ensures the reassortment of the genetic material and the emergence of new forms. It should be noted that many of the mechanisms ensuring outbreeding in the angiosperms depend upon the ovule being enclosed and accessible only through sporophytic tissue. The evolution of the carpel can therefore be regarded as an event of basic importance, making possible evolution of breeding systems which, by promoting genetic diversity, have probably contributed considerably to the rapid evolutionary advance of the flowering plants.

Physiological evolution

To conclude this outline of evolution within the angiosperms, mention should also be made of the striking physiological evolution that in some species and families has evidently accompanied the morphological. Occasionally this has extended to the abandonment of autotrophy, the species concerned becoming in consequence obligate heterotrophs, a transition already seen in other Divisions of the Plant Kingdom. Examples amongst the angiosperms are provided by *Epipogium*, an orchid which lacks chlorophyll and derives its nutrition from a highly developed mycorrhiza, and such parasites as *Lathraea* (toothwort), *Orobanche* (broomrape) and *Cuscuta* (dodder) (Fig. 8.43). A number of angiosperms, such as *Viscum* (mistletoe) and *Euphrasia* (eyebright), have become parasitic without losing the ability to photosynthesize, so they are not entirely dependent upon their hosts.

Physiological evolution has also extended to the appearance of some species or races tolerant to relatively high concentrations of normally toxic metals. The lead-tolerant forms of the grasses *Agrostis tenuis* and *Festuca ovina*, found on spoil heaps of old mines, are striking examples of recent evolution of this kind.

THE ORIGIN OF THE ANGIOSPERMS

The age of the angiosperms

There remains the problem of the origin of the angiosperms, and this inevitably raises that of the time of their first appearance in the earth's vegetation. Unfortunately, upon this cardinal point there is no general agreement. Although some have imagined the angiosperms to have appeared

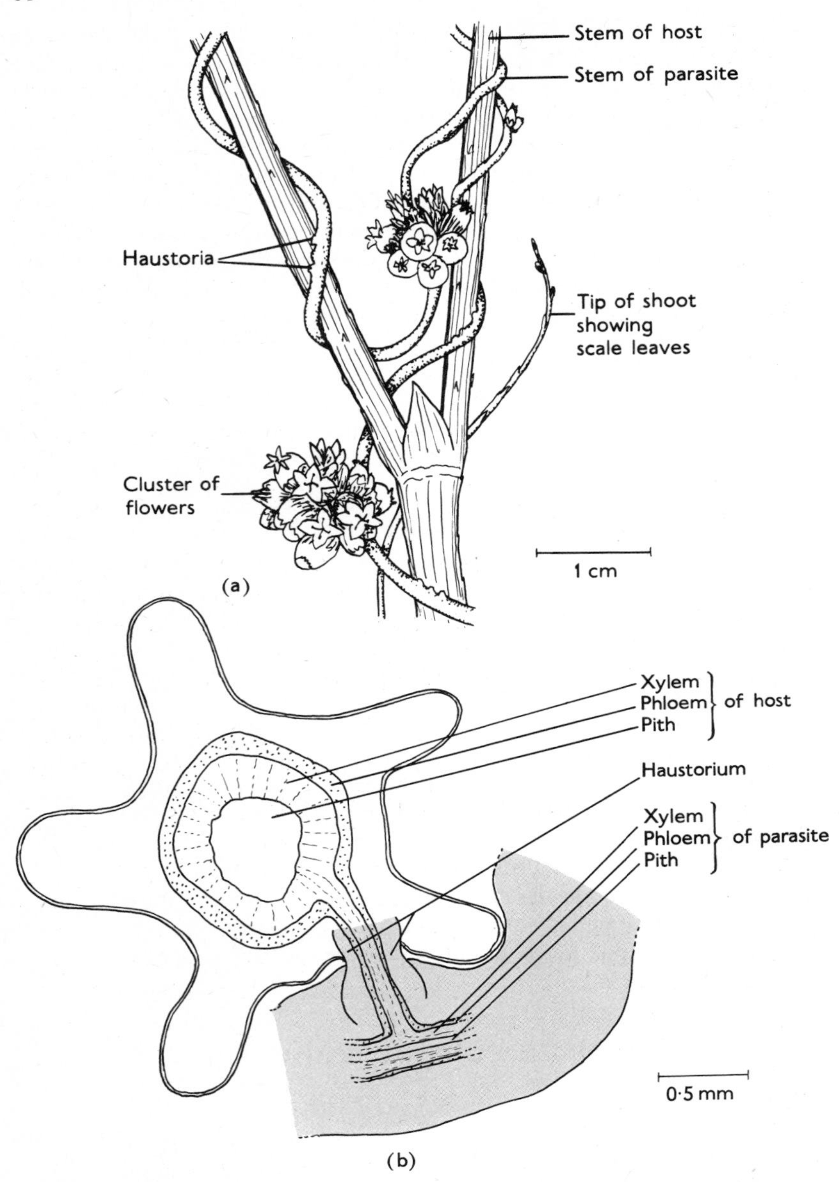

Fig. 8.43 *Cuscuta*. (**a**) Habit. (After Kerner von Marilaun and Oliver, *The Natural History of Plants*. Blackie, London, 1902.) (**b**) Transverse section of host stem showing the penetration of the vascular tissue of the parasite into that of the host.

in upland regions (and hence away from sites of fossilization) as early as the end of the Palaeozoic, this view owes more to inference than direct evidence, and has been disputed by many. There are admittedly several reports of vegetative remains of angiosperms from early Mesozoic rocks, but these should for the present be accepted with caution since the existence of angiosperms at this time has not yet received support from the study of fossil pollen. The identification of pollen, based upon the fine details of the structure of the wall and its ornamentation, has now reached a very high order of reliability, and the pollen of angiosperms can be readily distinguished from that of gymnosperms. So far as sampled, rocks of basal Cretaceous and greater age have yielded no well-established records of the angiosperm pollen,[48] and there are consequently few grounds for believing that the angiosperms were in existence before the Cretaceous period. In those few instances where remains whose angiospermous nature seems beyond question[77b] have been described as pre-Cretaceous, there is a variety of geological reasons why the age of the beds containing them cannot be taken as certain.

If the angiosperms arose at the end of the Jurassic or at the beginning of the Cretaceous period, this would imply that in about 25 million years (i.e. in about half the Cretaceous period) they had become the dominant component of the earth's vegetation. Provided, once the transition had been made, evolution within the angiosperms was comparatively rapid, this time scale is in no way unreasonable. The angiosperms may in fact have enjoyed from the time of their origin the advantage of relatively rapid sexual reproduction. In the angiosperms the interval between the production of the flower and the setting of the seed is usually a matter of weeks, sometimes only of days (e.g. in the so-called 'desert ephemerals'), whereas in most gymnosperms the equivalent process takes at least a year. The angiosperms thus have the possibility of more frequent sexual reproduction and consequently the opportunity for more rapid evolutionary advance (see p. 329).

Once initiated, if better suited to the environment than the existing gymnosperms, the angiosperms no doubt soon became dominant. The rapidity with which better adapted immigrant species will replace native is very striking. In parts of central Portugal, for example, the arboreal vegetation has been largely replaced in less than a century by introduced Australian species.

The nature of the immediate precursors of the angiosperms

The immediate ancestors of the angiosperms should therefore probably be sought in late Jurassic and early Cretaceous rocks. No fossils of this age have yet been generally accepted as intermediate between gymnosperms and angiosperms, although the Caytoniales (p. 246) were possibly close to the transitional forms. Despite this absence of direct evidence, however,

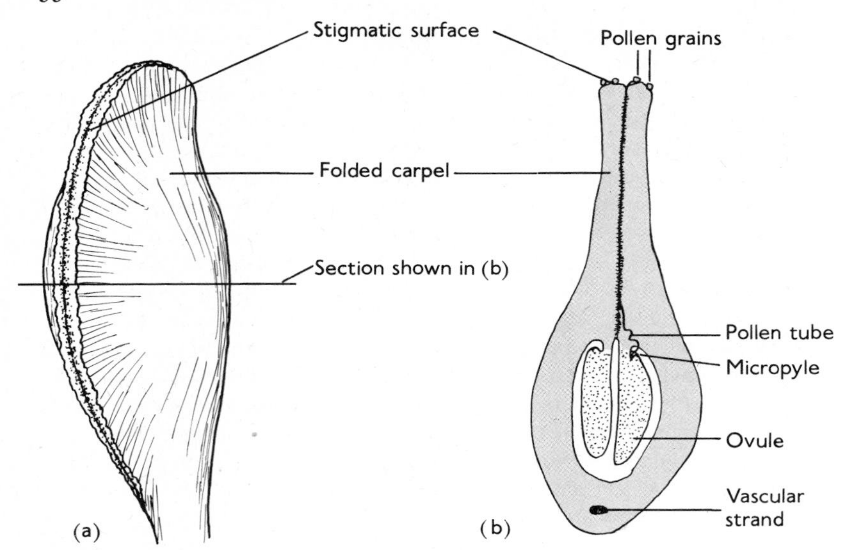

Fig. 8.44 *Drimys piperita*. (**a**) Carpel with stigmatic surface along the ventral suture. (**b**) Transverse section of carpel. (Both after Bailey and Swamy, from Foster and Gifford, *Comparative Morphology of Vascular Plants*. Freeman, San Francisco, Copyright © 1949)

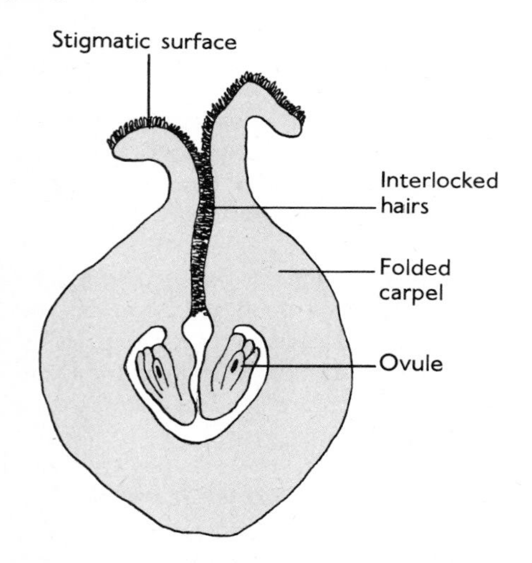

Fig. 8.45 *Degeneria* sp. Transverse section of conduplicate carpel with infused margins. (After Swamy, from Foster and Gifford, *Comparative Morphology of Vascular Plants*. Freeman, San Francisco, Copyright © 1949)

examination of the carpels of living angiosperms, particularly those possessing primitive wood (p. 329), gives strong indications of how angiospermy developed. In several of these genera the carpels are simple folded structures, with stigmatic surfaces lying along part or whole of the ventral suture. This is well shown in *Drimys piperita* (Fig. 8.44), a vessel-less dicotyledon. In *Degeneria* (Fig. 8.45), the carpel is similar, but is even more striking. The ventral margins of the carpel do not fuse, but become interlocked by papillae, between which the pollen tubes force their way. The principle of correlation thus indicates that the earliest angiospermous carpel was a conduplicate structure with a lateral stigma along the approximated margins. The carpel with a capitate stigma elevated on a style must therefore be regarded as a later development.

Fossils of possibly great significance in the evolution of the angiosperms have come from the Permian of South Africa.[65] They are associated with a reticulately veined leaf, *Glossopteris* (Fig. 8.46), characteristic of the Permo-Carboniferous of the southern hemisphere, and long believed to be the leaf of a seed-plant. Certain specimens of this leaf have been found in connection with reproductive organs. These are interpreted as consisting of two valves, enclosing both ovules and microsporangiophores, borne on a stalk apparently arising from the face of the leaf. This interpretation has aroused much controversy, an indication of the great importance of these fossils. Even though further investigation may require amendment of the reconstructions so far published, it seems clear that already by the end of the Palaeozoic era there was a clear tendency for ovules to become enclosed in possibly carpel-like structures. It was from the early gymnosperms showing this tendency that the angiosperms were presumably evolved.

Some have suggested that the angiosperms had more than one origin, i.e. that they, like the gymnosperms, are polyphyletic. This, however, seems unlikely in view of the general occurrence of the characteristic 'double fertilization' in the embryo sac. Nevertheless, the relationship between some monocotyledons, especially the palms, and dicotyledons certainly appears remote, and we must assume that dicotyledons and monocotyledons have followed more or less parallel lines of evolution for a considerable period.

THE MAIN TRENDS OF ANGIOSPERM EVOLUTION

To summarize, we can envisage the main evolutionary trends in the angiosperms to have been as follows. They emerged from the gymnosperms, probably, in view of the radial symmetry of the ovules and certain anatomical similarities, from a group evolving from the pteridosperms of the late Palaeozoic. The angiosperms thus inherited a seed-habit of long

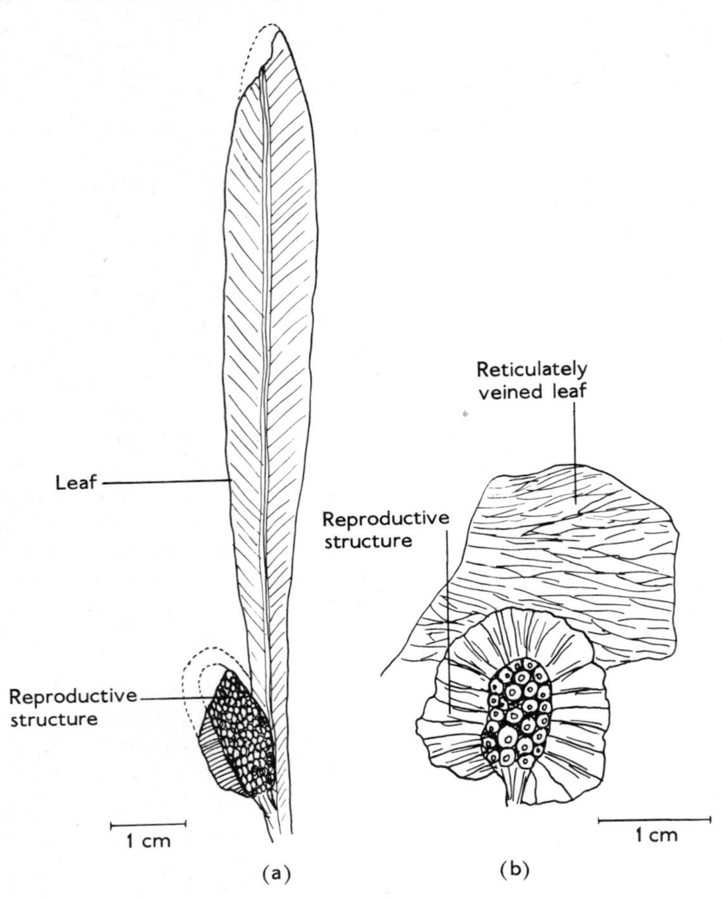

Fig. 8.46 *Glossopteris* sp. (**a**) Compression fossil of leaf and associated reproductive structure. (**b**) A further type of reproductive structure. (From photographs by Plumstead, *Paleontographica*, 100 B, 1 (1956))

standing, and a serviceable form of axial structure in which lateral branch systems had long ago become megaphylls, and in which megaphylls in turn were becoming differentiated into petioles and laminae with reticulate venation. They improved upon this structure, replacing coarse and angular ramification by elegant branching. Anatomically, strengthening tissues tended to become reduced to those required by mechanical laws, and stems and branches consequently acquired a springiness that is a protection against storms. Vessels were also developed from tracheids, and a more highly organized phloem containing sieve tubes and companion cells from simple sieve cells and parenchyma, both developments probably having advantageous physiological consequences. The reproductive regions became highly specialized, their evolution often being correlated with that of pollinating organisms.

We have considered the primitive angiosperms to be woody (p. 329), and the evolution of herbaceous angiosperms consequently a later phenomenon, certainly in the grasses correlated with that of the grazing animals. In tropical regions many herbaceous angiosperms are in fact epiphytes upon angiospermous trees and shrubs, and it seems clear that these followed rather than preceded the arboreal vegetation. Because of the rapidity with which the relatively succulent leaves of angiosperms decay in tropical conditions, humus is soon formed wherever they accumulate. Consequently, fertile substrata are found in abundance on horizontal surfaces and in the crotches of branches. This leads to an ecological situation different from that in gymnospermous forests, where the leaves mostly reach the ground and the humus, owing to the slow decay of the leaves, is covered by harsh and unconsolidated material unfavourable for the establishment of delicate herbaceous forms. It thus seems reasonable to suppose that the epiphytic angiosperms (and ferns) emerged in response to and are in the process of exploiting the new ecological habitat created by the rise of the angiosperm forests. Their evolution would then have been, in geological time, a relatively recent event.

Today the angiosperms in some ways resemble successful individuals reaching affluent middle age, determined to let no human experience escape them. They experiment freely with metabolic novelties, such as fragrant oils, peculiar alkaloids and unusual carbohydrates (such as inulin). In sexual reproduction, both as regards pollination and the cytology of the gametophytes, they show almost every conceivable variation. We probably witness the angiosperms at the height of their success. What they will be like in their senescence we can safely leave to our successors.

Glossary

Terms relating to special features defined in the text are not repeated here. These may be traced through the Index.

Abaxial, of lateral organs. The side away from the main axis (the same as DORSAL).

Abscission. The separation of structures from the stem of a plant after the formation of a layer of specialized cells (abscission layer).

Achene, of angiosperms. A dry, indehiscent, one-seeded FRUIT.

Acropetal. Developing from the base towards the apex, i.e. the youngest structures occur above, or, with regard to the movement of materials, towards the apex.

Actinomorphic. Radially symmetrical.

Adaxial, of lateral organs. The side towards the main axis (the same as VENTRAL).

Adventitious. Plant organs which appear in unusual positions, e.g. roots from stems, or buds not in the axils of leaves.

Adventive embryony. The development of EMBRYOS from cells other than EGG cells, and without FERTILIZATION.

Aerenchyma. A plant tissue composed of unthickened, often irregularly shaped, cells surrounding large air spaces.

Aerophore. A negatively geotropic root produced by certain plants growing in waterlogged conditions.

Akinete, of algae. A single-celled, non-motile, resting SPORE in which the original cell wall forms part of the SPORE wall.

Allopolyploidy. The type of POLYPLOIDY found in organisms of hybrid origin, at least two complete sets of CHROMOSOMES from each of the original parents being present.

Alternation of generations. A life cycle in which a SPOROPHYTE is succeeded by a GAMETOPHYTE and this in turn by a sporophyte.

Amphigastria, of liverworts. The lower leaves of a leafy liverwort, usually inserted transversely.

Amphiphloic. Having PHLOEM tissue on each side of the XYLEM.

Amyloplast. A PLASTID without CHLOROPHYLL, involved in the synthesis and storage of starch.

Androecium, of angiosperms. A collective term for the stamens of the FLOWER.

Aneuploid. Possessing a chromosome number not an exact multiple of the HAPLOID number.

Anisogamy. The production or fusion of GAMETES which are visibly different, usually in size.

Annual. A plant which completes its life cycle in one growing season and then dies.

Anther, of angiosperms. A group of fused MICROSPORANGIA.

Antheridium. A male sex organ producing ANTHEROZOIDS.

Antherocyte. The cell which differentiates to form an ANTHEROZOID (syn. SPERMATOCYTE).

Antherozoid. A motile male GAMETE (syn. SPERMATOZOID).

Anticlinal. Of cell divisions in which the new cell wall is perpendicular to the outer surface of the region in question.

Aphlebiae, of ferns. Small, leafy outgrowths at the base of the RACHIS of certain fossil and living ferns.

Aplanospore, of algae. A single-celled, non-motile SPORE which is liberated from the parent cell.

Apogamy. The ASEXUAL production of a SPOROPHYTE from a GAMETOPHYTE.

Apomixis, of seed plants. Reproduction by SEEDS which have not developed as a consequence of a sexual process.

Apophysis, of mosses. The sterile base of the SPORE CAPSULE.

Apospory. The production of a GAMETOPHYTE directly from the SPOROPHYTE without the intervention of MEIOSIS.

Archegonium. The flask-shaped, multicellular, female sex organ characteristic of certain land plants.

Archesporium. The group of cells giving rise to the SPORE mother cells (MEIOCYTES).

Areola (Areole). Area surrounded by VEINS in a RETICULATELY veined leaf. The tuft of spines indicating a lateral BUD in some succulents.

Aril. A fleshy outer covering of the OVULE of certain SEED plants.

Asexual. Reproduction not involving the fusion of GAMETES.

Autopolyploidy. The occurrence of three or more identical sets of CHROMOSOMES in the nucleus.

Autosome. A CHROMOSOME that is not a sex chromosome.

Autotrophic. Able to make its own food from inorganic raw materials.

Auxin. A general term for organic substances (naturally occurring or synthetic) which promote or regulate plant growth.

Auxospore, of diatoms. SPORES formed as the result of a rejuvenating process which may or may not include sexual fusion.

Axil. The upper angle which a leaf makes with a stem.

Basipetal. Developing from the apex towards the base, i.e. the youngest structures occur below, or, with regard to the movement of materials, towards the base.

Berry, of angiosperms. A many-seeded FRUIT with a fleshy PERICARP, produced from a single FLOWER.

Biennial. A plant in which the life cycle requires two growing seasons for completion, reproduction being followed by death.

Bifid. Divided deeply into two parts.

Bifurcate. To divide into two branches.

Bilateral symmetry. Symmetry with respect to a single plane; when divided by the plane, the two halves are mirror-images.

Blade. The broad, flattened portion of a leaf (the lamina).

Bordered pit. A PIT in which the thin area of primary wall is overhung by lignified secondary wall. The centre of the PIT is often thickened, forming the TORUS (see Fig. 7.20).

Bract, of angiosperms. A reduced leaf with a FLOWER in its AXIL.

Bracteole, of angiosperms. A reduced leaf produced on a FLOWER stalk.

Bract scale, of gymnosperms. The scale (in some species becoming woody) in the female CONE in the AXIL of which arises the OVULIFEROUS SCALE.

Bud. The apical MERISTEM of a stem, surrounded by leaf primordia and sometimes enclosed in scale leaves.

Bulb, of angiosperms. An underground perennating and storage organ, consisting of a reduced stem surrounded by fleshy leaves.

Callus. A PARENCHYMATOUS tissue formed over wounds, often capable of being cultured indefinitely in nutrient media.

Calyptra, of bryophytes. The protective covering of the CAPSULE, derived from the base of the neck of the ARCHEGONIUM.

Calyx, of angiosperms. A collective term for the SEPALS of a FLOWER.

Cambium. A secondary MERISTEMATIC tissue: the VASCULAR CAMBIUM produces secondary XYLEM and PHLOEM; the CORK CAMBIUM produces CORK and secondary CORTEX.

Capitulum, of angiosperms. An INFLORESCENCE in which the FLOWERS are closely grouped on a disc.

Capsule, of bryophytes. The SPORE-containing organ.

Carotenoids. A group of yellow pigments occurring in CHLOROPLASTS, CHROMOPLASTS and elsewhere.

Carpel, of angiosperms. The OVULE-bearing structure, usually regarded as a MEGASPOROPHYLL.

Carpogonium, of red algae. The female sex organ.

Carpospore, of red algae. A non-motile unicellular SPORE formed after division of the ZYGOTE.

Casparian band. A girdle of SUBERIZED thickening occurring on the radial and upper and lower walls of the cells of the ENDODERMIS.

Cauliflory, of angiosperms. The continued lateral production of FLOWERS on old woody stems.

Cauline. Appertaining to the stem.

Cell sap. The watery contents of the large VACUOLE usually present in plant cells.

Centrifugal. Developing from the centre towards the outside, i.e. the oldest structures are at the centre, and the youngest at the outside.

Centripetal. Developing towards the centre from the outside, i.e. the oldest structures are at the outside, and the youngest at the centre.

Chamaephyte. A plant in which the perennating BUDS occur within 20–30 cm of the soil surface (dwarf shrubs).

Chemotaxis. The movement of a motile organism, SPORE, or GAMETE in response to a chemical stimulus.

Chimaera. An organism with tissues of more than one GENOTYPE. It may be produced by mutation or by grafting.

Chlorophyll. A group of structurally similar green or blue-green PHOTOSYNTHETIC pigments common to all AUTOTROPHIC plants.

Chloroplast. An ORGANELLE bounded by a distinct membrane, containing the photosynthetic pigments, and found in all EUCARYOTIC photosynthesizing cells.

Chromatin. A complex of deoxyribonucleic acid and basic proteins present in CHROMOSOMES and defined by its staining properties.

Chromatophore. An intracellular body containing pigments, hence a general class containing CHLOROPLASTS, but not often used for the latter.

Chromoplast. A PLASTID lacking CHLOROPHYLL, but accumulating pigments, principally CAROTENOIDS.

Chromosomes. The organized structures in the NUCLEUS, carrying the Mendelian genes.

Cilium. A short, thread-like locomotory ORGANELLE which beats in a regular manner. Usually numerous.

Circumnutation. The sweeping movements made by the tip of a stem, particularly marked in twining plants, caused by unequal rates of growth around the axis.

Cladode, of angiosperms. A short, flattened stem of limited growth, replacing the leaves as the site of photosynthesis.

Coccoid, of algae. A unicellular, non-motile, spherical condition of the adult organism.

Coenobium. A COLONY of cells in which there is some degree of co-ordination.

Coenocyte. A multinucleate plant body or cell.

Coleoptile, of angiosperms. The sheath around the PLUMULE of certain MONOCOTYLEDONOUS genera.

Coleorhiza, of angiosperms. The sheath around the RADICLE of MONOCOTYLEDONOUS genera.

Collenchyma. A strengthening tissue composed of columnar cells with heavy deposits of cellulose at the angles of the primary walls.

Colloid. A system of minute, charged particles suspended in a liquid. The colloid itself may be in the form of a liquid (sol) or solid (gel).

Colony, of algae. A group of adhering unicellular organisms, all of which are equivalent and independent of one another.

Colpus. Each of the thin areas in the EXINE of a POLLEN grain, through one of which the pollen tube usually emerges.

Columella, of bryophytes. The central, sterile tissue found in certain SPORANGIA.

Comal. Occurring at the tip or apex.

Companion cells, of angiosperms. The small, nucleate cells with dense PROTOPLASM associated with SIEVE TUBES in the PHLOEM tissue.

Complanate. Arranged in a single plane.

Conceptacle, of brown algae. The flask-shaped cavity containing the sex organs in certain genera.

Concrescent. Grown together, coalesced.

Cone. A SHOOT bearing SPOROPHYLLS (usually closely packed).

Congenital (as used in ***Congenital fusion***). An inherited characteristic which becomes apparent during growth.

Coplanar. In the same plane.

Coradial. On the same radius.

Coralloid. Coral-like as a consequence of repeated branching.

Cordate. Heart-shaped.

Cork. The outer tissue formed by the activity of the CORK CAMBIUM and consisting of cells which become SUBERIZED and impervious to water.

Corm, of angiosperms. A subterranean storage and perennating organ consisting of a short swollen, stem base.

Corolla, of angiosperms. A collective term for the PETALS of a FLOWER.

Corona, of the Charales. A cluster of cells at the distal end of the OOGONIUM.

Corpus, of angiosperms. The central part of the apex of a stem in which cell divisions occur in a variety of planes.

Cortex. The tissue of ROOTS and SHOOTS which occurs between the VASCULAR TISSUE and EPIDERMIS.

Cotyledons, of seed plants. The first leaf or leaves of the EMBRYO.

Crenulate. Toothed, but the teeth small and rounded.

Cryptophyte. A plant in which the perennating BUDS are below the soil surface.

Cupule, of seed plants. An extra cup-shaped organ surrounding the ovule of certain fossil plants, or a cup formed by concrescent bracts at the base of the fruit of certain living genera (e.g. acorn).

Cuticle. The non-cellular waxy coating of the EPIDERMIS of all land plants.

Cymose branching. The type of branching in which the apices abort successively, growth being continued by laterals in each instance.

Cystocarp, of red algae. The complex SPORE-containing structure formed round the OOGONIUM after fertilization.

Cytoplasm. The PROTOPLAST of a cell, apart from the NUCLEUS.

Deciduous plant. One which sheds its leaves at intervals, in temperate regions usually at the end of the growing season, by means of an ABSCISSION layer.

Decumbent, of a shoot. Either initially, or soon becoming, prostrate.

Decussate. Arranged in opposite pairs, alternate pairs being at right angles to each other.

Dehiscence. Opening at maturity, by splitting, especially of SPORANGIA and FRUITS.

Diarch xylem. XYLEM tissue consisting of two strands, or having two PROTOXYLEM groups.

Dichotomous branching. Branching into two equal parts.

Diclinous, of angiosperms. Having STAMENS and CARPELS in separate FLOWERS, but on the same individual. Also used correspondingly of ARCHEGONIATE plants.

Dicotyledonous. Having two COTYLEDONS.

Dictyostele. A STELE composed of a network of MERISTELES.
Differentiation. The process of development of cells, tissues and organs by which differences in structure arise.
Dimorphic. Existing in two separate forms.
Dioecious. Bearing male and female sex organs on different individuals.
Diploid ($2n$). Having twice the basic (HAPLOID) number of CHROMOSOMES.
Distal. Distant from a point of reference or symmetry (ant. PROXIMAL).
Distichous. Arranged in two vertical ranks, diametrically opposed.
Dorsal, of lateral organs. The same as ABAXIAL.
——, of prostrate plants. The side away from the substratum.
Dorsiventral. Having distinct upper and lower sides.
Drupe, of angiosperms. A one-SEEDED FRUIT, formed from a single flower, in which the PERICARP is part fleshy and part stony.

Egg. A non-motile, female GAMETE.
Elaters, chiefly of bryophytes. Elongated cells with bands of thickening which assist SPORE dispersal, or bands of flexible material with a similar function (*Equisetum*).
Emarginate, of leaves. Having a small depression at the apex.
Embryo. The young, partly developed SPOROPHYTE.
Embryogeny. The process by which the EMBRYO develops from the ZYGOTE.
Embryo sac, chiefly of angiosperms. The mature female GAMETOPHYTE containing a number of cells with indistinct boundaries, one of those at the MICROPYLAR end of the sac being the EGG cell.
Endarch, of xylem. XYLEM tissue which differentiates CENTRIFUGALLY, leaving the PROTOXYLEM nearest the centre.
Endodermis. The inner layer of cells of the CORTEX of a stem or root, distinguished by the presence of CASPARIAN BANDS, and often by starch grains.
Endogenous. Developing or emanating from the inside.
Endomitosis. Duplication of the CHROMOSOMES without nuclear division, resulting in POLYPLOIDY.
Endophyte. A plant which lives inside another organism, but is not PARASITIC upon it.
Endopolyploidy. POLYPLOIDY which is the result of ENDOMITOSIS.
Endoscopic embryogeny. EMBRYOGENY in which the first division of the ZYGOTE is transverse, the apex of the EMBRYO developing from the inner cell.
Endosperm, of angiosperms. The nutritive tissue in SEEDS derived from the primary endosperm cell of the EMBRYO SAC. Notable for its frequently containing TRIPLOID nuclei and being a source of AUXINS.
Endosporic gametophyte. A GAMETOPHYTE which develops within the SPORE wall.
Endosymbiosis. SYMBIOSIS between two organisms, one of which lives inside the other.
Enzyme. A soluble protein, or protein complex, which catalyses a biochemical reaction.
Epidermis. The primary outer layer of cells of all plant organs.

Epigeal germination, of seed plants. Germination in which the COTYLEDONS are raised above the soil surface by elongation of the HYPOCOTYL.
Epigynous, of angiosperms. The insertion of the PERIANTH and ANDROECIUM above the GYNAECIUM of a flower.
Epiphyte. A plant which grows attached to another, but is not PARASITIC upon it.
Etiolation. The abnormal type of growth occurring in heavy shade or darkness, particularly the reduction in the amount of CHLOROPHYLL, and an exaggerated elongation of the INTERNODES.
Eucaryotic. Of organisms in which the NUCLEUS is enclosed in a distinct membranous envelope.
Eusporangiate. Of plants in which the SPORANGIA develop from a group of initial cells.
Eustele. A STELE consisting of a ring of collateral bundles.
Exarch. XYLEM tissue which differentiates CENTRIPETALLY, leaving the PROTOXYLEM furthest from the centre.
Exine. The outer layer of the wall of a SPORE or POLLEN grain.
Exogenous. Developing or emanating from the outside.
Exoscopic embryogeny. EMBRYOGENY in which the first division of the ZYGOTE is transverse, the apex of the EMBRYO developing from the outer cell.
Exoskeleton, of algae. A firm outer layer, especially calcium carbonate, secreted by certain species.

Facultative. Of a response which depends upon conditions; implying that the organism in other conditions is able to respond in another way.
Falcate. Sickle-shaped.
Fascicular. Relating to a cluster or bundle, usually implying association with VASCULAR bundles. Hence FASCICULAR CAMBIUM.
Fertilization. The fusion of GAMETES.
Fibre cell. An elongated cell with a thick lignified wall, tapering at each end.
Filament, chiefly of algae. A uniseriate row of cells.
Filiform. Thread-like.
Fission. The ASEXUAL division of a unicellular organism into two similar organisms.
Flagellum. A whip-like organ of locomotion, single, paired, or many, found in motile algae and GAMETES.
Flower, chiefly of angiosperms. The reproductive organs and associated structures.
Foot. The portion of the EMBRYO in the bryophytes and lower tracheophytes which absorbs nutrients from the GAMETOPHYTE.
Frond. The leaf of a fern, or of a cycad.
Fruit, of angiosperms. The SEED-containing structure derived from the OVARY, sometimes associated with other floral parts.
Fusion, in relation to organs. Used figuratively, implying CONCRESCENCE either during development or in the course of evolution.

Gametangium, of algae. An organ producing GAMETES.

Gamete. A mature sex cell, capable of fusing with another to form a ZYGOTE.

Gametophyte. The (normally) HAPLOID generation, producing GAMETES which fuse to form the DIPLOID SPOROPHYTE generation.

Gemma, chiefly of bryophytes. A multicellular, ASEXUAL, reproductive structure which becomes detached from the parent and is capable of growing into a new plant.

Generative cell, of seed plants. The cell of the male GAMETOPHYTE which by division gives rise to the two male GAMETES.

Genotype. The genetic constitution of an organism.

Geotropism. A growth movement of a plant dependent upon the direction of gravitation.

Glochidia, of aquatic ferns. Hooked processes developed on the MICROSPORANGIA of certain genera.

Guard cell. One of a pair of cells, containing chlorophyll, which occur in the EPIDERMIS and form a STOMA.

Gynaecium, of angiosperms. A collective term for the female reproductive structures.

Halophyte. A plant living in a habitat of high sodium chloride concentration, e.g. a saltmarsh.

Haploid (*n*). Possessing a single set of CHROMOSOMES.

Hapteron. Same as HOLDFAST. In *Equisetum*, a synonym for ELATERS.

Haustorium. An absorptive organ of PARASITES which penetrates the HOST tissue.

Hemicryptophyte. A plant whose perennating buds occur at or very close to the soil surface.

Heterocyst, of blue-green algae. A large cell, appearing empty, found in many filamentous genera.

Heteromorphic. Having more than one growth form.

Heterophyllous. Having more than one leaf form.

Heterospory. The production of spores of two sizes (MEGASPORES and MICROSPORES).

Heterothallic, of algae. The condition in which GAMETES from the same individual or strain of individuals cannot fuse.

Heterotrichous, of algae. Having two kinds of FILAMENTOUS growth, one prostrate and the other erect, in the same individual.

Heterotrophic. Unable to manufacture food from inorganic raw materials.

Histone. A highly basic protein often associated with deoxyribonucleic acid in the CHROMOSOMES.

Holdfast, of algae. The organ or cell which anchors the plant to the substratum.

Homologous. Said of organs believed to be identical in nature (but not necessarily in form), often (but not always) implying an evolutionary relationship.

Homospory. The production of only one type of SPORE.

Homothallic, of algae. The opposite of HETEROTHALLIC—the condition in which GAMETES from the same individual are able to fuse.

Hormocyst, of blue-green algae. A HORMOGONIUM which is enclosed in a thick wall.

Hormogonium, of blue-green algae. A portion of a FILAMENT which becomes detached and may grow into a new plant.

Host. The organism from which a PARASITE obtains its food.

Hyaline. Clear, translucent.

Hydathode. A water-secreting gland or pore found in the leaves of many plants.

Hygroscopic. Water absorbing.

Hypocotyl. The stem of a seedling between the ROOT and the COTYLEDONS.

Hypogeal germination, of seed plants. The type of germination in which the COTYLEDONS remain below the surface of the soil, the HYPOCOTYL not elongating.

Hypogynous, of angiosperms. A FLOWER in which the PERIANTH and ANDROECIUM are inserted below the GYNAECIUM.

Incompatibility. The inability of conspecific gametes to fuse and form a ZYGOTE, resulting from a number of different mechanisms in different kinds of plants.

Indusium, of ferns. The covering of the SORUS.

Inflorescence, A group of FLOWERS with a common stalk.

Integuments, of seed plants. The envelopes surrounding the NUCELLUS.

Intercalary. Of MERISTEMS that lie between two apices.

Internode. The portion of a stem occurring between two NODES.

Intine. The inner layer of the wall of a SPORE or POLLEN grain.

Involucre, of angiosperms. A whorl of BRACTS surrounding a flower or group of FLOWERS.

—, of bryophytes. Leaves or other structures surrounding sex organs.

Isodiametric. Of equal height, length and breadth.

Isogamy. The condition in which GAMETES are morphologically identical.

Isomorphic. Of similar form.

Isthmus, of desmids. The portion which separates the two halves of the cell.

Lacuna. A cavity.

Lamina. See BLADE.

Lanceolate. Spear-shaped; flattened, and tapering to a point, with the widest part near the centre.

Leaf gap. A gap in the VASCULAR cylinder of a stem immediately above the departure of a LEAF TRACE.

Leaf trace. The VASCULAR supply to a leaf, consisting of one or more strands.

Lenticel. A channel filled with loosely packed CORK cells permitting the diffusion of gases in stems (and occasionally in roots), formed by a characteristically shaped CORK CAMBIUM.

Leptosporangiate. Of plants in which the SPORANGIA develop from a single initial cell.

Leucoplast. A PLASTID lacking CHLOROPHYLL and other pigments, but often containing starch.

Lignin. A complex organic material deposited in the walls of certain cells.

Ligule, of grasses. A collar-like outgrowth at the inner junction of the leaf-sheath and BLADE.

—, of lycopods. A scale inserted into the upper surface of the leaf.

Littoral zone. The zone on the sea shore between the high- and low-tide marks.

Loculus, of angiosperms. The cavities in an OVARY, each containing one or more OVULES.

Lumen. The space bounded by a cell wall, more often used in relation to dead cells than living.

Lysigenous (of spaces). Formed by the dissolution of a cell or cells (cf. SCHIZOGENOUS).

Macrandrous, of the Oedogoniales. A species which does not produce dwarf male filaments.

Medulla. The central region of an axis or organ.

Megasporangiophore. A structure bearing MEGASPORANGIA.

Megasporangium. A sporangium producing only MEGASPORES.

Megaspore. A SPORE which on germination produces a female GAMETOPHYTE.

Megasporophyll. A structure bearing MEGASPORANGIA, and believed to be equivalent to a leaf.

Meiocyte. A cell in which MEIOSIS occurs.

Meiosis. The two consecutive nuclear divisions, in the first of which the CHROMOSOME number is halved, so that each NUCLEUS ultimately contains only one set of chromosomes.

Meristele. The unit of a DICTYOSTELE, composed of a central strand of XYLEM completely surrounded by PHLOEM and delimited by an ENDODERMIS.

Meristem. A group of undifferentiated cells remaining capable of division.

Mesophyll. The PARENCHYMATOUS tissue of leaves lying between the upper and lower epidermis; the site of most of the CHLOROPLASTS.

Metabolism. A collective term for the biochemical processes occurring in living cells.

Metaxylem. The portion of XYLEM tissue which is derived from the PROCAMBIUM (and is hence PRIMARY) and which DIFFERENTIATES after the PROTOXYLEM.

Micropyle, of seed plants. The minute pore in the distal end of the INTEGUMENTS of the OVULE.

Microsporangiophore. A structure bearing MICROSPORANGIA.

Microsporangium. A sporangium producing only MICROSPORES.

Microspore. A SPORE which on germination produces a male GAMETOPHYTE.

Microsporophyll. A structure bearing MICROSPORANGIA, and believed to be equivalent to a leaf.

Midrib. The central VEIN of a leaf.

Mitochondrion. An ORGANELLE of variable shape bounded by a distinct membrane, and occurring in all EUCARYOTIC cells. The site of the major part of the process of respiration.

Mitosis. Nuclear division in which two identical NUCLEI are formed.

Monoclinous, of angiosperms. Having functional STAMENS and CARPELS in the same FLOWER. Also used correspondingly of ARCHEGONIATE plants.

Monocolpate. Said of a SPORE or POLLEN grain which possesses only one COLPUS.

Monocotyledonous, of angiosperms. An EMBRYO possessing a single COTYLEDON.

Monoecious. The condition in which both male and female sex organs are produced on one plant.

Monolete spore. The kind of SPORE which, having been formed in a tetrad without tetrahedral symmetry, bears a linear scar on its PROXIMAL face.

Monopodial branching. The type of branching in which the main apex remains active, the shoots produced from AXILLARY BUDS remaining clearly lateral.

Monospores, of certain genera of brown algae. APLANOSPORES containing four NUCLEI.

Morphology. The study of the form and related anatomy of living organisms.

Mycorrhiza. A SYMBIOSIS between the roots of a plant and a fungus.

Myrmecophily. The association between certain plants and ants, of unknown significance.

Nanandrous, of the Oedogoniales. A species which produces dwarf male filaments.

Neck. The tube-like portion of an ARCHEGONIUM, through which the ANTHEROZOID reaches the EGG.

Nectar, of angiosperms. A sugary fluid produced by many species, often in FLOWERS, and collected by insects.

Nectary, of angiosperms. A gland which secretes nectar.

Nerve, of mosses. A group of elongated cells at the centre of the leaf.

Node. The position on a stem at which leaves and branches are attached.

Nucellus, of seed plants. The MEGASPORANGIUM.

Nucleolus. A body lying within the NUCLEUS, rich in ribonucleic acid.

Nucleus. A conspicuous body in the protoplasm containing most of the deoxyribonucleic acid of the cell, and the site of the Mendelian genes.

Oogamy. The condition in which the female GAMETE is non-motile and the male motile.

Oogonium, of algae. A unicellular structure which produces female GAMETES.

Operculum, of mosses. The circular lid of the CAPSULE.

Organelle. A part of a cell with certain definite functions and characteristic morphological features.

Organism. An individual animal or plant.

Ovary, of angiosperms. The portion of the GYNAECIUM containing the OVULES and composed of one or more CARPELS.

Ovule, of seed plants. The NUCELLUS (containing the EMBRYO SAC) and INTEGUMENTS, yielding the SEED after FERTILIZATION.

Ovuliferous scale, of gymnosperms. The scale on which the OVULES develop.

Palisade tissue. The portion of the leaf MESOPHYLL (usually the uppermost) which consists of regular, columnar cells, containing most of the CHLOROPLASTS.

Palmate. Shaped like a hand with finger-like lobes.

Palmelloid state, of algae. The aggregation of normally separate individuals into a gelatinous, more or less PALMATE, COLONY.

Paraphyses, of algae and bryophytes. Sterile hairs found among the reproductive organs in certain genera.

Parasite. An organism which lives at the expense of another, the HOST, upon which it confers no benefits.

Parenchyma. A tissue composed of VACUOLATED, thin-walled, more or less ISODIAMETRIC cells, functioning chiefly as a ground and storage tissue, often retaining the ability to divide.

Parthenogenesis. The development of a female GAMETE into a new individual without FERTILIZATION.

Parthenospore, of algae. A female GAMETE which develops into a resting SPORE without FERTILIZATION.

Peduncle, of angiosperms. A FLOWER stalk.

Pellicle, of algae. An elastic membrane surrounding a unicellular organism, found chiefly in the Euglenophyta.

Peltate. Umbrella-shaped.

Perianth, of angiosperms. Collective term for the PETALS and SEPALS of a FLOWER.

Pericarp, of angiosperms. The part of a FRUIT derived from the OVARY wall.

Perichaetium, of bryophytes. A curtain-like outgrowth of tissue surrounding groups of ARCHEGONIA in certain genera of liverworts, or the uppermost leaves surrounding the sex organs in mosses.

Pericycle. The tissue lying between the vascular tissue and the ENDODERMIS.

Periderm. The collective term for the PHELLEM, PHELLOGEN and PHELLODERM.

Perigynium, of liverworts. A cylindrical sheath enclosing an ARCHEGONIUM.

Perigynous, of angiosperms. A FLOWER in which the PERIANTH and ANDROECIUM are borne on the rim of a concave receptacle, separate from the central GYNAECIUM.

Perispore, of ferns. An irregular accretion around the spore, derived from the TAPETUM.

Peristome, of mosses. The ring of tooth-like structures encircling the mouth of the CAPSULE.

Petals, of angiosperms. The upper PERIANTH segments, especially when these are distinct by form or pigmentation from the lower.

Petiole. A leaf stalk.

Phanerophyte. A plant in which the perennating buds are borne high above ground level.

Phellem. See CORK.

Phelloderm. The layer of cells produced towards the inside by the PHELLOGEN (secondary CORTEX).

Phellogen. See CORK CAMBIUM.

Phenotype. The visible, or chemically or biologically detectable, manifestation of the GENOTYPE produced as a consequence of growth and development.

Phloem. VASCULAR tissue composed of SIEVE CELLS or SIEVE TUBES, and associated PARENCHYMATOUS and FIBROUS cells.

Phototaxis. The movement of a whole organism in response to light.

Phototropism. The change in the direction of a plant's growth in response to unequal illumination.

Phycocyanin, of blue-green algae. A blue pigment found in the cells.

Phycoerythrin, of red algae. A red pigment found in the CHROMATOPHORES.

Phyllode. A PETIOLE which is broad and leaf-like, the LAMINA often being little developed.

Phyllotaxis. The arrangement of the leaves on a stem.

Phylogeny. The evolutionary history of an organism, or group of organisms.

Physiology. The study of chemical and physical processes which occur in living organisms.

Pinna. The primary division of a PINNATE leaf.

Pinnate. Of leaves which have a series of leaflets on each side of a common MIDRIB.

Pinnule. The ultimate division of a PINNATE leaf.

Pistil, of angiosperms. The OVARY, STYLE and STIGMA.

Pit. A thin area in the cell wall through which substances may pass from cell to cell.

Pith. The core of PARENCHYMA which may occur at the centre of a STELE.

Placenta, of angiosperms. The region of attachment of the OVULE to the OVARY wall (CARPEL).

Plagiogeotropism. The orientation of growth at a definite angle to the gravitational field.

Plagiotropic. Tending to take up a position at a definite angle to an orientating influence (e.g. PLAGIOGEOTROPISM).

Plankton. Collective term for the microscopic organisms, both plant (phytoplankton) and animal (zooplankton), which occur at the surface of fresh and salt water.

Plasmalemma. The membrane which bounds the PROTOPLASM of a cell and adjoins the cell wall.

Plasma membrane. Same as PLASMALEMMA.

Plastid. Collective term for CHLOROPLASTS, LEUCOPLASTS, AMYLOPLASTS and CHROMOPLASTS.

Platyspermic, of seed plants. Having flattened, BILATERALLY SYMMETRICAL SEEDS.

Plumule. The embryonic shoot.

Polarity. DIFFERENTIATION in structure or function between two ends of an axis.

Pollen, of seed plants. The MICROSPORES, often showing ENDOSPORIC germination.

Pollination, of seed plants. The process by which POLLEN is transferred from the ANTHER to the CARPELS or OVARY.

Pollinium, of certain genera of angiosperms. A mass of POLLEN grains held together by a sticky secretion.

Polycyclic. Arranged in concentric circles.

Polyembryony, of seed plants. The development of more than one EMBRYO in an OVULE.

Polyploidy. The possession of three or more sets of CHROMOSOMES per NUCLEUS.

Polystelic. Having more than one STELE.

Primary tissues. Tissues produced as a direct consequence of the activity of apical MERISTEMS.

Procambium. A strand of relatively UNDIFFERENTIATED cells which later gives rise to the PRIMARY XYLEM and PRIMARY PHLOEM tissues.

Procarp, of red algae. The multicellular female sex organs.

Procaryotic. Of organisms in which the nuclear material is not enclosed in a membranous envelope.

Proembryo, of gymnosperms. The structure developing immediately from the ZYGOTE.

Prothallus. A little DIFFERENTIATED, free-living, GAMETOPHYTE generation.

Protocorm, chiefly of lycopods and orchids. A tuberous structure developing from a portion of the young EMBRYO.

Protonema, chiefly of mosses. The FILAMENT or small plate of cells produced by the germinating SPORE.

Protoplasm. The contents of a living cell, excluding the cell wall and vacuole.

Protoplast. The contents of a living cell, excluding the cell wall.

Protostele. The kind of STELE in which a solid cylinder of XYLEM is surrounded by PHLOEM.

Protoxylem. The portion of the PRIMARY XYLEM which DIFFERENTIATES first.

Proximal. Adjacent to a point of reference or symmetry (ant. DISTAL).

Pseudo-endosperm, of gymnosperms. The nutritive tissue in SEEDS derived from the female GAMETOPHYTE (and hence HAPLOID).

Pseudogamy. The development of a female GAMETE into a new individual after stimulation by a male GAMETE, but without actual FERTILIZATION.

Pseudoparenchyma, in algae and fungi. A tissue consisting of closely adpressed, often interwoven FILAMENTS, giving the impression of PARENCHYMA.

Pseudopodium, of mosses. A stalk-like outgrowth of the GAMETOPHYTE of certain genera which elevates the SPOROPHYTE.

Pyrenoid, of algae. An intracellular particle associated with the CHLOROPLAST, around which starch forms.

Raceme, of angiosperms. An INFLORESCENCE in which FLOWERS are borne on branches of the main axis, those at the base maturing first.

Radial symmetry. Symmetry about a central axis; when divided longitudinally along any diameter, the two halves are mirror-images.

Radicle. The embryonic root.
Radiospermic, of seed plants. Bearing RADIALLY SYMMETRICAL SEEDS.
Ray. A vertical sheet of PARENCHYMATOUS cells traversing the STELE radially.
Receptacle, of angiosperms. The enlarged tip of the FLOWER stalk to which the floral parts are attached.
Reniform. Kidney-shaped.
Reticulate. In the form of a network.
Rhizoids. Thread-like anchoring and absorbing organs produced by plants lacking roots.
Rhizome. A PLAGIOTROPIC underground stem.
Root. That part of a VASCULAR plant which in branching produces only simple axes like itself, usually subterranean.
Root cap. A cap of tissue over the root apex.
Root hair. Hair-like outgrowths of the EPIDERMAL cells of the young root which absorb water and minerals.
Rootstock. A short, vertical, underground stem, bearing roots; in horticulture the plant providing the root system on the stem of which another kind of plant is grafted.

Saprophyte. An organism obtaining its food in the form of complex molecules from dead organic matter.
Scalariform. In the form of a ladder.
Scandent. Climbing.
Scarious. Thin, dry and membranous.
Schizogenous (of spaces). Formed by the separation of cells (cf. LYSIGENOUS).
Sclerenchyma. A strengthening tissue composed of FIBRES or STONE CELLS.
Secondary tissues. Produced from MERISTEMS which arise after the DIFFERENTIATION of the PRIMARY TISSUES.
Seed, of seed plants. The product of the OVULE after fertilization, comprising the EMBRYO, with its surrounding food reserves and protective coverings.
Sepals, of angiosperms. The lowermost perianth segments, especially when green, and thus distinguished from PETALS.
Septum. A partition.
Sessile. Without a stalk.
Seta, of bryophytes. The stalk bearing the CAPSULE.
Shoot. The leafy part of a plant, usually aerial.
Shrub. A woody plant with no distinct trunk, the main branches arising near ground level.
Sieve cell. A cell with a PROTOPLAST but lacking a distinct NUCLEUS, concerned with the transport of organic materials. Found in the PHLOEM.
Sieve plate. The perforate area of the cell wall between two SIEVE CELLS or SIEVE TUBE elements.
Sieve tube, of angiosperms. A column of SIEVE CELLS (here usually called SIEVE TUBE elements) with more or less transverse end walls bearing SIEVE PLATES.

Sinuose. Undulating.

Siphon(ac)eous. Tubular.

Siphonostele. A STELE comprising a hollow cylinder of VASCULAR tissue with PITH at the centre, the PHLOEM solely abaxial.

Solenostele. A SIPHONOSTELE in which PHLOEM occurs on both abaxial and adaxial sides of the XYLEM.

Sorus, chiefly of ferns. A group of SPORANGIA.

Sperm (-atozooid). See ANTHEROZOID.

Spermatium, of red algae. A non-motile male GAMETE.

Spermatocyte. See ANTHEROCYTE.

Spike, chiefly of angiosperms. A RACEMOSE INFLORESCENCE in which the FLOWERS are SESSILE.

Spongy tissue. The lower portion of the MESOPHYLL of a leaf, consisting of irregular cells with large air spaces.

Sporangiophore. A structure bearing one or more SPORANGIA.

Sporangium. An organ which produces SPORES.

Spore. A unicellular, ASEXUAL reproductive cell, usually uninucleate.

Sporophyll. A structure bearing SPORANGIA, and believed to be equivalent to a leaf.

Sporophyte. The (normally) DIPLOID generation, producing HAPLOID SPORES which germinate to give the GAMETOPHYTE generation.

Stamen, of angiosperms. The MICROSPORANGIA (ANTHER) together with their common stalk (FILAMENT), usually regarded as a MICROSPOROPHYLL.

Staminode, of angiosperms. An infertile STAMEN, often highly modified or reduced.

Stele. The VASCULAR tissue of a root or stem, extending to the ENDODERMIS if present.

Stigma, of algae. A small pigmented region, possibly photo-sensitive (eye-spot).

—, of angiosperms. The special region of the CARPEL on which the POLLEN grains germinate, often borne on a STYLE.

Stipule, of angiosperms. Outgrowths (usually two) at the base of the PETIOLE.

Stolon. A short, PLAGIOTROPIC shoot which develops a new plant at the tip, eventually severing connection with the parent.

Stoma. A pore in the EPIDERMIS bounded by a pair of GUARD CELLS.

Stomium, of ferns. The point of rupture of the SPORANGIUM.

Stone cells. More or less isodiametric, heavily LIGNIFIED cells.

Strain. A mating group within a species; or a variety within a species with distinctive PHYSIOLOGICAL or MORPHOLOGICAL features.

Strobilus. See CONE.

Style, of angiosperms. An elongation of the distal end of the CARPEL, bearing the STIGMA.

Suberin. An impervious waxy substance with which CORK and other cells are impregnated.

Subsidiary cells. Cells occurring adjacent to the GUARD CELLS of STOMATA, which are usually of constant shape and position, and distinguishable from other EPIDERMAL cells.

Suspensor, chiefly of seed plants. The cell or group of cells arising with the EMBRYO proper from the ZYGOTE, and pushing the young EMBRYO into the nutritive material of the SEED.

Swarmer. See ZOOSPORE.

Symbiosis. An association between two organisms in which there is mutual benefit.

Sympodial branching. The type of branching in which the terminal bud ceases to grow, growth being continued in each instance by the uppermost lateral branch. A form of CYMOSE BRANCHING.

Synangium. A composite structure formed by the CONCRESCENCE of SPORANGIA.

Syngamy. The fusion of GAMETES.

Tangential. Perpendicular to a radius. Used principally to describe the orientation of longitudinal sections of axes, or of cell division.

Tapetum. A layer of nutritive cells surrounding the SPORE mother cells in the SPORANGIUM. Conspicuous only in plants with VASCULAR TISSUE.

Taproot. A persistent primary root, often swollen with food reserves.

Tendril, of angiosperms. A modified leaf, leaflet or stem, often sensitive to contact and able to coil round objects it touches, thus supporting and anchoring the plant.

Testa, of angiosperms. The covering of the SEED, derived from the INTEGUMENTS.

Tetrad. The group of four cells produced by a MEIOTIC division, frequently genetically dissimilar.

Thallus. A relatively undifferentiated plant body.

Therophyte, of angiosperms. An ANNUAL plant, the perennating organs being the SEEDS.

Tonoplast. The membrane surrounding the VACUOLE, often also called the vacuolar membrane.

Trabeculum. A bar, or elongated cell, traversing a cavity.

Tracheid. An element of the XYLEM tissue consisting of an elongated, elaborately pitted, lignified cell with oblique end walls.

Tree. A woody plant with a definite trunk; the main branching well above ground level.

Trichogyne, of algae. A hair-like projection from the female sex organ.

Trichothallic growth, of algae. The type of growth in which cell divisions are confined to a region at the base of the FILAMENT.

Trilete spore. The kind of SPORE which, having been formed in a tetrahedral TETRAD, bears a tri-radiate scar on its PROXIMAL face.

Triploid. Having three times the basic (HAPLOID) number of chromosomes.

Trisomy. The presence of one chromosome in triplicate in the diploid nucleus. A form of ANEUPLOIDY.

Tuber. A rounded storage organ formed from a root (root tuber) or a stem (stem tuber).

Tunica, of angiosperms. The outer layer or layers of cells at the apex of a stem in which the divisions are principally ANTICLINAL.

Vacuole. The portion of the cell which contains the CELL SAP, bounded by the TONOPLAST.

Vascular tissue. A collective term for the XYLEM and PHLOEM.

Veins. The VASCULAR strands of leaves.

Velum, of the fertile leaves of Isoetales. The membranous covering of the SPORANGIUM.

Venation. The pattern of VEINS in a leaf.

Ventral, of lateral organs. The same as ADAXIAL.

--, of prostrate plants. The side towards the substratum.

Vessels, of angiosperms. Tubes occurring in the XYLEM and concerned with the conduction of water. Formed from columns of broad, short cells, the end walls of which break down during DIFFERENTIATION.

Wood. The XYLEM, principally the SECONDARY.

Xanthophyll. A class of yellow, CAROTENOID pigments associated with CHLOROPHYLL in the CHLOROPLAST.

Xeromorph. A plant possessing the features often found in XEROPHYTES, but not necessarily confined to dry places.

Xerophyte. A plant growing characteristically in dry situations, and often possessing anatomical and physiological features enabling it to withstand prolonged drought.

Xylem. VASCULAR tissue which may comprise VESSELS, TRACHEIDS, FIBRES, together with some PARENCHYMA.

Zoosporangium. An organ producing ZOOSPORES.

Zoospore. A motile SPORE.

Zygomorphic. BILATERALLY SYMMETRICAL.

Zygote. The cell formed by the fusion of two GAMETES.

References

1. ALLEN, C. E. (1917). *Science, N.Y.*, **46**, 466.
2. ANDREWS, H. N. (1960). *Cold Spr. Harb. Sym. quant. Biol.*, **24**, 217.
3. ANDREWS, H. N. (1961). *Studies in Paleobotany*. Wiley, New York and London.
4. ANDREWS, H. N., JR. and MURDY, W. H. (1958). *Am. J. Bot.*, **45**, 552.
5. ARBER, A. (1937). *Biol. Rev.*, **12**, 157.
6. ARBER, A. (1941). *Biol. Rev.*, **16**, 81.
6a. ARNOLD, C. A. (1968). *Earth Sci. Rev.*, **4**, 245.
7. BARGHOORN, E. S. and SCHOPF, J. W. (1965). *Science, N.Y.*, **150**, 337.
7a. BECK, C. B. (1970). *Biol. Rev.*, **45**, 379.
8. BELL, P. R. (1956). *Ann. Bot., N.S.*, **20**, 69.
9. BENNET-CLARK, T. A. and BALL, N. G. (1951). *J. exp. Bot.*, **2**, 169.
10. BEWS, J. W. (1925). *Plant Forms*. Longmans, Green, London.
11. BIERHORST, D. W. (1953). *Am. J. Bot.*, **40**, 649.
11a. BIERHORST, D. W. (1967). *Am. J. Bot.*, **54**, 538.
11b. BIERHORST, D. W. (1968). *Phytomorphology*, **18**, 232.
12. BOPP, M. (1954). *Z. Bot.*, **42**, 331.
13. BOPP, M. (1963). *J. Linn. Soc. (Bot.)*, **58**, 305.
14. BOWER, F. O. (1923–8). *The Ferns*. 3 vols. University Press, Cambridge.
15. BOWER, F. O. (1930). *Size and Form in Plants*. Macmillan, London.
16. BRYAN, G. S. and EVANS, R. I. (1956). *Am. J. Bot.*, **43**, 640.
17. BURGEFF, H. (1943). *Genetische Studien an Marchantia*. Fischer, Jena.
18. CAMPBELL, D. H. (1905). *Mosses and Ferns*. 2nd ed. Macmillan, New York.
19. CHALONER, W. G. (1958). *Ann. Bot., N.S.*, **22**, 197.
20. CHURCH, A. H. (1919). *Thalassiophyta and the Subaerial Transmigration*. Oxford University Press, London.
21. CLOWES, F. A. L. (1961). *Apical Meristems*. Oxford University Press, London.
22. CORNER, E. J. H. (1964). *The Life of Plants*. Weidenfeld and Nicolson, London.
23. CROCKER, W. and BARTON, L. V. (1957). *Physiology of Seeds*. Chronica Botanica, Waltham, Mass.
23a. D'AMATO, F. and AVANZI, S. (1968). *Caryologia*, **21**, 83.

24. DAVY, J. B. (1922). *J. Ecol.*, **10**, 211.
25. DE LA RUE, C. D. and NARAYANASWAMI, S. (1957). *New Phytol.*, **56**, 61.
25a. DICKINSON, H. G. (1970). *Cytobiologie*, **1**, 437.
25b. DOUGHERTY, E. C. (1957). *J. Protozool.*, **4** (suppl.), 14.
26. DREW, K. M. (1955). *Biol. Rev.*, **30**, 343.
26a. DUCKETT, J. G. (1970). *New Phytol.*, **69**, 333.
27. EAMES, A. J. (1936). *Morphology of Vascular Plants*. McGraw-Hill, New York.
28. ECHLIN, P. and MORRIS I. (1965). *Biol. Rev.*, **40**, 143.
29. EGGERT, D. A. (1961). *Palaeontographica*, **108B**, 43.
30. EICHLER, A. W. (1878). *Blüthendiagramme*, II. Engelmann, Leipzig.
30a. FERGUSON, C. W. (1968). *Science, N.Y.*, **159**, 839.
31. FLORIN, R. (1951). *Acta Hort. berg.*, **15**, 285.
32. FOGG, G. E. (1951). *Ann. Bot., N.S.*, **15**, 23.
33. FOGG, G. E. (1953). *The Metabolism of Algae*. Methuen, London.
34. FOGG, G. E. (1965). *Proc. R. Soc., ser. B*, **162**, 517.
35. FREEBERG, J. A. and WETMORE, R. H. (1957). *Phytomorphology*, **7**, 204.
36. FREI, E. and PRESTON, R. D. (1961). *Proc. R. Soc., ser. B*, **154**, 70.
37. FRITSCH, F. E. (1935, 1945). *Structure and Reproduction of the Algae*, 1 and 2. University Press, Cambridge.
38. FRITSCH, F. E. (1945). *Ann. Bot., N.S.*, **9**, 1.
39. FROST, F. H. (1930). *Bot. Gaz.*, **89**, 67 and **90**, 198.
40. GANTT, E. and CONTI, S. F. (1965). *J. Cell Biol.*, **26**, 365.
41. GLAESSNER, M. F. (1962). *Biol. Rev.*, **37**, 467.
42. GOEBEL, K. (1886). *Ber. dt. bot. Ges.*, **4**, 184.
43. GOLUB, S. J. and WETMORE, R. H. (1948). *Am. J. Bot.*, **35**, 755 and 767.
44. GRAUSTEIN, J. E. (1930). *Bot. Gaz.*, **90**, 46.
45. HARRIS, T. M. (1939). *Annls bryol.*, **12**, 57.
46. HESLOP HARRISON, J. (1962). *Nature, Lond.*, **195**, 1069.
47. HUEBER, F. M. (1964). *Mem. Torrey bot. Club*, **21**, no. 5, 5.
48. HUGHES, N. F. (1961). *Sci. Progr.*, **49**, 84.
49. JACOT-GUILLARMOD, A. (1958). *Nature, Lond.*, **182**, 474.
50. JAYASEKERA, R. D. E. and BELL, P. R. (1959). *Planta*, **54**, 1.
50a. KNAPP, E. and SCHREIBER, H. (1939). *Proc. 7th Gen. Congr. Edinburgh*, 175.
51. KNUTH, P. (1898–1905). *Handbuch der Blütenbiologie*. Engelmann, Leipzig.
52. LEE, C. L. (1954). *Am. J. Bot.*, **41**, 545.
52a. LEEDALE, G. F., LEADBEATER, B. S. C. and MASSALSKI, A. (1970). *J. Cell Sci.*, **6**, 109.
52b. LEWIN, R. A. (Ed.) (1962). *Physiology and Biochemistry of Algae*. Academic Press, New York.
53. LYON, A. C. (1964). *Nature, Lond.*, **203**, 1082.
53a. LØVLIE, A. and BRÅTEN, T. (1970). *J. Cell Sci.*, **6**, 109.
54. MACALLUM, A. B. (1905). *J. Physiol.*, **32**, 95.
55. MAHESHWARI, P. (1950). *An Introduction to the Embryology of Angiosperms*. McGraw-Hill, New York.
56. MAHESHWARI, P. and VASIL, V. (1961). *Gnetum*. C.S.I.R., New Delhi.
56a. MAMAY, S. H. (1969). *Science, N.Y.*, **164**, 295.
57. MANTON, I. (1950). *Problems of Cytology and Evolution in the Pteridophyta*. University Press, Cambridge.

58. MANTON, I. (1959). *J. mar. biol. Ass. U.K.*, **38**, 319.
59. MANTON, I. (1965). *Adv. bot. Res.*, **2**, 1.
60. MARTENS, P. and WATERKEYN, L. (1964). *La Cellule*, **65**, 1.
61. MASCARENHAS, J. P. and MACHLIS, L. (1964). *Plant Physiol.*, **39**, 70.
61a. MAYER, A. M. and POLJAKOFF-MAYBER, A. (1963). *The Germination of Seeds*. Pergamon, Oxford.
62. MERESCHKOWSKY, C. (1905). *Biol. Zbl.*, **25**, 593.
63. PANT, D. D. (1962). *Proc. Summer School in Botany*, Darjeeling, 276.
64. PARKE, M. (1961). *Br. phycol. Bull.*, **2**, 47.
65. PLUMSTEAD, E. P. (1956). *Palaeontographica*, **100B**, 1.
66. PRINGSHEIM, E. G. (1963). *Farblose Algen*. Fischer, Stuttgart.
67. RAUH, W. and FALK, H. (1959). *Sber. heidelb. Akad. Wiss. Math. naturw. Kl.*, **1**.
68. RAUNKIAER, C. (1934). *The Life Forms of Plants*. Oxford University Press, London.
69. REID, G. K. (1961). *Ecology of Inland Waters and Estuaries*. Reinhold, New York.
70. RICHARDS, P. W. (1952). *The Tropical Rain Forest*. University Press, Cambridge.
70a. RIDLEY, H. N. (1930). *The Dispersal of Plants*. Reeve, Ashford.
71. ROUND, F. E. (1965). *The Biology of the Algae*. Edward Arnold, London.
72. SAX, K. (1935). *J. Arnold Arbor.*, **16**, 301.
72a. SCHOPF, J. W. (1970). *Biol. Rev.*, **45**, 319.
73. SPORNE, K. (1962). *The Morphology of Pteridophytes*. Hutchinson, London.
74. SPORNE, K. (1965). *The Morphology of Gymnosperms*. Hutchinson, London.
75. STAINER, R. Y. and VAN NIEL, C. B. (1962). *Arch. Mikrobiol.*, **42**, 17.
76. STEBBINS, G. L. (1965). *Ann. Mo. bot. Gdn*, **52**, 457.
76a. STEWARD, F. C. (1970). *Proc. R. Soc., ser. B*, **175**, 1.
76b. STEWART, W. D. P., HAYSTEAD, A. and PEARSON, H. W. (1969). *Nature, Lond.*, **224**, 226.
77. SUMMERHAYES, V. S. (1951). *Wild Orchids of Britain*. Collins, London.
77a. TAYLOR, T. N. (1969). *Science, N.Y.*, **164**, 294.
77b. TIDWELL, W. D., RUSHFORTH, S. R., REVEAL, J. L. and BEHUNIN, H. (1970). *Science, N.Y.*, **168**, 834.
78. WARDLAW, C. W. (1952). *Phylogeny and Morphogenesis*. Macmillan, London.
79. WARDLAW, C. W. (1955). *Embryogenesis in Plants*. Methuen, London.
80. WARDLAW, C. W. (1965). In Ruhland, W. ed. *Handbuch der Pflanzenphysiologie*, **15**, 1, 1008.
80a. WAREING, P. F. and PHILLIPS, I. D. J. (1970). *The Control of Growth and Differentiation in Plants*. Pergamon, Oxford.
81. WATSON, E. V. (1964). *The Structure and Life of Bryophytes*. Hutchinson, London.
82. WETMORE, R. H., DE MAGGIO, A. E. and MOREL, G. (1963). *J. Indian bot. Soc.*, **42A**, 306.
83. WHITEHOUSE, H. L. K. (1950). *Ann. Bot., N.S.*, **14**, 199.
84. WILLIAMS, S. (1950). *Trans. Br. bryol. Soc.*, **1**, 357.
85. WILLIS, A. J. (1957). *Nature, Lond.*, **179**, 380.
86. WOLLERSHEIM, M. (1957). *Z. Bot.*, **45**, 245.

Index

Bold page numbers indicate definitions